NON-ACCELERATOR ASTROPARTICLE PHYSICS

The Abdus Salam
International Centre for Theoretical Physics

United Nations
Educational, Scientific
and Cultural Organization

International Atomic
Energy Agency

Proceedings of the 7th School

NON-ACCELERATOR ASTROPARTICLE PHYSICS

ICTP, Trieste, Italy
26 July – 6 August 2004

Editors

R. A. Carrigan, Jr.
Fermilab, USA

G. Giacomelli
University of Bologna, Italy

N. Paver
University of Trieste, Italy

World Scientific

NEW JERSEY • LONDON • SINGAPORE • BEIJING • SHANGHAI • HONG KONG • TAIPEI • CHENNAI

Published by

World Scientific Publishing Co. Pte. Ltd.
5 Toh Tuck Link, Singapore 596224
USA office: 27 Warren Street, Suite 401-402, Hackensack, NJ 07601
UK office: 57 Shelton Street, Covent Garden, London WC2H 9HE

British Library Cataloguing-in-Publication Data
A catalogue record for this book is available from the British Library.

NON-ACCELERATOR ASTROPARTICLE PHYSICS
Proceedings of the 7th School

ISBN 981-256-316-4

Printed by FuIsland Offset Printing (S) Pte Ltd, Singapore

FOREWORD

The proceedings of the 7th School on Non-Accelerator Astroparticle Physics present a timely coverage of this intere and rapidly expanding subject. They enlarge and complement the earlier volumes prepared for the 4th, 5th and 6th Schools*. We have endeavored to maintain an informative pedagogical tone so that the book can serve as the basis for a modern course on the subject and as useful reference book.

The first section introduces the fundamentals of particle physics with a review of the standard model and beyond. The comprehensive section on neutrino physics and astrophysics covers neutrino masses and oscillations, short and long baseline neutrino experiments, atmospheric and solar neutrinos, and neutrino telescopes. The section on dark matter includes a theoretical presentation and a review of existing and potential dark matter searches. Searches for axions, magnetic monopoles, and nuclearites are also discussed. Cosmic rays and astrophysics are covered with reviews on experiments in space, extreme energy cosmic rays, and photons and antimatter in space. The theory of gravitational waves and searches for gravitational waves are discussed. A section deals with the LEP legacy and future accelerators and superbeams. Large scale facilities, detectors, data acquisition and large scale computing are reviewed. The final section concerns the World of Science, with reviews on 100 years of science, Science and Society and the Universe: Today, Yesterday and Tomorrow. Abstracts of the many posters presented by participants at the school give a broad picture of world-wide activities in the field.

The school was conducted under the auspices of the Abdus Salam ICTP, Trieste, and was also sponsored by the INFN.

We gratefully acknowledge the kind and efficient help of Ms. G. De Meo, Ms. A. Triolo, the staff of ICTP, Ms. A. Casoni, and the technical cooperation of Drs. R. Giacomelli, Y. Becherini and M. Giorgini (Bologna).

The school's lecture notes and recordings can be found at http://cdsagenda5.ictp.trieste.it/full_display.php?ida=a0355#.

Trieste, January 2005

The Editors:

R.A. Carrigan Jr., G. Giacomelli, N. Paver

*Proceedings of the Fourth School on Non-Accelerator Particle Astrophysics, Eds. E. Bellotti, R.A. Carrigan, Jr., G. Giacomelli and N. Paver (World Scientific, Singapore, ISBN 981022688, 1996); Proceedings of the Fifth School on Non-Accelerator Particle Astrophysics, Eds. R.A. Carrigan, Jr., G. Giacomelli and N. Paver (INFN, Edizioni Università di Trieste, Trieste, 1999); Proceedings of the Sixth School on Non-Accelerator Astroparticle Physics, Eds. R.A. Carrigan, Jr., G. Giacomelli, A. Masiero and N. Paver (World Scientific, Singapore, ISBN 9810249438, 2002).

CONTENTS

Cosmic Rays and Astrophysics

Gravitational Waves

Accelerator-Based Physics

Large Scale Facilities

The World of Science

STANDARD MODEL AND BEYOND

A. BARTL AND S. HESSELBACH

Institut für Theoretische Physik, Universität Wien, A-1090 Vienna, Austria

We first discuss the basic features of electroweak 1-loop corrections in the Standard Model. We also give a short and elementary review on Higgs boson searches, grand unification, supersymmetry and extra dimensions.

1. Introduction

The Standard Model (SM) is our present theory of the fundamental interactions of the elementary particles. It includes quantum chromodynamics (QCD) as the theory of the strong interactions and the Glashow-Salam-Weinberg (GSW) theory as the unified theory of the electromagnetic and weak interactions. Both QCD and GSW theory are non-Abelian gauge theories, based on the principle of local gauge invariance. The gauge symmetry group of QCD is SU(3) with colour as the corresponding quantum number, that of the GSW theory is SU(2) $\times$ U(1) with the quantum numbers weak isospin and hypercharge. The gauge symmetry group of the GSW theory is spontaneously broken by the Higgs mechanism from SU(2) $\times$ U(1) to the electromagnetic U(1).[1] According to the gauge symmetry groups there are eight massless gluons mediating the strong interactions, one massless photon for the electromagnetic interaction and three vector bosons $W^{\pm}$ and Z^0 for the charged and neutral weak interactions. The weak vector bosons aquire their masses by the spontaneous breaking of the electroweak symmetry group.

The matter particles have spin $\frac{1}{2}$ and are grouped into three families of quarks and leptons. The fermions appear as left-handed and right-handed states, except for the neutrinos which in the SM are only left-handed and massless. The left-handed fermions are grouped in isodoublets, the right-handed fermions are isosinglets. The quark generations are mixed by the charged weak currents. This quark mixing is described by the Cabibbo-Kobayashi-Maskawa (CKM) matrix. The Glashow-Iliopoulos-Maiani (GIM) mechanism guarantees the absence of flavour changing neutral currents (FCNC) at tree level. In QCD the coupling of the gluon to the quarks is flavour independent ("flavour-blind"). The flavour dependence in the SM is essentially due to the quark mixing. To emphasize this aspect of the flavour dependence this part of the GSW theory is also called quantum flavour dynamics (QFD). Note that in the present formulation of the SM there is no mixing between lepton families. While this is true to a high accuracy for the charged leptons, we know that it is not true for the neutrino sector because neutrino oscillations occur.[2]

The spontaneous breaking of the electroweak symmetry is achieved by introducing one doublet of complex scalar Higgs fields. This is the minimum number of Higgs fields necessary to spontaneously break the $SU(2) \times U(1)$ symmetry and to introduce the mass terms for all particles apart from the neutrinos. After spontaneous symmetry breaking there remains one neutral scalar Higgs particle as physical state. The other three scalar fields become the longitudinal components of the massive $W^{\pm}$ and Z^0 bosons.

The SM is phenomenologically very successful. Highlights of the experimental development were the discoveries of the $W^{\pm}$ and Z^0 bosons, the τ lepton, the heavy quarks and the gluon at the large accelerator centres CERN and DESY in Europe, BNL, FNAL and SLAC in the USA. At present the SM can reproduce all accelerator-based experimental data. The gauge sector of the SM has been extremely well tested. If radiative corrections are included, the theoretical predictions are in very good agreement with the data of LEP, SLC, Tevatron and HERA.[3] Some observables have been measured with an error of less than one per mille, the theoretical predictions have a similar accuracy. However, the Higgs sector has up to now not been sufficiently well tested. In particular, the Higgs boson has not been found yet. Our theoretical ideas about the spontaneous electroweak symmetry breaking have still to be verified. If the Higgs mechanism of the SM is the right way of electroweak symmetry breaking, then we know from the direct searches at LEP that the mass of the Higgs boson has the lower bound $m_h > 114.4$ GeV.[3,4]

Despite its phenomenological success it is generally believed that the SM is just the low-energy limit of a more fundamental theory. Obviously, the SM in its present form cannot describe the recent experimental results on neutrino oscillations, which are only possible if the neutrinos have mass. Several theoretical ideas have been proposed for introducing neutrino mass terms. For a review we refer to Ref. 2.

We have also theoretical arguments for our believe that the SM has to be extended. One attempt is to embed the SM into a grand unified theory (GUT) where all gauge interactions become unified at a high scale $M_{\rm GUT} \approx 10^{16}$ GeV. Another extension of the SM is provided by supersymmetry (SUSY), which is probably the most intensively studied one so far. Other modifications are composite models, technicolour, strong electroweak symmetry breaking, little Higgs etc. In recent years the idea of "large extra dimensions" has been proposed and intensively studied, which could also provide a solution of some of the theoretical flaws of the SM. All these extensions of the SM will be probed at the Large Hadron Collider LHC,[5] which is presently under construction at CERN and will start operating in the year 2007. In the last decade the design of an e^+e^- linear collider has been intensively studied.[6] At such a machine all extensions of the SM could even be more precisely tested.

In this series of lectures we will first review the basics of electroweak radiative corrections in the SM and then present a short comparison with the experimental data. Then we will briefly discuss how to search for the Higgs boson in e^+e^- collisions and at $p\bar{p}$ and pp colliders. In the following sections we will discuss some aspects of physics beyond the SM. We will shortly treat GUTs, then give a phenomenological introduction to SUSY and close with some remarks about large extra dimensions.

2. Standard Model Physics

The SM is a renormalizable quantum field theory because QCD and the GSW theory are gauge theories. This enables us to calculate the theoretical predictions for the various observables with high accuracy. In the last years both the QCD and the electroweak 1-loop corrections for all important observables have been calculated. For some observables even the leading terms of the higher order corrections are known. Moreover, also the QCD corrections to a number of electroweak processes as well as the electroweak corrections to some QCD reactions have been calculated. Comparison with the precision data of LEP, SLC, Tevatron and HERA allows us to test the SM with high accuracy. In the following subsection we give a short review of the electroweak 1-loop corrections, essentially following the treatments of Refs. 7, 8.

2.1. *Electroweak Radiative Corrections*

The Lagrangian of the SM follows from the construction principles for gauge theories. It consists of the gauge field part, the fermion kinetic terms, the gauge interaction terms of the fermion fields, the kinetic and potential terms of the Higgs doublet, the gauge interaction of the Higgs doublet, and the terms for the Yukawa interaction between the fermion and the Higgs field. Their explicit form will not be given here, but can be found, e. g., in Ref. 1.

The Higgs sector of the SM, after spontaneous symmetry breaking, gets the following shape: the Higgs fields H^+, H^{+*} and $\mathrm{Im}\, H^0$ become the longitudinal components of $W^\pm$ and Z^0. After the shift $\mathrm{Re}\, H^0(x) = \frac{1}{\sqrt{2}}(v+h(x))$ the real scalar field $h(x)$ becomes the physical Higgs field. Its Lagrangian can be brought into the form[9]

$$\mathcal{L}_{\text{Higgs}} = \tfrac{1}{2}(\partial_\mu h)(\partial^\mu h) - \tfrac{1}{2}m_h^2 h^2 \left[1 + \frac{h}{v} + \frac{1}{4}\left(\frac{h}{v}\right)^2\right] - \sum_f \frac{m_f}{v}\bar{f} f\, h$$
$$+ \left(2\frac{h}{v} + \frac{h^2}{v^2}\right)\left[m_W^2 W_\mu^+ W^{-\mu} + \tfrac{1}{2}m_Z^2 Z_\mu Z^\mu\right] \ , \tag{1}$$

where the physical Higgs boson mass at tree level is $m_h^2 = 2\lambda v^2$, with λ being the quartic coupling constant in the original Higgs potential. Eq. (1) determines all properties of the SM Higgs boson. It has cubic and quartic self–interactions whose strengths are proportional to m_h^2. Its couplings to the vector bosons $W^\pm$, Z^0, and to fermions f are proportional to m_W^2, m_Z^2, and m_f, respectively. Therefore, the Higgs boson couples dominantly to the heavy particles. The Higgs boson mass m_h is experimentally not known. In the analyses it is usually treated as a free parameter of the SM.

The weak vector bosons $W^\pm$ and Z^0 get masses by the Higgs mechanism, which are

$$m_W^2 = \frac{1}{4}g^2 v^2 \ , \qquad m_Z^2 = \frac{1}{4}(g^2 + g'^2)v^2 = \frac{m_W^2}{\cos^2\theta_W} \ , \tag{2}$$

where g and g' are the $SU(2)$ and $U(1)$ coupling constants, θ_W is the electroweak mixing or Weinberg angle, $g'/g = \tan\theta_W$, and v is the vacuum expectation value (vev) of the H^0 component of the Higgs field. The photon and Z^0 are linear combinations of the neutral $SU(2)$ and $U(1)$ vector bosons with mixing angle θ_W, and the electromagnetic coupling is $e = g\sin\theta_W$. Comparison with the muon decay $\mu^+ \to e^+\nu_e\bar{\nu}_\mu$ leads to the relation

$$\frac{1}{\sqrt{2}}G_\mu = \frac{g^2}{8m_W^2} = \frac{e^2}{8m_W^2\sin^2\theta_W} \ , \tag{3}$$

where G_μ is the Fermi coupling constant. Inserting the experimental values[3,10]

$$G_\mu = (1.16637 \pm 0.00001) \times 10^{-5}\ \text{GeV}^{-2} \ , \tag{4}$$
$$\sin^2\theta_W = 0.23149 \pm 0.00015 \tag{5}$$

together with the fine structure constant $\alpha = \frac{e^2}{4\pi} = 1/137.03599911(46)$ into Eqs. (2) and (3) gives $m_W \approx 77.5$ GeV, $m_Z \approx 88.4$ GeV, and $v \approx 246$ GeV. These results for the vector boson masses are already very close to their experimental values, and historically this was one of the first triumphs of the SM. However, when compared with the recent experimental values with very small errors,[3,10]

$$m_W = 80.425 \pm 0.038 \text{ GeV} \ , \qquad m_Z = 91.1876 \pm 0.0021 \text{ GeV} \ , \qquad (6)$$

the theoretical values disagree by several standard deviations. This shows that the tree–level relations Eqs. (2) and (3) are not accurate enough, and that the electroweak loop–corrections have to be taken into account.

The high precision experiments at LEP, SLAC, and Tevatron have measured some of the electroweak observables with a very high accuracy.[3,10,11] For example, the Z^0 mass is known to 0.002%, the $W^\pm$ mass, the Z^0 width Γ_Z, and some of the partial widths $\Gamma(Z^0 \to f\bar{f})$, are known to about 0.1%. Some of the forward–backward asymmetries A_{FB} and left–right polarisation asymmetries A_{LR} for $e^+e^- \to f\bar{f}$ are also measured with very high experimental accuracy. In comparison, the electroweak radiative corrections are usually of the order of 1%, with a numerical accuracy of about 0.1%. This means that we can only get a theoretical accuracy comparable to the experimental one by taking into account the electroweak radiative corrections.

The analysis of the electroweak radiative corrections provides very accurate tests of the SM and leads to substantial restrictions on the allowed range of the Higgs boson mass. The $\mathcal{O}(\alpha)$ electroweak corrections at 1–loop level arise from self–energy diagrams, vertex corrections and box diagrams. They affect the basic SM parameters in characteristic ways. The self–energy diagrams of the vector bosons play a special role. The vacuum polarisation diagrams with charged lepton pairs and light quark pairs in the loops lead to a logarithmic q^2 dependence of the electromagnetic coupling. The bulk of the 1–loop corrections can be taken into account by including this q^2–dependence in an effective $\alpha(q^2)$. At $q^2 = m_Z^2$ this gives $\alpha(m_Z^2) = 1/(128.939 \pm 0.024)$, where the error is mainly due to the uncertainty in the hadronic contribution to the vacuum polarisation.[12]

The self–energy diagrams of the vector bosons $W^\pm$ and Z^0 lead to shifts of their renormalised masses $m_Z^2 \to m_Z^2 + \delta m_Z^2$, and $m_W^2 \to m_W^2 + \delta m_W^2$. At tree–level we have the relation $\sin^2\theta_W^{(0)} = 1 - m_W^2/m_Z^2$. At higher orders it is useful to define the effective electroweak mixing angle

$$\sin^2 \theta_W = \frac{1}{4|q_f|} \left(1 - \frac{g_{Vf}}{g_{Af}}\right), \tag{7}$$

where q_f is the electric charge and g_{Vf}/g_{Af} the ratio of the vector and the axial vector couplings of the fermion f.

The parameter ρ is introduced for comparing the SM predictions with the weak charged and neutral current data. It is defined as the ratio between the neutral and charged current amplitudes. In the SM at tree–level $\rho = m_W^2/(m_Z \cos\theta_W)^2 = 1$. If higher order corrections are taken into account, or in modifications of the SM, we may have $\rho \neq 1$. The deviation from 1, $\Delta\rho$ is a measure of the influence of heavy particles. $\Delta\rho$ can be expressed in terms of the vector boson self–energy contributions. The main SM 1–loop contribution was calculated in Ref. 13 and is

$$\Delta\rho \simeq \frac{3G_\mu m_t^2}{8\pi^2\sqrt{2}} + \mathcal{O}\left(\frac{m_b^2}{m_t^2}\right), \tag{8}$$

where m_t and m_b are the top and bottom quark masses. Some of the 2–loop corrections to $\Delta\rho$ have also been calculated.[7,8] They can be of the order of 10% of the 1–loop contribution Eq. (8). Experimentally we have[14] $\Delta\rho = (5.4 \pm 1.0) \times 10^{-3}$. As can be seen from Eq. (8), the main contribution to $\Delta\rho$ comes from heavy particle loops.

The analysis of the decay $\mu \to e\nu_e\nu_\mu$ leads to a relation between m_Z, m_W, and the Fermi coupling constant G_μ, which at tree–level is given in Eq. (3). This relation is modified when the electroweak radiative corrections to $\mu \to e\nu_e\nu_\mu$ are taken into account:

$$m_W^2 \left(1 - \frac{m_W^2}{m_Z^2}\right) = \frac{\pi\alpha}{\sqrt{2}\,G_\mu} \cdot \frac{1}{1 - \Delta r} \tag{9}$$

with

$$\Delta r = \frac{1}{2}\frac{\Delta\alpha}{\alpha} - \frac{m_W^2}{m_Z^2(1 - \frac{m_W^2}{m_Z^2})}\Delta\rho\,. \tag{10}$$

This shows that the main part of the radiative corrections to the tree-level relation (3) is contained in the quantity Δr. There are additional contributions from vertex corrections and box diagrams. Numerically Δr is in the range 0.04 to 0.07. Analogous to $\Delta\rho$, the main contributions to Δr come from heavy particle loops.

In the analysis of the precision data of LEP one usually proceeds in the following way:[7,8,14] The SM parameters α, G_μ, m_Z, (see Eqs. (4), (5) and

(6)), the strong coupling $\alpha_s(m_Z) = 0.1187 \pm 0.0020$, and m_h are taken as the main input parameters, and the quantities m_W and $\sin^2\theta_W$ are calculated with the help of Eqs. (7) to (10). Also the other Z–boson observables are calculated including the electroweak radiative corrections. The Higgs boson mass m_h is not known and it is taken as a free parameter and varied in the allowed range $m_h < 1$ TeV. In general, very good agreement between theory and experiment is obtained. This can also be illustrated in Fig. 1 from Ref. 15, where the theoretical relation between the $W^\pm$ mass m_W and the top quark mass m_t in the SM (lower band) together with the experimental error ellipses from LEP/Tevatron, Tevatron/LHC and the GigaZ option of an e^+e^- linear collider are shown. This theoretical relation between m_W and m_t is due to the radiative corrections to the $W^\pm$ boson mass, where the loops involving the top quark play a special role. The leading corrections depend quadratically on m_t and logarithmically on the Higgs boson mass m_h. While in this calculation essentially all basic electroweak parameters enter, m_W depends very significantly on m_t and on m_h. The width of the SM band is mainly due to the variation of the Higgs boson mass in the range 113 GeV $\lesssim m_h \lesssim$ 400 GeV. If the Higgs boson mass is left as a free parameter and a global fit to the precision data is performed, the best fit is obtained for the value $m_h = 114^{+69}_{-45}$ GeV, or equivalently, $m_h < 260$ GeV at 95% confidence level.[16] This result is consistent with the present experimental lower bound from LEP2, $m_h > 114.4$ GeV.[3,4]

From this analysis we learn that a heavy particle can be probed when it appears in the virtual loops of the quantum corrections. As these loop effects influence some of the measurable observables, it may be possible to derive limits on the allowed mass range of this heavy particle. This is possible although the energy is not high enough to directly produce the heavy particle in experiment. The prize to be paid is, of course, high precision in experiment as well as in the theoretical calculation.

2.2. *Higgs Boson Searches*

For a complete verification of the SM and its electroweak symmetry breaking mechanism we have to find the Higgs boson. In the preceding section we have seen that consistency of the SM with the existing precision data requires $m_h < 260$ GeV. Combining the final results from LEP2 of the four experiments ALEPH, DEPLPHI, L3 and OPAL a lower bound for the SM Higgs boson mass of 114.4 GeV at 95% confidence level arises.[3,4] A considerable part of the allowed mass range for the Higgs boson is within the

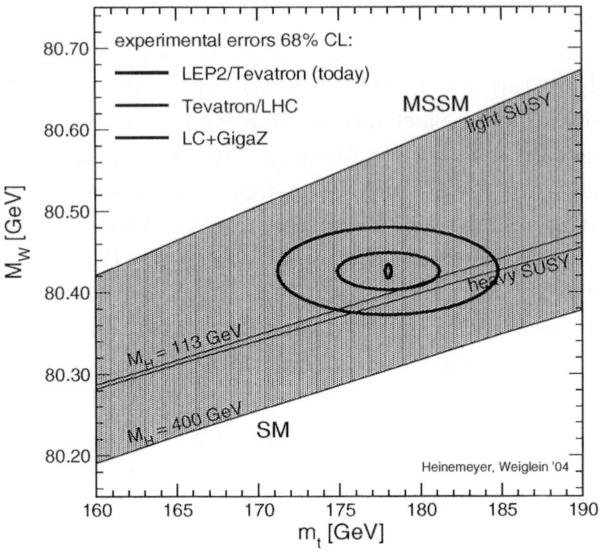

Figure 1. The present experimental accuracy for m_W and m_t after the experiments at LEP and Tevatron (large ellipse) and the expected accuracies at Tevatron + LHC (medium size ellipse) and LC + GigaZ (small ellipse). The lower and upper bands show the predictions of SM and MSSM, respectively, where the small intermediate band denotes the overlap between the predictions of SM and MSSM. From Ref. 15.

reach of Tevatron. A full coverage of this mass range will be provided by LHC and a future e^+e^- linear collider or muon collider. The search for the Higgs boson, therefore, has high priority at all present and future colliders. In this section we will discuss the principle ideas of Higgs boson searches at the Tevatron, LHC and a future e^+e^- linear collider.

The main production mechanisms at hadron colliders are gluon–gluon fusion, WW or ZZ fusion, associated production with W or Z and associated production with $t\bar{t}$ or $b\bar{b}$.[17] At the $p\bar{p}$ collider Tevatron with $\sqrt{s} = 2$ TeV the most relevant production mechanism is the associated production with W or Z bosons, where a detectable rate of Higgs events is expected for $m_h = 120$ GeV and an integrated luminosity $\int \mathcal{L} = 2\ \text{fb}^{-1}$. For example, a clear signature is expected for the reaction[18]

$$p + \bar{p} \to W^{\pm} + h \to \ell^{\pm} + (q\bar{q}) + p_{T\text{miss}} \tag{11}$$

where 85% of the $q\bar{q}$ pairs are $b\bar{b}$, and the ν_ℓ from $W^{\pm} \to \ell^{\pm}\nu_\ell, \ell = e, \mu$ is reconstructed from the missing transverse momentum $p_{T\text{miss}}$. The WW or ZZ fusion cross sections are slightly smaller for $m_h \lesssim 150$ GeV. The cross sections for associated production with $t\bar{t}$ or $b\bar{b}$ are rather low.

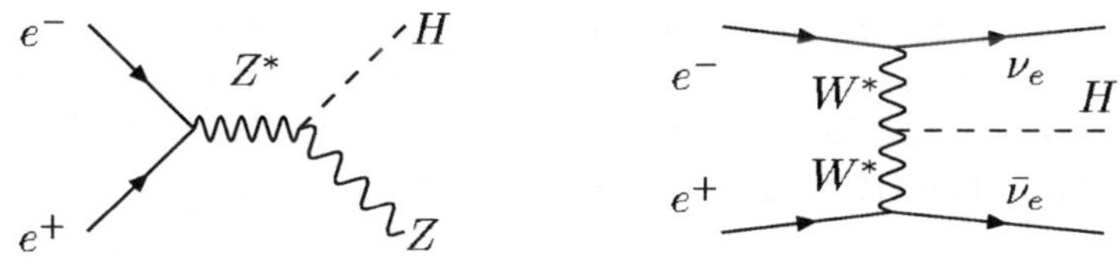

Figure 2. Feynman diagrams for the Higgsstrahlung and WW fusion production mechanisms of SM Higgs bosons in e^+e^+ annihilation.

The dominant production mechanism at the LHC with $\sqrt{s} = 14$ TeV is gluon–gluon fusion with a cross section $\gtrsim 10$ pb for $m_h < 260$ GeV. The cross section for WW or ZZ fusion is of the order of a few pb, whereas the cross sections for the associated productions with gauge bosons or $t\bar{t}$, $b\bar{b}$ may contribute for lower Higgs masses.

The search for the Higgs boson will also have a very high priority at a future linear collider.[6] The production of a SM Higgs boson in e^+e^- annihilation can proceed via "Higgsstrahlung" $e^+e^- \to Zh$, WW fusion $e^+e^- \to \nu_e\bar{\nu}_e h$, and ZZ fusion $e^+e^- \to e^+e^-h$. At $\sqrt{s} = 500$ GeV the Higgsstrahlung process dominates for $m_h \gtrsim 160$ GeV, whereas for $m_H \lesssim 160$ GeV the WW fusion process gives the largest contribution. The higher $\sqrt{s}$ the more important is the WW fusion process. The Feynman diagrams for the Higgsstrahlung and the WW fusion processes are shown in Fig. 2. Only for $\sqrt{s} \gtrsim 800$ GeV the ZZ fusion process can contribute about $\gtrsim 10\,\%$ of the total production rate. If $m_h \lesssim 260$ GeV as suggested by the electroweak precision data, an optimal choice for the c.m.s. energy is $\sqrt{s} \approx 350 - 500$ GeV.

In conclusion one can say that there are some prospects of finding the Higgs boson at the Tevatron. The LHC will cover the full mass range up to $m_h \approx 1$ TeV. Precise determinations of all important Higgs boson couplings will be possible at a future e^+e^- linear collider or muon collider.[6,18]

3. Grand Unification

We can study the scale dependence of the three gauge coupling "constants" with the help of the renormalization group equations (RGE). If we evolve the strong, electromagnetic and weak coupling constants to higher energy scales, they become approximately equal at $M_U \approx 10^{14}$ GeV to 10^{16} GeV, the grand unification scale. This behaviour of the gauge coupling constants suggests that the SM is embedded into an underlying grand unified theory (GUT). If we assume that this GUT is also a gauge theory, its symmetry group has to be semi-simple and it has to contain $SU(3) \times SU(2) \times U(1)$ as a subgroup. The GUT gauge group is unbroken at energies higher than the GUT scale M_U, and is spontaneously broken to the SM gauge group at lower energies. The smallest semi-simple GUT group with rank 4 is $SU(5)$. Other possible choices are $SO(10)$, $E(6)$ etc. In this section we will shortly mention the basic features of $SU(5)$ and $SO(10)$ grand unification (for other examples see Refs. 19, 20, 21).

In the $SU(5)$ GUT model the 15 helicity states of each family of quarks and leptons are put into the $\bar{\mathbf{5}}$ (e_L, ν_{eL}, d^C_{iL}, $i = 1, 2, 3$) and **10** (e^C_L, u_{iL}, u^C_{iL}, d_{iL}, $i = 1, 2, 3$) representations. Here the right–handed states f_R are written as the charge conjugate left–handed states f^C_L, and $i = 1, 2, 3$ denotes the three colours of the quarks. Furthermore the $SU(5)$ GUT model contains 24 vector bosons corresponding to the 24 generators of the Lie group $SU(5)$, i.e. the gluons, the electroweak gauge bosons and 12 new coloured and charged gauge bosons called X and Y which are leptoquarks and diquarks. The spontaneous breaking of $SU(5)$ can be achieved in a two-step procedure. In a first step a **24** multiplet of scalar Higgs fields with masses $\mathcal{O}(M_U)$ breaks $SU(5)$ to the SM group $SU(3) \times SU(2) \times U(1)$. In a second step the SM group is broken to $SU(3) \times U(1)$ by a **5** multiplet of Higgs fields with masses $\mathcal{O}(m_Z)$.

The gauge bosons X have couplings of the form $X\ell q$ with leptons and quarks and can therefore induce proton decay, for example, $p \to \pi^0 e^+$. The mass of the vector bosons X has to be of the order $m_X \approx M_U$. The order of magnitude for the proton lifetime can be estimated as $\tau_p^{-1} \approx \alpha_U^2 m_p^2 / m_X^2$, where m_p is the proton mass. In the non-supersymmetric $SU(5)$ GUT model with $M_U \approx 10^{14}$ GeV one obtains $\tau_p \approx 10^{30}$ years, whereas the present experimental lower bound for the proton lifetime is $\tau_p > 1.9 \times 10^{33}$ years. In the supersymmetric $SU(5)$ GUT model the unification scale turns out to be $M_U \approx 10^{16}$ GeV. This leads to a larger value for the proton lifetime, which is in agreement with the experimental lower bound although the parameter space of the supersymmetric $SU(5)$ GUT is tightly constrained.[22]

The supersymmetric $SO(10)$ GUT model has a number of additional desirable features compared to $SU(5)$.[21,23] For example, the 15 helicity states of quarks and leptons together with a SM gauge singlet right-handed neutrino state leading to nonzero neutrino masses are included in one **16** representation of $SO(10)$. Furthermore the $SO(10)$ GUT model can solve the SUSY CP and R-parity problems because it is left-right symmetric. There are many ways to break $SO(10)$ down to the SM, details can be found in Ref. 21.

4. Supersymmetry

Supersymmetry (SUSY) is a new symmetry relating bosons and fermions. The particles combined in a SUSY multiplet have spins which differ by $\frac{1}{2}$. This is different from the symmetries of the SM or a GUT where all particles in a multiplet have the same spin (for an introduction to SUSY see e.g. Ref. 24).

SUSY is at present one of the most attractive and best studied extensions of the SM. The most important motivation for that is the fact that SUSY quantum field theories have in general better high–energy behaviour than non–SUSY ones. This is due to the cancellation of the divergent bosonic and fermionic contributions to the 1-loop radiative corrections. A particularly important example is the cancellation of the quadratic divergencies in the loop corrections to the Higgs mass. This cancellation mechanism provides one of the best ways we know to stabilize the mass of an elementary scalar Higgs field against radiative corrections and keep it "naturally" of the order $\mathcal{O}(m_Z)$.

Practically all SUSY modifications of the SM are based on local $N = 1$ SUSY. In the "minimal" SUSY extension of the SM a hypothetical SUSY partner is introduced for every known SM particle. The SUSY partners of the neutrinos, leptons, and quarks are called scalar neutrinos $\tilde{\nu}$, left and right scalar leptons $\tilde{\ell}_L$, $\tilde{\ell}_R$, and left and right scalar quarks $\tilde{q}_L$, $\tilde{q}_R$, respectively. They have spin 0. The SUSY partners of the gauge vector bosons have spin $\frac{1}{2}$ and are called gauginos. The photino $\tilde{\gamma}$, $W^{\pm}$-ino $\tilde{W}^{\pm}$, Z-ino $\tilde{Z}$, and gluino $\tilde{g}$ are the partners of γ, $W^{\pm}$, Z^0, and the gluon, respectively. In the local version of SUSY the graviton gets a spin–$\frac{3}{2}$ SUSY partner, called gravitino. Furthermore, at least two isodoublets of Higgs fields $H_i, i = 1, 2$, have to be introduced, together with their SUSY partners, the higgsinos $\tilde{H}_i, i = 1, 2$, which have spin $\frac{1}{2}$. In this way the anomalies in the triangular loops cancel. The model obtained in this way is the Minimal Supersymmetric Standard Model (MSSM).[25] In the "next-to-minimal" SUSY extension of the SM (NMSSM) an additional Higgs singlet and the corresponding higgsino are introduced (see for example Ref. 26 and References therein).

The gauginos and higgsinos form quantum mechanically mixed states. The charged and neutral mass eigenstates are the charginos $\tilde{\chi}_i^{\pm}$, $i = 1, 2$, and neutralinos $\tilde{\chi}_i^0$, $i = 1, \ldots 4$, respectively. The left and right states of the scalar fermions are also mixed, with a mixing term proportional to the corresponding fermion mass. Therefore, the mass eigenstates of the first and second generation scalar fermions are to a good approximation the left and right states. However, there may be strong left–right mixing in the sector of the scalar top and bottom quarks and the scalar tau lepton.

If SUSY was an exact symmetry, then the masses of the SUSY partners would be the same as those of the corresponding SM particles. This is evidently not observed in nature, therefore, SUSY must be broken. Essentially, the idea is to break local SUSY spontaneously at a high energy scale.[27] The result is the global SUSY Lagrangian plus additional "soft SUSY–breaking terms", which are mass terms for the SUSY partners, and additional trilinear coupling terms for the scalar fields.[19,20,25] Further assumptions are necessary to fix the additional soft–breaking parameters. For example, we can assume that at the GUT scale M_U all scalar SUSY partners have the same mass M_0, all gauginos have a common mass $M_{1/2}$, and all trilinear couplings of the scalar fields have a common strength A_0. We obtain their values at the weak scale by evolving them with the RGEs from $Q = M_U$ to $Q \approx M_Z$.[28] The model obtained in this way is called constrained MSSM (CMSSM) or minimal supergravity-inspired model (mSUGRA). In Fig. 3 we show an example where we plot the gaugino mass parameters M_1, M_2, M_3 as a function of the scale Q.

Radiative electroweak symmetry breaking is a further attractive feature of SUSY. This can be achieved by exploiting the logarithmic scale dependence of the squares of the masses of the Higgs fields H_1^0 and H_2^0. Starting at the scale $Q = M_U$ with the mass values $M_{H_1}^2 = M_{H_2}^2 = M_0^2$ and evolving to lower energies, it turns out that $M_{H_2}^2$ can become negative at $Q \approx M_Z$. The reason is that $M_{H_2}^2$ gets large negative contributions from the top–quark loops. In this way spontaneous breaking of the electroweak symmetry is induced. H_1^0 and H_2^0 get vev's $\langle H_i^0 \rangle = \frac{1}{\sqrt{2}} v_i, i = 1, 2$, and the vector bosons get masses $m_W^2 = \frac{1}{4} g^2 (v_1^2 + v_2^2)$ and $m_Z^2 = \frac{1}{4}(g^2 + g'^2)(v_1^2 + v_2^2)$. This mechanism works because the top–quark mass is much larger than the other quark masses (as one can show $m_t > 60$ GeV must be fulfilled). Furthermore, the mass difference between the SM particles and their SUSY partners must be less than about 1 TeV.

The Higgs sector of the MSSM contains five Higgs bosons, the CP–even h^0 and H^0, the CP–odd A^0, and a pair of charged ones, $H^\pm$.[30] An important prediction of the MSSM is that the mass of the lighter CP–even state h^0 is always $m_{h^0} < m_Z$ at tree–level. There are large radiative

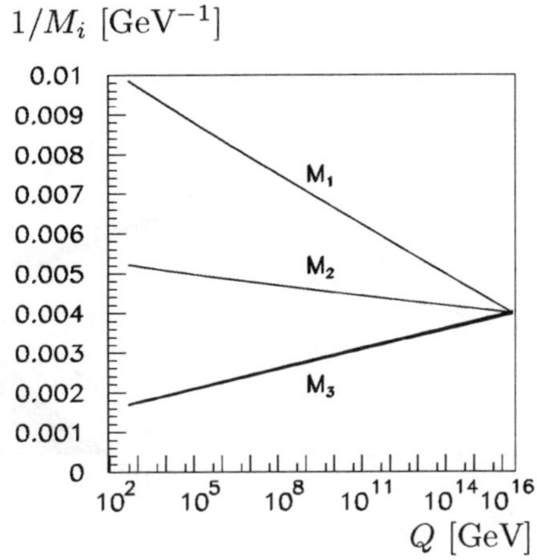

Figure 3. Evolution of the gaugino mass parameters M_i from low to high scales. From Ref. 29.

corrections which change this prediction to $m_{h^0} \lesssim 140$ GeV.[31,32] Comparing with the discussion in subsection 2.1 we see that this prediction for m_{h^0} lies within the allowed range for the Higgs boson mass obtained in the analysis of the electroweak precision data. We note in passing that some SUSY parameters may be complex and induce CP-violating effects, for example, mixing between the CP-odd A^0 and the CP-even h^0 and H^0.[32,33]

It turns out that the unification of the three gauge couplings works better in the MSSM than without SUSY.[28] We illustrate this in Fig. 4, where we plot the gauge couplings in the MSSM as a function of the energy scale. The evolution in the MSSM is different from that in the SM, because the RGEs of the MSSM contain also the contributions from the SUSY particles. In the MSSM the unification scale turns out to be of the order $M_U \approx 2 \times 10^{16}$ GeV, provided the masses of the SUSY particles are not much larger than approximately 1 TeV.

The experimental search for SUSY particles has a high priority at present colliders and will become even more important at LHC and the future e^+e^- linear collider ILC. In the discussion of the possible signatures one has to distinguish the two cases whether the multiplicative quantum number R-parity $R_P = (-1)^{3B+L-2S}$ is conserved or violated. SUSY particles have $R_P = -1$ and ordinary particles have $R_P = +1$. If R_P is conserved, then there exists a lightest SUSY particle (LSP) which is stable. Cosmological arguments suggest that it is neutral and only weakly inter-

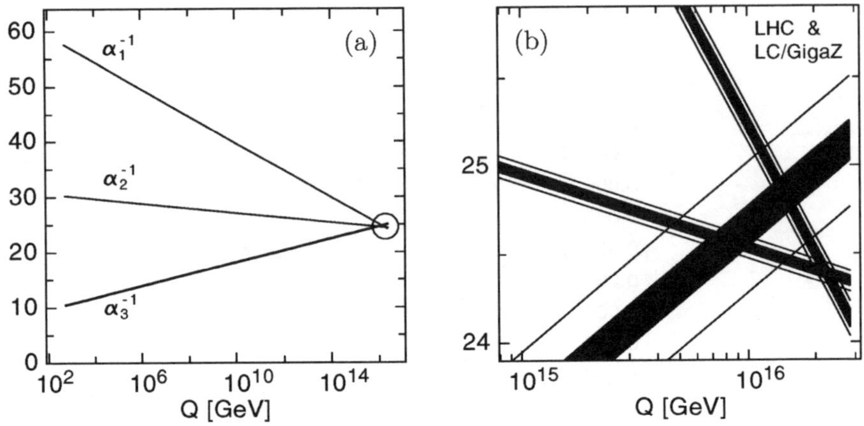

Figure 4. (a) Running of the inverse gauge couplings from low to high energies. (b) Expansion of the area around $Q = 10^{16}$ GeV. The wide error bands are based on present data, and the spectrum of supersymmetric particles from LHC measurements within mSUGRA. The narrow bands demonstrate the improvement expected by future GigaZ analyses and the measurement of the complete spectrum at "LHC+LC". From Ref. 29.

acting. It is an excellent candidate for dark matter. We assume that the lightest neutralino $\tilde{\chi}_1^0$ is the LSP, which in the CMSSM holds in most of the parameter space. In experiment the LSP behaves like a neutrino and its energy and momentum are not observable. Therefore, in the R_P conserving case the characteristic experimental signatures for SUSY particles are events with missing energy E_{miss} and missing momentum p_{miss}.

In the R_P violating case the SUSY Lagrangian contains additional terms which are allowed by SUSY and the gauge symmetry, but are lepton number violating and/or baryon number violating. Consequently, the LSP is not stable and decays into SM particles. Therefore, in the R_P violating case the E_{miss} and p_{miss} signature is in general not applicable. However, due to the decay of the LSP there are more leptons and/or jets in the final state. At an e^+e^- collider the main signature for R_P violation is, therefore, an enhanced rate of multi-lepton and/or multi-jet final states. If the mean decay length of the LSP is too large and it decays outside the detector, then its energy and momentum remain invisible and the E_{miss} and p_{miss} signature is again applicable. If the LSP decays within the detector and the decay length is long enough, then displaced vertices may occur, which then provide a further important observable for R_P violating SUSY. At a hadron collider the situation may be more involved. If the lepton number violating terms dominate over the baryon number violating ones, then the enhanced number of multi-lepton final states is again a good signature.

R_P violating SUSY can also provide a viable framework for non-vanishing neutrino masses and a quantitative description of the present data on neutrino oscillations. This is a very attractive feature of R_P violating SUSY, which in its bilinear formulation can be shortly described in the following way (for a review see Ref. 34): As lepton number is not conserved, the neutrinos mix with the neutralinos and the charged leptons mix with the charginos, where the amount of mixing depends on the R_P violating parameters. In bilinear R_P violating SUSY one neutrino gets a non-vanishing mass already at tree level, while the other two neutrinos get their masses at 1-loop level. In this way "small" neutrino masses are obtained and the data on neutrino oscillations can be quantitatively described. After fixing the R_P violating parameters by the solar and atmospheric neutrino data, the R_P violating decay widths of the SUSY particles can be predicted. This means that the low energy phenomena in the neutrino sector are linked to the SUSY particle sector, which we expect to probe at high energy colliders.

At LEP no supersymmetric particles have been found.[35] This implies lower mass bounds which are $m_{\tilde{\chi}_1^\pm} > 103.5$ GeV (for $m_{\tilde{\nu}_e} > 300$ GeV), $m_{\tilde{e}} > 99.9$ GeV, $m_{\tilde{\mu}} > 94.9$ GeV, $m_{\tilde{\tau}} > 86.6$ GeV, $m_{\tilde{t}} > 95$ GeV and $m_{\tilde{b}} > 94$ GeV. The limit on the mass of $\tilde{\chi}_1^0$ is model dependent. Within the CMSSM the non–observation of charginos and neutralinos excludes certain CMSSM parameter regions. From these follows the limit on the $\tilde{\chi}_1^0$ mass $m_{\tilde{\chi}_1^0} > 50.3$ GeV. The non–observation of Higgs bosons leads to the mass limits $m_{h^0} > 92.9$ GeV and $m_{A^0} > 93.3$ GeV in the CP-conserving MSSM with real parameters. In the CP-violating MSSM with complex parameters no universal lower bound for the masses of the neutral Higgs bosons can be defined.[36]

At the Tevatron the strong interaction processes of gluino and squark production, $p\bar{p} \to \tilde{g}\tilde{g}, \tilde{g}\tilde{q}, \tilde{q}\bar{\tilde{q}}$, are the SUSY reactions with the highest cross sections. Gluinos and squarks may have cascade decays which start with $\tilde{g} \to q\bar{q}\tilde{\chi}_i^0$, $q\bar{q}'\tilde{\chi}_i^\pm$, $\tilde{q} \to q\tilde{\chi}_i^0$, $\tilde{q} \to q'\tilde{\chi}_i^\pm$, and continue until the LSP $\tilde{\chi}_1^0$ is reached. Suitable kinematical cuts are necessary to distinguish a possible signal from the huge SM background. The present gluino and squark mass limits are $m_{\tilde{g}} \gtrsim 400$ GeV, and $m_{\tilde{q}} \gtrsim 250$ GeV if $m_{\tilde{q}} \approx m_{\tilde{g}}$, $m_{\tilde{q}} \gtrsim 200$ GeV if $m_{\tilde{g}} \approx 500$ GeV whereas for $m_{\tilde{g}} \gtrsim 560$ GeV no limit on the squark mass can be obtained from measurements at Tevatron.[37] For $\tilde{t}_1$ and $\tilde{b}_1$ the mass limits are different: $m_{\tilde{t}_1} \gtrsim 115$ GeV provided $m_{\tilde{\chi}_1^0} \lesssim 50$ GeV and $m_{\tilde{b}_1} \gtrsim 140$ GeV provided $m_{\tilde{\chi}_1^0} \lesssim 70$ GeV, respectively.[35] Another interesting SUSY reaction which can be studied at the hadron colliders is $p\bar{p} \to \tilde{\chi}_1^\pm \tilde{\chi}_2^0$. It leads to the very clean signature $3\ell + p_{\rm miss}$, $\ell = e, \mu$.[38] The Tevatron mass limit for $\tilde{\chi}_1^\pm$, following from the non-observation of this reaction, is close to the LEP limit. At the upgraded Tevatron the expected SUSY mass reach will be $m_{\tilde{g}} \approx m_{\tilde{q}} \approx 390$ GeV, $m_{\tilde{t}_1} \approx 180$ GeV, $m_{\tilde{\chi}_1^\pm} \approx 250$ GeV, for an integrated luminosity of 2 fb^{-1}.

At LHC gluinos and squarks will be detectable up to masses of approximately $1-2$ TeV, as is illustrated in Fig. 5. The cascade decays of these particles will play an important role.[39] On the one hand they will give rise to characteristic signatures, for example the same–sign dilepton signature of gluinos. On the other hand, in the cascade decays the weakly interacting charginos and neutralinos will appear whose properties can also be studied. If weak–scale SUSY is not found at the Tevatron, then the LHC is the collider where it will be either discovered or definitely disproved.

The reach in the mSUGRA parameter space of an e^+e^- linear collider with $\sqrt{s} = 0.5$ to 1 TeV will be be somewhat smaller than that of the LHC (Fig. 5 (b)). However, due to the high luminosity and good energy resolution expected an e^+e^- linear collider will be inevitable for precision measurements, especially in the neutralino and chargino sectors. This will enable us to determine very precisely the SUSY parameters and to reconstruct the underlying theory.[6,28,42] However, the signatures will be more complicated than those at LEP, because also the heavier SUSY particles will be produced which have cascade decays. This will lead to characteristic events with several leptons and/or jets, and missing energy and momentum.

Inspecting again Fig. 1 it can be seen that already the precision data obtained at the GigaZ mode of the linear collider (small ellipse) will presumably allow us to discriminate between the SM and the MSSM or another extension of the SM. The present experimental errors (large ellipse) do not allow to discriminate between the two models. In this figure the MSSM band is obtained by varying the SUSY parameters in the range allowed by the experimental and theoretical constraints. There is a small overlap of the SM and MSSM bands (small intermediate band) for a light Higgs boson ($m_h = 113$ GeV) and a heavy SUSY spectrum.

5. Extra dimensions

A solution to the hierarchy problem can in principle be obtained by formulating gravity in $4+\delta$ dimensions, where $\delta = 1, 2, 3, \ldots$ are the so-called "extra" dimensions,[43,44] which are assumed to be compactified with a ra-

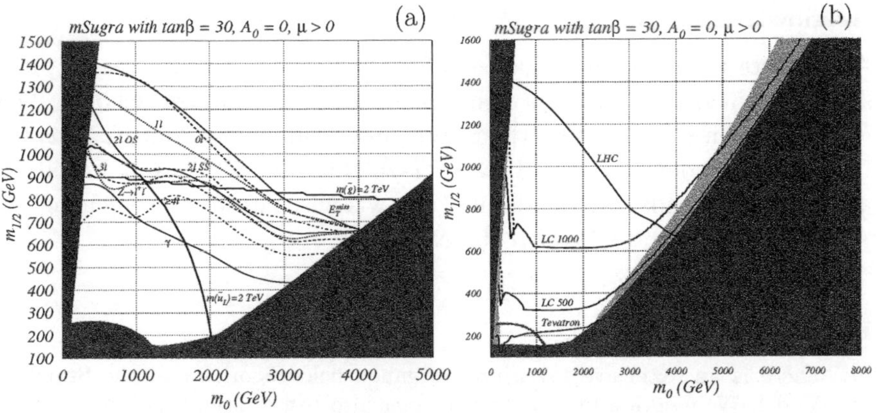

Figure 5. (a) The reach of the LHC for various production channels of SUSY particles in the the mSUGRA model for $\tan\beta = 30$, $A_0 = 0$ and $\mu > 0$, assuming 100 fb^{-1} of integrated luminosity. The shaded region is excluded by theoretical and experimental constraints. From Ref. 40. (b) Reach of an e^+e^- linear collider with $\sqrt{s} = 0.5$ and 1 TeV in the mSUGRA model for $\tan\beta = 30$, $A_0 = 0$ and $\mu > 0$. For comparison the reach of the Tevatron assuming 10 fb^{-1} of integrated luminosity (for isolated trileptons) and the reach of the LHC (in the "inclusive" $\not{E}_T$ channel) assuming 100 fb^{-1} of integrated luminosity is shown. The dark shaded region is excluded by theoretical and experimental constraints. The light shaded region shows points where the relic density $\Omega h^2 < 0.129$ as preferred by WMAP. From Ref. 41.

dius R (for a review see Ref. 45). In the model of Ref. 43 it is assumed that SM physics is restricted to the 4-dimensional brane, whereas gravity acts in the $4 + \delta$ dimensional bulk. In 4-dimensional space-time the Planck mass is $M_{\rm Pl} = 1.2 \cdot 10^{19}$ GeV. In the $(4 + \delta)$-dimensional space the corresponding Planck mass M_D is given by $M_D^{2+\delta} = M_{\rm Pl}^2/R^\delta$. Assuming further that the compactification radius R is many orders of magnitude larger than the Planck length, $R \gg M_{\rm Pl}^{-1}$, R and δ may be adjusted such that $M_D \approx \mathcal{O}(1$ TeV$)$. In this way the Planck scale is close to the electroweak scale and there is no hierarchy problem.

As a consequence of the compactification Kaluza-Klein towers of the gravitons can be excited. This leads to two possible signatures at an e^+e^- linear collider. The first one is $e^+e^- \to \gamma/Z + G_n$ where G_n means the graviton and its Kaluza-Klein excitations, which appear as missing energy in the detector. The main background to this process is $e^+e^- \to \nu\bar{\nu}\gamma$, which strongly depends on the e^- beam polarisation. The second signature is due to graviton exchange in $e^+e^- \to f\bar{f}$, which leads to a modification of cross sections and asymmetries compared to the SM prediction.

Acknowledgements

A.B. wants to thank the organizers for inviting him to give the lectures, and for providing a pleasant atmosphere at their summer school. This work has been supported by the European Community's Human Potential Programme under contract HPRN-CT-2000-00149 "Physics at Colliders" and by the "Fonds zur Förderung der wissenschaftlichen Forschung" of Austria, FWF Project No. P16592-N02.

References

1. There is an extensive literature on gauge field theories and the Standard Model. We mention two books and refer also to the literature cited therein: O. Nachtmann, Elementary Particle Physics, Springer–Verlag (1990); M. E. Peskin and D. V. Schroeder, An Introduction to Quantum Field Theory, Addison-Wesley (1995).
2. S. Petcov, lecture at the *VII School on Non-Accelerator Astroparticle Physics,* July 26 – August 6, 2004, Trieste, Italy.
3. G. Giacomelli, lecture at the *VII School on Non-Accelerator Astroparticle Physics,* July 26 – August 6, 2004, Trieste, Italy.
4. R. Barate *et al.* [The LEP Collaborations ALEPH, DELPHI, L3, OPAL and the LEP Electroweak Working Group], Phys. Lett. B **565** (2003) 61 [arXiv:hep-ex/0306033].
5. J. G. Branson, D. Denegri, I. Hinchliffe, F. Gianotti, F. E. Paige and P. Sphicas [ATLAS and CMS Collaborations], Eur. Phys. J. directC **4** (2002) N1.
6. E. Accomando *et al.* [ECFA/DESY LC Physics Working Group Collaboration], Phys. Rept. **299** (1998) 1 [arXiv:hep-ph/9705442]; T. Abe *et al.* [American Linear Collider Working Group Collaboration], in *Proc. of the APS/DPF/DPB Summer Study on the Future of Particle Physics (Snowmass 2001)* ed. N. Graf, arXiv:hep-ex/0106056; J. A. Aguilar-Saavedra *et al.* [ECFA/DESY LC Physics Working Group Collaboration], arXiv:hep-ph/0106315; K. Abe *et al.* [ACFA Linear Collider Working Group Collaboration], arXiv:hep-ph/0109166.
7. W. Hollik *et al.*, Acta Phys. Polon. B **35** (2004) 2533 and References therein.
8. S. Heinemeyer, W. Hollik and G. Weiglein, arXiv:hep-ph/0412214.
9. J. F. Gunion, H. E. Haber, G. L. Kane, and S. Dawson, The Higgs Hunter's Guide, Addison-Wesley (1990).
10. S. Eidelman *et al.* [Particle Data Group Collaboration], Phys. Lett. B **592** (2004) 1, see `http://pdg.lbl.gov/`.
11. The LEP Collaborations ALEPH, DELPHI, L3, OPAL, the LEP Electroweak Working Group, the SLD Electroweak and Heavy Flavour Groups, arXiv:hep-ex/0312023.
12. F. Jegerlehner, arXiv:hep-ph/0312372.
13. M. J. G. Veltman, Nucl. Phys. B **123** (1977) 89.

14. G. Altarelli, arXiv:hep-ph/0406270.
15. G. Weiglein, talk at the *International Conference on Linear Colliders, LCWS 04,* April 19–22, 2004;
    S. Heinemeyer and G. Weiglein, arXiv:hep-ph/0012364.
16. The LEP Electroweak Working Group,
    `http://lepewwg.web.cern.ch/LEPEWWG/`.
17. A. Djouadi, Pramana **60** (2003) 215 [arXiv:hep-ph/0205248].
18. J. Gunion *et al.*, Proc. of the 1996 DPF/DPB Summer Study on High–Energy Physics, Snowmass, Colorado, Vol. 2, p. 541, eds. D. G. Cassel, L. Trindle Gennari, R. H. Siemann.
19. G. G. Ross, Grand Unified Theories, Addison–Wesley (1985).
20. R. N. Mohapatra, Unification and Supersymmetry, Springer–Verlag (1986).
21. R. N. Mohapatra, arXiv:hep-ph/9911272.
22. B. Bajc, P. Fileviez Perez and G. Senjanovic, Phys. Rev. D **66** (2002) 075005 [arXiv:hep-ph/0204311].
23. H. Baer, M. Brhlik, M. A. Diaz, J. Ferrandis, P. Mercadante, P. Quintana and X. Tata, Phys. Rev. D **63** (2001) 015007 [arXiv:hep-ph/0005027].
24. J. Wess and J. Bagger, Supersymmetry and Supergravity, Princeton University Press (1983);
    D. Bailin and A. Love, Supersymmetric Gauge Field Theory and String Theory, IOP Publishing (1994).
25. H. P. Nilles, Phys. Rept. **110** (1984) 1;
    H. E. Haber and G. L. Kane, Phys. Rept. **117** (1985) 75.
26. F. Franke and H. Fraas, Int. J. Mod. Phys. A **12** (1997) 479 [arXiv:hep-ph/9512366];
    U. Ellwanger, J. F. Gunion and C. Hugonie, arXiv:hep-ph/0406215.
27. R. Arnowitt, A. Chamseddine, and P. Nath, Applied N=1 Supergravity, World Scientific (1984).
28. G. A. Blair, W. Porod and P. M. Zerwas, Eur. Phys. J. C **27** (2003) 263 [arXiv:hep-ph/0210058].
29. B. C. Allanach *et al.*, arXiv:hep-ph/0407067.
30. J. F. Gunion and H. E. Haber, Nucl. Phys. B **272** (1986) 1 [Erratum-ibid. B **402** (1993) 567].
31. M. Carena and H. E. Haber, Prog. Part. Nucl. Phys. **50** (2003) 63 [arXiv:hep-ph/0208209].
32. S. Heinemeyer, arXiv:hep-ph/0407244.
33. M. Carena, J. R. Ellis, A. Pilaftsis and C. E. M. Wagner, Nucl. Phys. B **586** (2000) 92 [arXiv:hep-ph/0003180].
34. M. Hirsch and J. W. F. Valle, New J. Phys. **6** (2004) 76 [arXiv:hep-ph/0405015].
35. The LEP SUSY Working Group, ALEPH, DELPHI, L3 and OPAL Collaborations, `http://lepsusy.web.cern.ch/lepsusy/Welcome.html`.
36. The LEP Working Group for Higgs Boson Searches, ALEPH, DELPHI, L3 and OPAL Collaborations,
    `http://lephiggs.web.cern.ch/LEPHIGGS/www/Welcome.html`.
37. A. Meyer, talk at the *32nd International Conference on High Energy Physics,*

ICHEP'04, August 16–22, 2004, Beijing, China.
38. H. Baer and X. Tata, Phys. Rev. D **47** (1993) 2739.
39. H. Baer, C. h. Chen, F. Paige and X. Tata, Phys. Rev. D **49** (1994) 3283 [arXiv:hep-ph/9311248]; Phys. Rev. D **50** (1994) 4508 [arXiv:hep-ph/9404212];
D. Denegri, W. Majerotto and L. Rurua, Phys. Rev. D **58** (1998) 095010 [arXiv:hep-ph/9711357];
A. Bartl et al. Proc. of the 1996 DPF/DPB Summer Study on High–Energy Physics, Snowmass, Colorado, Vol. 2, p. 693, eds. D. G. Cassel, L. Trindle Gennari, R. H. Siemann.
40. H. Baer, C. Balazs, A. Belyaev, T. Krupovnickas and X. Tata, JHEP **0306** (2003) 054 [arXiv:hep-ph/0304303].
41. H. Baer, A. Belyaev, T. Krupovnickas and X. Tata, JHEP **0402** (2004) 007 [arXiv:hep-ph/0311351].
42. G. Weiglein *et al.* [The LHC / LC Study Group], arXiv:hep-ph/0410364.
43. N. Arkani-Hamed, S. Dimopoulos and G. R. Dvali, Phys. Lett. B **429** (1998) 263 [arXiv:hep-ph/9803315]; Phys. Rev. D **59** (1999) 086004 [arXiv:hep-ph/9807344];
I. Antoniadis, N. Arkani-Hamed, S. Dimopoulos and G. R. Dvali, Phys. Lett. B **436** (1998) 257 [arXiv:hep-ph/9804398].
44. L. Randall and R. Sundrum, Phys. Rev. Lett. **83** (1999) 3370 [arXiv:hep-ph/9905221].
45. T. G. Rizzo, arXiv:hep-ph/0409309 and References therein.

NEUTRINO MASSES, MIXING AND OSCILLATIONS

S.T. PETCOV

SISSA/INFN - Sezione di Trieste, I-34014 Trieste, Italy *

There exist at present compelling experimental evidences for oscillations of solar, atmospheric and reactor neutrinos. They imply the existence of 3-neutrino mixing in vacuum. We review the theory of neutrino oscillations, the phenomenology of 3-neutrino mixing, and the current data on the 3-neutrino mixing parameters. The opened questions and the main goals of future research in the field of neutrino mixing and oscillations are outlined.

1. Introduction

The hypothesis of neutrino oscillations was formulated in [1]. In [2] it was suggested that the solar ν_e can take part in oscillations involving another active or sterile neutrino. The evidences of solar neutrino ($\nu_{\odot}$-) oscillations obtained first in the Homestake experiment and strengthened by the results of Kamiokande, SAGE and GALLEX/GNO experiments [3,4], were made compelling in the last several years by the data of Super-Kamiokande (SK), SNO and KamLAND (KL) experiments [5,6,7]. Under the plausible assumption of CPT-invariance, the results of the KL reactor neutrino experiment [7] established the large mixing angle (LMA) MSW oscillations/transitions [8,9] as the dominant mechanism at the origin of the observed solar ν_e deficit. The Kamiokande experiment [4] provided the first evidences for oscillations of atmospheric ν_μ and $\bar{\nu}_\mu$, while the data of the Super-Kamiokande experiment made the case of atmospheric neutrino oscillations convincing [10,11]. Evidences for oscillations of neutrinos were obtained also in the first long baseline accelerator neutrino experiment K2K [12]. Indications for ν-oscillations were reported by the LSND collaboration [13].

The recent new SK data on the L/E-dependence of multi-GeV μ-like atmospheric neutrino events [11], L and E being the distance traveled by neutrinos and the ν energy, and the new spectrum data of KL and K2K experiments [14,15] are the latest significant contributions to the remarkable

*Also at: Institute of Nuclear Research and Nuclear Energy, Bulgarian Academy of Sciences, BG–1784 Sofia, Bulgaria.

progress made in the last several years in the studies of ν-oscillations. For the first time the data exhibit directly the effects of the oscillatory dependence on L/E and E of the probabilities of ν-oscillations in vacuum [16]. We begin to "see" the oscillations of neutrinos. As a result of these magnificent developments, the oscillations of solar ν_e, atmospheric ν_μ and $\bar{\nu}_\mu$, accelerator ν_μ (at $L \sim$250 km) and reactor $\bar{\nu}_e$ (at $L \sim$180 km), driven by nonzero ν-masses and ν-mixing, can be considered as practically established.

The neutrino oscillation data imply the existence of 3-neutrino mixing in vacuum. In the present lectures we review the theory of neutrino oscillations, the phenomenology of 3-ν mixing, and the current data on the 3-ν mixing parameters. We discuss also the opened questions and the main goals of future research in the field of neutrino mixing and oscillations.

2. Neutrino Oscillations in Vacuum

We shall consider first the simplest possibility of two-neutrino oscillation in vacuum (see, e.g., [17,18,19]). Let us assume that the state vector of the electron neutrino, $|\nu_e\rangle$, produced in vacuum with momentum $\vec{p}$ in some weak interaction process, is a coherent superposition of the state vectors $|\nu_i\rangle$ of two neutrinos ν_i, i=1,2, having the same momentum $\vec{p}$ and definite masses in vacuum, $\mathrm{m_i}$, $\mathrm{m_1} \neq \mathrm{m_2}$, while the linear combination of $|\nu_1\rangle$ and $|\nu_2\rangle$, which is orthogonal to $|\nu_e\rangle$, represents the state vector $|\nu_x\rangle$ of another weak-eigenstate neutrino, $|\nu_x\rangle = |\nu_{\mu(\tau)}\rangle$ or $|\nu_s\rangle$, ν_s being a sterile neutrino:

$$\begin{aligned} |\nu_e\rangle &= |\nu_1\rangle \cos\theta + |\nu_2\rangle \sin\theta \ , \\ |\nu_\mu\rangle &= -|\nu_1\rangle \sin\theta + |\nu_2\rangle \cos\theta \ , \end{aligned} \tag{1}$$

where θ is the neutrino mixing angle in vacuum and we have chosen (for concreteness) $\nu_x \equiv \nu_\mu$. Obviously, $|\nu_{1,2}\rangle$ are eigenstates of the Hamiltonian of the ν-system in vacuum, H_0:

$$H_0 \ |\nu_i\rangle \ = E_i \ |\nu_i\rangle, \quad E_i = \sqrt{\vec{p}^{\,2} + m_i^2}, \ i = 1, 2. \tag{2}$$

If ν_e is produced at time $t = 0$ in the state given by (1), after a time t the latter will evolve (in vacuum) into the state

$$|\nu_e(t)\rangle = e^{-iE_1 t} \ |\nu_1\rangle \cos\theta + e^{-iE_2 t} \ |\nu_2\rangle \sin\theta = A_{ee}(t) \ |\nu_e\rangle + A_{\mu e}(t) \ |\nu_\mu\rangle \ , \tag{3}$$

where we have ignored the overall space coordinate dependent factor $\exp(i\,\vec{p}\vec{r})$ in the right-hand side of (3) and used (1). Here

$$A_{ee} = e^{-iE_1 t} \cos^2\theta + e^{-iE_2 t} \sin^2\theta, \ A_{\mu e} = \frac{1}{2} \sin 2\theta (e^{-iE_2 t} - e^{-iE_1 t}) \tag{4}$$

are the probability amplitudes to find respectively ν_e and ν_μ at time t of the evolution of the ν-system if neutrino ν_e has been produced at time $t = 0$. Thus, if $m_1 \neq m_2$ and if neutrino mixing exists in vacuum, $\theta \neq n\pi/2$, $n = 0, 1, 2, ...$, we have $|A_{\mu e}(t)|^2 \neq 0$ and transitions in flight between ν_e and ν_μ are possible. Assuming that ν_1 and ν_2 are stable and relativistic, we obtain from (4) the probabilities that a ν_e will not change into ν_μ, $P(\nu_e \to \nu_e)$, or will transform into ν_μ, $P(\nu_e \to \nu_\mu)$:

$$
\begin{aligned}
P(\nu_e \to \nu_e; t) = |A_{ee}(t)|^2 = 1 - \tfrac{1}{2}\sin^2 2\theta \left(1 - \cos 2\pi \tfrac{L}{L_v}\right) , \\
P(\nu_e \to \nu_\mu; t) = |A_{\mu e}(t)|^2 = \tfrac{1}{2}\sin^2 2\theta \left(1 - \cos 2\pi \tfrac{L}{L_v}\right) ,
\end{aligned}
\tag{5}
$$

where $\Delta m^2 = m_2^2 - m_1^2$, $L \cong t$ is the distance traveled by neutrinos and

$$
L_v = 4\pi \frac{E}{\Delta m^2} \cong 2.48 \text{ m } \frac{E[MeV]}{\Delta m^2[eV^2]}
\tag{6}
$$

is the oscillation length in vacuum. In deriving (5) and we have used the equality $E_2 - E_1 \cong E + \Delta m^2/(2E)$, $E \cong |\vec{p}|$, valid for relativistic $\nu_{1,2}$. The quantities Δm^2 and $\sin^2 2\theta$ are typically considered as free parameters to be determined by the analysis of the neutrino oscillation data.

It should be clear from the above discussion that the neutrino oscillations are a purely quantum mechanical phenomenon. The requirements of coherence between the states $|\nu_1\rangle$ and $|\nu_2\rangle$ in the superposition (1) representing the ν_e (or $\nu_{\mu(\tau)}$) at the production point, and that the coherence be maintained during the evolution of the neutrino system up to the moment of neutrino detection, are crucial for the neutrino oscillations to occur. The subtleties and the implications of the coherence condition for neutrino oscillations continue to be discussed (see, e.g., [17,20,21]).

It follows from CPT-invariance, which we will assume to hold, that

$$
P(\nu_e \to \nu_e; t) = P(\bar{\nu}_e \to \bar{\nu}_e; t) \; , \quad P(\nu_e \to \nu_\mu; t) = P(\bar{\nu}_\mu \to \bar{\nu}_e; t) \; . \tag{7}
$$

Combined with the probability conservation, $P(\nu_e \to \nu_e; t) + P(\nu_e \to \nu_\mu; t) = 1$, $P(\bar{\nu}_e \to \bar{\nu}_e; t) + P(\bar{\nu}_e \to \bar{\nu}_\mu; t) = 1$, eq. (7) implies that in the simple case of two-neutrino oscillations we are considering one has

$$
P(\nu_e \to \nu_\mu; t) = P(\bar{\nu}_e \to \bar{\nu}_\mu; t) = P(\nu_\mu \to \nu_e; t) = P(\bar{\nu}_\mu \to \bar{\nu}_e; t) \; . \tag{8}
$$

As it follows from (5), $P(\nu_e \to \nu_\mu; t)$ depends on two factors: on $(1 - \cos 2\pi L/L_v)$, which exhibits oscillatory dependence on the distance L and on the ν energy E (hence the name "neutrino oscillations"), and on $\sin^2 2\theta$ which determines the amplitude of the oscillations. In order to have $P(\nu_e \to \nu_\mu; t) \cong 1$, two conditions have to be fulfilled: the neutrino

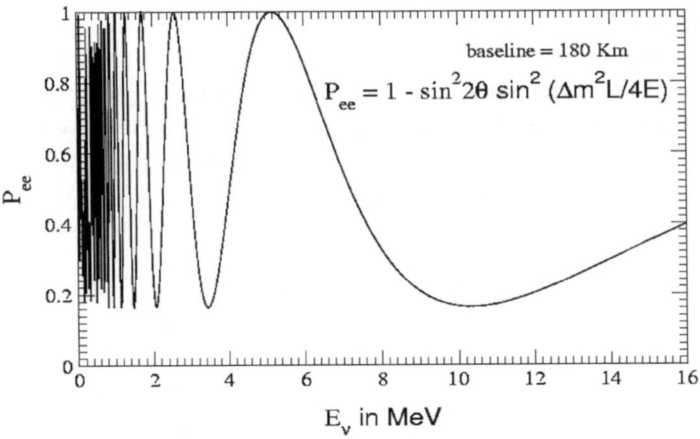

Figure 1. The probability of ν_e ($\bar{\nu}_e$) survival, $P(\nu_e \to \nu_e; t) = P(\bar{\nu}_e \to \bar{\nu}_e; t)$, as a function of the neutrino energy for $L = 180$ km and $\Delta m^2 = 8.0 \times 10^{-5}$ eV^2 (from [22]).

mixing in vacuum must be large, $\sin^2 2\theta \cong 1$, and the oscillation length in vacuum L_v has to be of the order of or smaller than the distance traveled by the neutrinos, $L_v \lesssim 2\pi L$. If $L_v \gg 2\pi L$, the oscillations do not have enough time to develop on the way to the neutrino detector and one has $P(\nu_e \to \nu_\mu; t) \cong 0$. This is illustrated in Fig. 1 showing the dependence of the probability $P(\nu_e \to \nu_e; t) = P(\bar{\nu}_e \to \bar{\nu}_e; t)$ on the neutrino energy.

A given experiment searching for ν-oscillations, is specified, in particular, by the average energy of the neutrinos being studied, $\bar{E}$, and by the distance traveled by the neutrinos to the detector L. The requirement $L_v \lesssim 2\pi L$ determines the minimal value of Δm^2 to which the experiment is sensitive (figure of merit of the experiment): $\min(\Delta m^2) \sim 2\bar{E}/L$. Because of the interference nature of ν- oscillations, the ν- oscillation experiments can probe, in general, rather small values of Δm^2 (see, e.g., [17,18]). Values of $\min(\Delta m^2)$, characterizing qualitatively the sensitivity of different experiments are given in Table 1. They correspond to the reactor experiments CHOOZ ($L \sim 1$ km) and KamLAND ($L \sim 100$ km), to accelerator experiments - past ($L \sim 1$ km), current and future (K2K, MINOS, OPERA)), to Super-Kamiokande experiment studying atmospheric and solar neutrino oscillations, and to the solar neutrino experiments. Due to the large Sun - Earth distance the relatively low energies of the solar ν_e, the experiments with solar neutrinos have a remarkable sensitivity to Δm^2.

In certain cases the dimensions of the neutrino source, ΔR, are not negligible in comparison with the oscillation length. Similarly, when analyzing

Source	Type of ν	$\bar{E}$[MeV]	L[km]	min(Δm^2)[eV^2]
Reactor	$\bar{\nu}_e$	~ 1	1	$\sim 10^{-3}$
Reactor	$\bar{\nu}_e$	$\sim$1	100	$\sim 10^{-5}$
Accelerator	$\nu_\mu, \bar{\nu}_\mu$	$\sim 10^3$	1	~ 1
Accelerator	$\nu_\mu, \bar{\nu}_\mu$	$\sim 10^3$	1000	$\sim 10^{-3}$
Atmospheric ν's	$\nu_{\mu,e}, \bar{\nu}_{\mu,e}$	$\sim 10^3$	10^4	$\sim 10^{-4}$
Sun	ν_e	~ 1	1.5×10^8	$\sim 10^{-11}$

neutrino oscillation data one has to include the energy resolution of the detector, ΔE, etc. in the analysis. As can be shown [18], if $2\pi\Delta R/(L_v) \gg 1$, and/or $L\Delta m^2 \Delta E/(E^2) \gg 1$, the oscillating term in the neutrino oscillation probability will be strongly suppressed. In this case the effects of ν-oscillations will be effectively determined by the average probabilities:

$$\bar{P}(\nu_e \to \nu_e) \cong 1 - \frac{1}{2}\sin^2 2\theta \ , \quad \bar{P}(\nu_e \to \nu_\mu) \cong \frac{1}{2}\sin^2 2\theta \ . \tag{9}$$

As we have seen, if (1) is realized and $\Delta m^2 L/(2E) \gtrsim 1$ for reactor $\bar{\nu}_e$, for instance, they can take part in vacuum oscillations on the way to the detector (see eqs. (8) and (7). In this case the flavour content of the $\bar{\nu}_e$ state vector will change periodically on the way to the detector due to the different time evolution of the vector's massive neutrino components. If $\sin^2 2\theta$ is sufficiently large, the neutrinos that are being detected at distance L will be in states representing, in general, certain superpositions of the states of [a] $\bar{\nu}_e$ and $\bar{\nu}_\mu$. The reactor $\bar{\nu}_e$ are detected through the charged current (CC) reaction $\bar{\nu}_e + p \to e^+ + n$. Obviously, the $\bar{\nu}_\mu$ component of the state being detected will not give a contribution to the signal in the detector. As a result, the measured signal in the reactor $\bar{\nu}_e$ oscillation experiment should be noticeably smaller than the predicted one in the absence of oscillations. This is what is observed in the KamLAND experiment [7,14]. Similar considerations apply to the case of mixing and oscillations between ν_μ ($\bar{\nu}_\mu$) and ν_τ ($\bar{\nu}_\tau$), which is relevant for the interpretation of the Super-Kamiokande atmospheric neutrino data in terms of ν-oscillations [10], etc.

3. Matter-Enhanced Transitions

The presence of matter can drastically change the pattern of neutrino oscillations: neutrinos can interact with the particles forming the matter.

[a] Obviously, if ν_e mixes with ν_μ and/or ν_τ, these states will be superpositions of the states of $\bar{\nu}_\mu$ and/or $\bar{\nu}_\tau$.

Accordingly, the Hamiltonian of the neutrino system in matter differs from the Hamiltonian of the neutrino system in vacuum H_0,

$$H_m = H_0 + H_{int} \; , \tag{10}$$

where H_{int} describes the interaction of neutrinos with the particles of matter. When, e.g., ν_e propagate in matter, they can scatter (due to the H_{int}) on the electrons (e^-), protons (p) and neutrons (n) present in matter. The incoherent elastic and the quasi-elastic scattering, in which the states of the initial particles change in the process (destroying the coherence between the neutrino states), are not of interest - they have a negligible effect on the solar neutrino propagation in the Sun and on the solar, atmospheric and reactor neutrino propagation in the Earth [b] : even in the center of the Sun, where the matter density is relatively high (~ 150 g/cm^3), an ν_e with energy of 1 MeV has a mean free path with respect to the indicated scattering processes, which exceeds 10^{10} km (recall that the solar radius is much smaller: $R_\odot = 6.96 \times 10^5$ km). The oscillating ν_e and ν_μ can scatter also elastically in the forward direction on the e^-, p and n, with the momenta and the spin states of the particles remaining unchanged. In such a process the coherence of the neutrino states is being preserved.

The ν_e and ν_μ coherent elastic scattering on the particles of matter generates nontrivial indices of refraction of the ν_e and ν_μ in matter [8]: $\kappa(\nu_e) \neq 1$, $\kappa(\nu_\mu) \neq 1$. Most importantly, we have $\kappa(\nu_e) \neq \kappa(\nu_\mu)$. The difference $\kappa(\nu_e) - \kappa(\nu_\mu)$ is determined essentially by the difference of the real parts of the forward $\nu_e - e^-$ and $\nu_\mu - e^-$ elastic scattering amplitudes [8] [c] and can be calculated in the Standard Theory. One finds [8,25,26]:

$$\kappa(\nu_e) - \kappa(\nu_\mu) = -\,\frac{1}{p}\sqrt{2}G_F N_e \; , \tag{11}$$

where G_F is the Fermi constant and N_e is the e^- number density in matter. Knowing $\kappa(\nu_e) - \kappa(\nu_\mu)$, it is possible to write the system of evolution equations which describes the $\nu_e \leftrightarrow \nu_\mu$ oscillations in matter [8,25,26]:

$$i\frac{d}{dt}\begin{pmatrix} A_e(t,t_0) \\ A_\mu(t,t_0) \end{pmatrix} = \begin{pmatrix} -\epsilon(t) & \epsilon' \\ \epsilon' & \epsilon(t) \end{pmatrix} \begin{pmatrix} A_e(t,t_0) \\ A_\mu(t,t_0) \end{pmatrix} \tag{12}$$

[b] These processes are important, however, for the supernova neutrinos (see, e.g., [23]).

[c] We standardly assume that the weak interaction of the flavour neutrinos ν_e, ν_μ and ν_τ and antineutrinos $\bar{\nu}_e$, $\bar{\nu}_\mu$ and $\bar{\nu}_\tau$ is described by the standard (Glashow-Salam-Weinberg) theory of electroweak interaction (for an alternative possibility see, e.g., [24]). Let us add that the imaginary parts of the forward scattering amplitudes (responsible, in particular, for decoherence effects) are proportional to the corresponding total scattering cross-sections and in the case of interest are negligible in comparison with the real parts.

where $A_e(t,t_0)$ ($A_\mu(t,t_0)$) is the amplitude of the probability to find neutrino ν_e (ν_μ) at time t of the evolution of the neutrino system if at time t_0 the neutrino ν_e or ν_μ has been produced, $t \geq t_0$, and

$$\epsilon(t) = \frac{1}{2}\,[\frac{\Delta m^2}{2E}\cos 2\theta - \sqrt{2}G_F N_e(t)],\ \epsilon' = \frac{\Delta m^2}{4E}\sin 2\theta. \qquad (13)$$

The term $\sqrt{2}G_F N_e(t)$ in the parameter $\epsilon(t)$ accounts for the effects of matter on neutrino oscillations. The system of evolution equations describing the oscillations of antineutrinos $\bar{\nu}_e \leftrightarrow \bar{\nu}_\mu$ in matter has exactly the same form except for the matter term in $\epsilon(t)$ which changes sign.

Consider first the case of $\nu_e \leftrightarrow \nu_\mu$ oscillations in matter with constant density: $N_e(t) = N_e = const.$ Due to the interaction term H_{int} in H_m, the eigenstates of the Hamiltonian of the neutrino system in vacuum, $|\nu_1\rangle$ and $|\nu_2\rangle$, are not eigenstates of H_m. It proves convenient to find the states $|\nu^m_{1,2}\rangle$, which diagonalize the evolution matrix in the r.h.s. of the system (12) or equivalently, the Hamiltonian H_m. We have:

$$\begin{aligned} |\nu_e\rangle &= \ \ |\nu_1^m\rangle\cos\theta_m + |\nu_2^m\rangle\sin\theta_m\ , \\ |\nu_\mu\rangle &= -|\nu_1^m\rangle\sin\theta_m\ + |\nu_2^m\rangle\cos\theta_m\ . \end{aligned} \qquad (14)$$

Here θ_m is the neutrino mixing angle in matter [8],

$$\sin 2\theta_m = \frac{\epsilon'}{\sqrt{\epsilon^2+\epsilon'^2}} = \frac{\tan 2\theta}{\sqrt{(1-\frac{N_e}{N_e^{res}})^2 + \tan^2 2\theta}}, \qquad (15)$$

where the quantity

$$N_e^{res} = \frac{\Delta m^2\cos 2\theta}{2E\sqrt{2}G_F} \qquad (16)$$

is called "resonance density" [25]. The matter-eigenstates $|\nu^m_{1,2}\rangle$ (which are also called "adiabatic") have energies $E^m_{1,2}$ whose difference is given by

$$E_2^m - E_1^m = 2\sqrt{\epsilon^2+\epsilon'^2} = \frac{\Delta m^2}{2E}\ \left((1-\frac{N_e}{N_e^{res}})^2\cos^2 2\theta + \sin^2 2\theta\right)^{\frac{1}{2}}. \qquad (17)$$

It should be clear from (14) and (17)) that the probability of $\nu_e \to \nu_\mu$ transition in matter with $N_e = const.$ is given by [9]

$$P_m(\nu_e \to \nu_\mu; t) = |A_\mu(t)|^2 = \frac{1}{2}\sin^2 2\theta_m\ [1 - \cos 2\pi\frac{L}{L_m}\ ]\ , \qquad (18)$$

where $L_m = (E_2^m - E_1^m)/(2\pi)$ is the oscillation length in matter. As (15) indicates, the dependence of the amplitude of $\nu_e \leftrightarrow \nu_\mu$ oscillations in matter,

$\sin^2 2\theta_m$, on N_e has a resonance character [9]. Indeed, if $\Delta m^2 \cos^2 2\theta > 0$, for any $\sin^2 2\theta \neq 0$ there exists a value of N_e equal to N_e^{res}, such that

$$\sin^2 2\theta_m = 1\,, \quad for\ N_e = N_e^{res}\ , \tag{19}$$

even if the mixing angle in vacuum is small, i.e., if $\sin^2 2\theta \ll 1$. This implies that the presence of matter can lead to a strong enhancement of the oscillation probability $P_m(\nu_e \to \nu_\mu; t)$ even when the $\nu_e \leftrightarrow \nu_\mu$ oscillations in vacuum are strongly suppressed due to a small value of $\sin^2 2\theta$.

The oscillation length at resonance is given by $L_m^{res} = L_v / \sin 2\theta$, while the width in N_e of the resonance (i.e., the "distance" in N_e between the points at which $\sin^2 2\theta_m = 1/2$) reads $\Delta N_e^{res} = 2N_e^{res} \tan 2\theta$. Thus, if the mixing angle in vacuum is small the resonance is narrow, $\Delta N_e^{res} \ll N_e^{res}$, and L_m at resonance is relatively large, $L_m^{res} \gg L_v$. As it follows from (17), the energy difference $E_2^m - E_1^m$ has a minimum at the resonance: $(E_2^m - E_1^m)^{res} = \min\,(E_2^m - E_1^m) = (\Delta m^2/(2E)) \sin 2\theta$.

It is instructive to consider two limiting case. If $N_e \ll N_e^{res}$, as it follows from (15) and (17), $\theta_m \cong \theta$, $\mathrm{L_m} \cong \mathrm{L_v}$ and the neutrinos oscillate practically as in vacuum. In the opposite limit, $N_e \gg N_e^{res}$, $N_e^{res} \tan^2 2\theta$, $\theta_m \cong \pi/2$ ($\cos 2\theta_m \cong -1$) and the presence of matter suppresses the $\nu_e \leftrightarrow \nu_\mu$ oscillations. In this case we get from (14) and (15): $|\nu_e\rangle \cong |\nu_2^m\rangle$, $|\nu_\mu\rangle = -|\nu_1^m\rangle$, i.e., ν_e practically coincides with the heavier of the two matter-eigenstate ν_2^m, while the ν_μ coincides with the lighter one ν_1^m.

Since the neutral current weak interaction of neutrinos in the Standard Theory is flavour symmetric, the formulae and results we have obtained are valid for the case of $\nu_e - \nu_\tau$ mixing and $\nu_e \leftrightarrow \nu_\tau$ oscillations in matter as well. The case of $\nu_\mu - \nu_\tau$ mixing, however, is different. It is possible to show that to a relatively good precision we have for the ν_μ and ν_τ indeces of refraction $\kappa(\nu_\mu) \cong \kappa(\nu_\tau)$. As a consequence, the $\nu_\mu \leftrightarrow \nu_\tau$ oscillations in matter (e.g., in the Earth) proceed as in vacuum [d].

The analogs of eqs. (15) - (18) for oscillations of antineutrinos, $\bar{\nu}_e \leftrightarrow \bar{\nu}_\mu$, in matter can formally be obtained by replacing N_e with $(-N_e)$ in the indicated equations. It should be clear that depending on the sign of $\Delta m^2 \cos 2\theta$, the presence of matter can lead to resonance enhancement either of the $\nu_e \leftrightarrow \nu_\mu$ or of the $\bar{\nu}_e \leftrightarrow \bar{\nu}_\mu$ oscillations, but not of the both types of oscillations. This is a consequence of the fact that the matter in

[d] In what concerns the possibility of mixing and oscillations between the ν_e and a sterile neutrino ν_s, $\nu_e \leftrightarrow \nu_s$, the relevant formulae can be obtained from the formulae derived for the case of $\nu_e \leftrightarrow \nu_{\mu(\tau)}$ oscillations by [27] replacing N_e with $(N_e - 1/2N_n)$, where N_n is the number density of neutrons in matter.

the Sun or in the Earth we are interested in, is not charge-symmetric (it contains e^-, p and n, but does not contain their antiparticles) and therefore the oscillations in matter are neither CP- nor CPT- invariant [27] [e].

The formalism we have developed can be applied, e.g., to the study of the matter effects in the $\nu_e \leftrightarrow \nu_{\mu(\tau)}$ ($\nu_{\mu(\tau)} \leftrightarrow \nu_e$) oscillations of neutrinos which traverse the Earth mantle [f] (but do not traverse the Earth core). N_e changes little around the mean value of $\bar{N}_e \cong 2.3\ cm^{-3}\ N_A$, along the trajectories of neutrinos which cross a substantial part of the Earth mantle and the $N_e = const.$ approximation was shown to be remarkably accurate in what concerns the calculation of ν-oscillation probabilities. If, for example, $\Delta m^2 = 10^{-3}$ eV^2, $E = 1$ GeV and $\sin^2 2\theta \cong 0.5$, we have: $N_e^{res} \cong 4.6\ cm^{-3}\ N_A$, $\sin^2 2\theta_m \cong 0.8$ and the oscillation length in matter, $L_m \cong 3 \times 10^3$ km, is of the order of the depth of the Earth mantle, so that one can have $2\pi L \gtrsim L_m$.

In the case of neutrinos crossing the Earth core, new resonant effects become apparent. For $\sin^2\theta < 0.05$ and $\Delta m^2 > 0$, we can have $P_m(\nu_e \to \nu_\mu \equiv P^m_{e\mu} \cong 1$ *only due to the effect of maximal constructive interference between the amplitudes of the the $\nu_e \to \nu_\mu$ transitions in the Earth mantle and in the Earth core* [30,31]. The effect differs from the MSW one [30] and the enhancement happens in the case of interest at a value of the energy between the resonance energies corresponding to the density in the mantle and that of the core. The *mantle-core enhancement effect* is caused by the existence (for a given ν-trajectory through the Earth core) of *points of resonance-like total neutrino conversion*, $P^m_{e\mu} = 1$, in the corresponding space of ν-oscillation parameters [31]. The points where $P_{2\nu} = 1$ are determined by the conditions [31]:

$$\tan\phi' \pm \sqrt{\frac{-\cos 2\theta''_m}{\cos(2\theta''_m - 4\theta'_m)}},\quad \tan\phi'' = \pm\frac{\cos 2\theta'_m}{\sqrt{-\cos 2\theta''_m \cos(2\theta''_m - 4\theta'_m)}} \qquad (20)$$

where the signs are correlated and $\cos 2\theta''_m \cos(2\theta''_m - 4\theta'_m) \leq 0$. In eq. (20)

---

[e] The matter effects in the $\nu_e \leftrightarrow \nu_\mu$ ($\bar{\nu}_e \leftrightarrow \bar{\nu}_\mu$) oscillations will be invariant with respect to the operation of time reversal if the N_e distribution along the neutrino path is symmetric with respect to this operation. The latter condition is fulfilled for the N_e distribution along a path of a neutrino crossing the Earth [28].

[f] The Earth density distribution in the existing Earth models [29] is assumed to be spherically symmetric and there are two major density structures - the core and the mantle, and a certain number of substructures (shells or layers). The Earth radius is 6371 km; the Earth core has a radius of 3486 km, so the Earth mantle depth is 2885 km. The mean electron number densities in the mantle and in the core read [29]: $\bar{N}_e^{man} \cong 2.2\ N_A cm^{-3}$, $\bar{N}_e^c \cong 5.4\ N_A cm^{-3}$, N_A being the Avogadro number.

$2\phi'$ and $2\phi''$ are the oscillation phases (phase differences) accumulated by the (two) neutrino states after crossing respectively the first mantle layer and the core, and θ'_m and θ''_m are the ν-mixing angles in the mantle and in the core. A rather complete set of values of $\Delta m^2/E$ and $\sin^2 2\theta$ for which both conditions in eq. (20) hold and $P^m_{e\mu} = 1$ was found in [31]. The location of these points determines the regions where $P^m_{e\mu}$ is large, $P^m_{e\mu} \gtrsim 0.5$. For $\sin^2\theta < 0.05$, there are two sets of values of Δm^2 and $\sin^2\theta$ for which eq. (20) is fulfilled and $P_{2\nu} = 1$. These two solutions of eq. (20) occur for, e.g., values of the Nadir angle $\theta_n = 0;\ 13^0; 23^0$, at 1) $\sin^2 2\theta = 0.034;\ 0.039;\ 0.051$, and at 2) $\sin^2 2\theta = 0.15;\ 0.17;\ 0.22$ (see Table 2 in the last article quoted in [31]). For $\Delta m^2 = 2.0\ (3.0) \times 10^{-3}$ eV^2, for instance, $P^m_{e\mu} = 1$ occurs in the case of the first solution [g] at $E \cong (2.8 - 3.1)$ GeV ($E \cong (4.2 - 4.7)$ GeV).

The effects of the mantle-core enhancement of $P^m_{e\mu}$ are relevant, in particular, for the searches of subdominant $\nu_{e(\mu)} \to \nu_{\mu(e)}$ oscillations of atmospheric neutrinos (see, e.g., [32]).

4. Analytic Description of the Solar Neutrino Oscillations

Consider next the oscillations of solar ν_e while they propagate from the central part, where they are produced [33], to the surface of the Sun. For details concerning the production, spectrum, magnitude and particularities of the solar neutrino flux, the methods of detection of solar neutrinos, description of solar neutrino experiments and of the data they provided we refer the reader to [33,34] and to the lectures of D. Cowen [35]. The electron number density N_e changes considerably along the neutrino path in the Sun: it decreases monotonically from the value of ~ 100 cm^{-3} N_A in the center of the Sun to 0 at the surface of the Sun. According to the contemporary solar models (see, e.g., [33,36]), N_e decreases approximately exponentially in the radial direction towards the surface of the Sun:

$$N_e(t) = N_e(t_0) \exp\left\{-\frac{t - t_0}{r_0}\right\}, \tag{21}$$

where $(t - t_0) \cong d$ is the distance traveled by the neutrino in the Sun, $N_e(t_0)$ is the electron number density in the point of ν_e production in the Sun, r_0 is the scale-height of the change of $N_e(t)$ and one has [36] $r_0 \sim 0.1 R_\odot$.

[g] The first solution corresponds to $\cos 2\phi' \cong -1$, $\cos 2\phi'' \cong -1$ and $\sin^2(2\theta''_m - 4\theta'_m) = 1$. The enhancement effect in this case was called "neutrino oscillation length resonance" (NOLR) in [30].

The system of evolution equations (12) does not admit, in general, exact solutions. However, there are few notable exceptions in which the evolution equations can be solved exactly (see, e.g., [37,38]). Remarkably, these include the case of exponentially varying N_e [39,40], eq. (21), relevant for the description of the solar neutrino oscillations in the Sun. Perhaps even more remarkable is the fact that [41] the system of evolution equations (12), with N_e given by eq. (21), describing the solar neutrino oscillations in the Sun, is equivalent to a second order differential equation - the confluent hypergeometric equation [42], which coincides in form with the Schrödinger (energy eigenvalue) equation obeyed by the radial part of the non-relativistic wave function of the hydrogen atom [43]. On the basis of the corresponding exact solutions expressed in terms of confluent hypergeometric functions, using the asymptotic series expansions of the latter [42], a simple expression for the solar neutrino survival probability, $P_{\odot}(\nu_e \to \nu_e)$, containing only elementary functions, has been derived [39,44] (see also [45]). It was also demonstrated that the expression for $P_{\odot}(\nu_e \to \nu_e)$ thus found provides a very precise (and actually, the most precise) analytic description of the MSW oscillations and transitions of the solar neutrinos in the Sun [46,47,48]. The expression of interest for $P_{\odot}(\nu_e \to \nu_e)$ has the form [39,44]:

$$P_{\odot}(\nu_e \to \nu_e) = \bar{P}_{\odot} + P_1^{osc}, \tag{22}$$

where $\bar{P}_{\odot}$ is the average probability of solar ν_e survival,

$$\bar{P}_{\odot} = \frac{1}{2} + \left(\frac{1}{2} - \mathrm{P}_c\right) \cos 2\theta_m^0 \cos 2\theta, \tag{23}$$

and P_1^{osc} is an oscillating term

$$P_1^{osc} = -\sqrt{P_c(1 - P_c)} \cos 2\theta_m^0 \sin 2\theta \cos(\Phi_{21} - \Phi_{22}) \ . \tag{24}$$

In eqs. (23) and (24)

$$P_c = \frac{\exp\left[-2\pi \mathrm{r}_0 \frac{\Delta m^2}{2E} \sin^2\theta\right] - \exp\left[-2\pi \mathrm{r}_0 \frac{\Delta m^2}{2E}\right]}{1 - \exp\left[-2\pi \mathrm{r}_0 \frac{\Delta m^2}{2E}\right]} \tag{25}$$

is [39] the "jump" or "level-crossing" probability for exponentially varying electron number density N_e [h], and θ_m^0 is the neutrino mixing angle in

[h]An expression for the "jump" probability corresponding to the case of density (N_e) varying linearly along the neutrino path was derived a long time ago by Landau and Zener [49]. An analytic description of the average probability of solar neutrino transitions based on the linear approximation for the change of N_e in the Sun and on the Landau-Zener result was proposed in [50]. The drawbacks of this description, which in certain

matter [8] in the point of ν_e production in the Sun. The phases Φ_{21} and Φ_{22} in the oscillating term, eq. (24), have a simple physical interpretation [44,41]. In the exponential density approximation one finds [44]:

$$\Phi_{21} - \Phi_{22} = -2\arg\Gamma(1-c) - \arg\Gamma(a-1) + \arg\Gamma(a-c)$$
$$-r_0 \frac{\Delta m^2}{2E} \ln[r_0\sqrt{2}G_F N_e(x_0)] + \frac{\Delta m^2}{2E}(L - x_0) \qquad (26)$$

where $a = 1 + ir_0\Delta m^2/(2E)\sin^2\theta$, $c = 1 + ir_0\Delta m^2/(2E)$, $\Gamma(y)$ is the Gamma function and $L = 1$ A.U. The part of the phase $(\Phi_{21} - \Phi_{22})$ given by $\Delta m^2(L - R_\odot)/(2E)$, is accumulated on the path of neutrinos in vacuum from the solar surface to the surface of the Earth; the rest is generated in the Sun. Numerical studies have shown that $(\Phi_{21} - \Phi_{22})$ does not depend on the value of $N_e(x_0)$, i.e., on the point of ν_e production in the Sun [48].

Few comments are in order. Both eqs. (25) and (26) are valid for any value of Δm^2 (or $\Delta m^2/(2E)$) and for any θ, including $\theta \geq \pi/4$ [44]. The solar neutrino transitions are called "adiabatic" [9] if $P_c \cong 0$; otherwise they are called "non-adiabatic" [i]. As was shown in [45], the oscillating term P_1 can be relevant in the solar neutrino transitions, i.e., can give a non-negligible contribution in $P_\odot(\nu_e \to \nu_e)$, only for $\Delta m^2/(2E) \lesssim 10^{-8}$ eV^2/MeV: at $\Delta m^2/(2E) \gtrsim 5 \times 10^{-8}$ eV^2/MeV we have effectively $P_\odot(\nu_e \to \nu_e) \cong \bar{P}_\odot$. In the latter case one speaks about solar neutrino transitions. At $\Delta m^2/(2E) \lesssim 10^{-8}$ eV^2/MeV a very precise and easy to use expression for the phase $(\Phi_{21} - \Phi_{22})$ was found in [48]:

$$\Phi_{21} - \Phi_{22} \cong 0.130\,(\frac{\Delta m^2}{2E}\,R_\odot) + 1.67\times 10^{-3}(\frac{\Delta m^2}{2E}\,R_\odot)^2\cos 2\theta + \frac{\Delta m^2}{2E}(L - R_\odot). \qquad (27)$$

The effects of solar matter in the $\nu_e \to \nu_{\mu(\tau)}$ oscillations or transitions of solar neutrinos become negligible at sufficiently large [9] and sufficiently small [39,44,45] Δm^2. For solar neutrinos we have at $\Delta m^2 \gtrsim 6 \times 10^{-4}$ eV^2: $P_c \cong 0$, $P_1 \cong 0$, $\cos 2\theta_m^0 \cong \cos 2\theta$, and $P_\odot(\nu_e \to \nu_e) \cong 1 - 1/2\ \sin^2 2\theta$, which coincides with the average probability of survival of ν_e when the oscillations take place in vacuum. At $\Delta m^2 \lesssim 5\times 10^{-10}$ eV^2 one finds [39,44,45] $P_c \cong \cos^2\theta$, $\cos 2\theta_m^0 \cong -1$, $(\Phi_{21} - \Phi_{22}) \cong \Delta m^2(L - x_0)/(2E)$, and correspondingly $P_\odot(\nu_e \to \nu_e) \cong 1 - 1/2\ \sin^2 2\theta\ [1 - \cos(\Delta m^2(L - x_0)/(2E))]$, i.e., the solar

cases (e.g., non-adiabatic transitions with relatively large $\sin^2 2\theta$) is considerably less accurate [46] than the description based on the results obtained in the exponential density approximation, were discussed in [37,39,46].

[i]For a more rigorous definition of the adiabatic and non-adiabatic neutrino transitions see [51,46,48].

neutrinos oscillate as in vacuum. For 5×10^{-10} eV$^2 \lesssim \Delta m^2 \lesssim 2 \times 10^{-8}$ eV^2 the solar matter effects are still not negligible and solar neutrinos take part is the so-called "quasi-vacuum oscillations (QVO)". The analytic expression for $P_{\odot}(\nu_e \rightarrow \nu_e)$ given by eqs. (22) - (26) and (27) provides a remarkably precise analytic description of the solar ν_e oscillations/transitions if one uses for the scale height r_0, entering into the expression for P_c, eq. (25), not a constant, but a "running" value [46,48].

5. The Neutrino Mixing Parameters

The formalism of neutrino oscillations in vacuum and in matter we have developed is used in the analyses of the neutrino oscillation data provided by the solar, atmospheric and reactor neutrino experiments as well as by the experiments with accelerator neutrinos.

The Super-Kamiokande atmospheric neutrino data (see, e.g., [52]) and the K2K data (see, e.g., [53]) are best described in terms of dominant 2-neutrino $\nu_\mu \rightarrow \nu_\tau$ ($\bar{\nu}_\mu \rightarrow \bar{\nu}_\tau$) vacuum oscillations. The corresponding $\nu_\mu \rightarrow \nu_\tau$ oscillation probability is given by:

$$
\begin{aligned}
P(\nu_\mu \rightarrow \nu_\mu) &\cong 1 - \sin^2 2\theta_{\rm A} \sin^2 \frac{\Delta m^2_{\rm A} {\rm L}}{4{\rm E}}, \\
P(\nu_\mu \rightarrow \nu_\mu) = 1 - P(\nu_\mu \rightarrow \nu_\tau) &= P(\bar{\nu}_\mu \rightarrow \bar{\nu}_\mu) = 1 - P(\bar{\nu}_\mu \rightarrow \bar{\nu}_\tau) \, .
\end{aligned}
\tag{28}
$$

The best fit values and the 99.73% C.L. allowed ranges of the atmospheric neutrino ($\nu_{\rm A}$-) oscillation parameters read [10]:

$$
\begin{aligned}
|\Delta m^2_{\rm A}| &= 2.1 \times 10^{-3} \ {\rm eV}^2 \, , \quad \sin^2 2\theta_{\rm A} = 1.0 \, , \\
|\Delta m^2_{\rm A}| &= (1.3 - 4.2) \times 10^{-3} \ {\rm eV}^2 \, , \quad \sin^2 2\theta_{\rm A} \geq 0.85 \, .
\end{aligned}
\tag{29}
$$

The sign of $\Delta m^2_{\rm A}$ and of $\cos 2\theta_{\rm A}$, if $\sin^2 2\theta_{\rm A} \neq 1.0$, cannot be determined using the existing data. The latter implies that when, e.g., $\sin^2 2\theta_{\rm A} = 0.92$, one has $\sin^2 \theta_{\rm A} \cong 0.64$ or 0.36.

Recently, the SK collaboration presented the first evidence for an "oscillation dip" in the $L/E-$dependence, L and E being the distance traveled by neutrinos and the neutrino energy, of a particularly selected sample of (essentially milti-GeV) $\mu-$like events [j] [11]. Such a dip is predicted due to the oscillatory dependence of the $\nu_\mu \rightarrow \nu_\tau$ ($\bar{\nu}_\mu \rightarrow \bar{\nu}_\tau$) oscillation probability on L/E: the $\nu_\mu \rightarrow \nu_\tau$ ($\bar{\nu}_\mu \rightarrow \bar{\nu}_\tau$) transitions of atmospheric neutrinos are predominantly two-neutrino transitions governed by vacuum

[j]These are $\mu-$like events for which the relative uncertainty in the experimental determination of the L/E ratio does not exceed 70%.

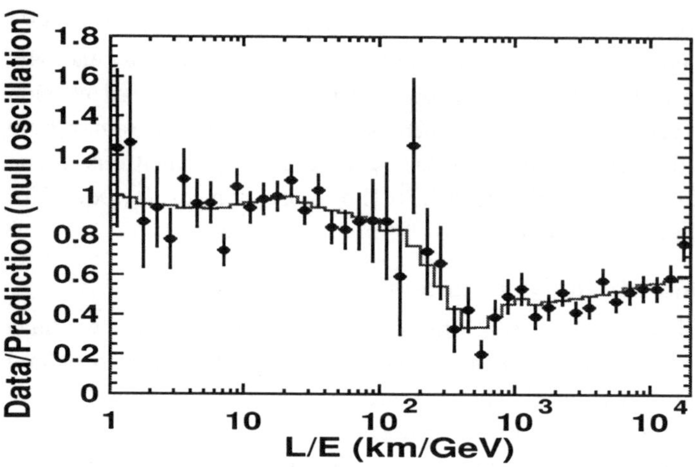

Figure 2. The L/E dependence of the μ-like atmospheric neutrino event rate observed in the Super-Kamiokande experiment [11].

oscillation probability. The dip in the observed L/E distribution corresponds to the first oscillation minimum of the ν_μ ($\bar{\nu}_\mu$) survival probability, $P(\nu_\mu \rightarrow \nu_\mu)$ ($P(\bar{\nu}_\mu \rightarrow \bar{\nu}_\mu)$), as L/E increases starting from values for which $|\Delta m^2_{\rm A}|L/(2E) \ll 1$ and $P(\nu_\mu \rightarrow \nu_\mu) \cong 1$. This beautiful result represents the first ever observation of a direct effect of the oscillatory dependence on L and E of the probability of neutrino oscillations in vacuum.

The combined 2-neutrino oscillation analysis of the solar neutrino and the new KL 766.3 Ty spectrum data (see, e.g., [35]) shows [14,54] that the $\nu_\odot$-oscillation parameters lie in the low-LMA region :

$\Delta m^2_\odot = (7.9^{+0.6}_{-0.5}) \times 10^{-5}$ eV^2, $\tan^2\theta_\odot = (0.40^{+0.09}_{-0.07})$.

The value of $\Delta m^2_\odot$ is determined with a remarkably high precision. The high-LMA solution (see, e.g, [55]) is excluded at $\sim 3.3\sigma$. Maximal $\nu_\odot$-mixing is ruled out at $\sim 6\sigma$. One also has: $\Delta m^2_\odot / |\Delta m^2_{\rm A}| \sim 0.04 \ll 1$.

The interpretation of the solar and atmospheric neutrino, and of K2K and KL data in terms of ν-oscillations requires the existence of 3-ν mixing in the weak charged lepton current:

$$\nu_{l\rm L} = \sum_{j=1}^{3} U_{lj}\, \nu_{j\rm L}, \quad l = e, \mu, \tau, \tag{30}$$

where ν_{lL} are the flavour neutrino fields, $\nu_{j\rm L}$ is the left-handed field of neutrino ν_j having a mass m_j and U is the Pontecorvo-Maki-Nakagawa-Sakata (PMNS) ν-mixing matrix [1,56]. All existing ν-oscillation data, except the data of LSND experiment [k] [13], can be described assuming 3-ν mixing in vacuum and we will consider only this possibility. The minimal 4-ν mixing scheme which could incorporate the LSND indications for ν-oscillations is strongly disfavored by the data [58]. The ν-oscillation explanation of the LSND results is possible assuming 5-ν mixing [59].

The PMNS matrix can be parametrized by 3 angles and, depending on whether the massive neutrinos ν_j are Dirac or Majorana particles, by 1 or 3 CP-violation (CPV) phases [60,61]. In the standard parameterization [62]

$$U_{\rm PMNS} = V(\theta_{12}, \theta_{13}, \theta_{23}, \delta)\ {\rm diag}(1, e^{i\alpha}, e^{i\beta}),$$

$$V = \begin{pmatrix} c_{12}c_{13} & s_{12}c_{13} & s_{13} \\ -s_{12}c_{23} - c_{12}s_{23}s_{13}e^{i\delta} & c_{12}c_{23} - s_{12}s_{23}s_{13}e^{i\delta} & s_{23}c_{13}e^{i\delta} \\ s_{12}s_{23} - c_{12}c_{23}s_{13}e^{i\delta} & -c_{12}s_{23} - s_{12}c_{23}s_{13}e^{i\delta} & c_{23}c_{13}e^{i\delta} \end{pmatrix}, \quad (31)$$

where $c_{ij} = \cos\theta_{ij}$, $s_{ij} = \sin\theta_{ij}$, the angles $\theta_{ij} = [0, \pi/2]$, $\delta = [0, 2\pi]$ is the Dirac CPV phase and α, β are two Majorana CPV phases [60,61]. One can identify $\Delta m^2_\odot = \Delta m^2_{21} > 0$. In this case $|\Delta m^2_{\rm A}|=|\Delta m^2_{31}| \cong |\Delta m^2_{32}|$, $\theta_{12} = \theta_\odot$, $\theta_{23} = \theta_{\rm A}$. The angle θ_{13} is limited by the data from the CHOOZ and Palo Verde experiments [63]. The existing $\nu_{\rm A}$-data is essentially insensitive to θ_{13} obeying the CHOOZ limit [10]. The probabilities of survival of reactor $\bar{\nu}_e$ and solar ν_e, relevant for the interpretation of the KL, CHOOZ and $\nu_\odot$- data, depend on θ_{13}:

$$P^{3\nu}_{\rm KL} \cong \sin^4\theta_{13} + \cos^4\theta_{13}\left[1 - \sin^2 2\theta_{12}\sin^2\frac{\Delta m^2_{21}{\rm L}}{4E}\right],$$

$$P^{3\nu}_{\rm CHOOZ} \cong 1 - \sin^2 2\theta_{13}\sin^2\frac{\Delta m^2_{31}{\rm L}}{4{\rm E}},$$

$$P^{3\nu}_\odot \cong \sin^4\theta_{13} + \cos^4\theta_{13}\ P^{2\nu}_\odot(\Delta m^2_{21}, \theta_{12}; \theta_{13}),$$

where $P^{2\nu}_\odot$ is the 2-ν mixing solar ν_e survival probability, eq. (22), in the case of transitions driven by Δm^2_{21} and θ_{12}, in which (the solar e^- number density) N_e is replaced by $N_e\cos^2\theta_{13}$ [64], $P^{2\nu}_\odot = \bar{P}_\odot + P^{osc}_1$ (see eqs. (23) and (24)). In the LMA solution region one has [45] $P^{2\nu}_{\odot\ \rm osc} \cong 0$. Using the 3σ allowed range of $|\Delta m^2_{\rm A}|=|\Delta m^2_{31}|$ [10] and performing a combined analysis of the solar neutrino, CHOOZ and KL data, one finds [54]:

$$\sin^2\theta_{13} < 0.055, \qquad 99.73\%\ {\rm C.L.}$$

---

[k]In the LSND experiment indications for oscillations $\bar{\nu}_\mu \to \bar{\nu}_e$ with $(\Delta m^2)_{\rm LSND} \simeq 1\ {\rm eV}^2$ were obtained. The LSND results are being tested in the MiniBooNE experiment [57].

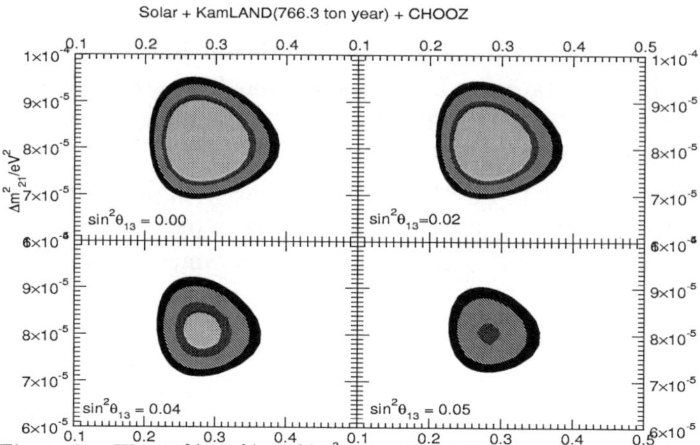

Figure 3. The 90%, 95%, 99% and 99.73% C.L. allowed regions in the Δm^2_{21} -$\sin^2\theta_{12}$ plane, obtained in a three-neutrino oscillation analysis of the solar neutrino, KamLAND and CHOOZ data [54].

Similar constraint is obtained from a global 3-ν oscillation analysis of the data [58,65]. In Fig. (3) we show the allowed regions in the $\Delta m^2_{21} - \sin^2\theta_{12}$ plane for few fixed values of $\sin^2\theta_{13}$ [54].

Thus, the fundamental parameters characterizing the 3-neutrino mixing are: i) the 3 angles θ_{12}, θ_{23}, θ_{13}, ii) depending on the nature of ν_j - 1 Dirac (δ), or 1 Dirac + 2 Majorana (δ, α, β), CPV phases, and iii) the 3 neutrino masses, m_1, m_2, m_3. It is convenient to express the two larger masses in terms of the third mass and the measured $\Delta m^2_\odot = \Delta m^2_{21} > 0$ and $\Delta m^2_{\rm A}$. In the convention we are using, the two possible signs of $\Delta m^2_{\rm A}$ correspond to two types of ν-mass spectrum:

- with normal hierarchy, $m_1 < m_2 < m_3$,
  $\Delta m^2_{\rm A} = \Delta m^2_{31} > 0$, $m_{2(3)} = (m_1^2 + \Delta m^2_{21(31)})^{\frac{1}{2}}$, and
- with inverted hierarchy, $m_3 < m_1 < m_2$,
  $\Delta m^2_{\rm A} = \Delta m^2_{32} < 0$, $m_2 = (m_3^2 - \Delta m^2_{32})^{\frac{1}{2}}$, etc.

The spectrum can also be

- *normal hierarchical (NH)*: $m_1 \ll m_2 \ll m_3$,
  $m_2 \cong (\Delta m^2_\odot)^{\frac{1}{2}} \sim 0.009$ eV, $m_3 \cong |\Delta m^2_{\rm A}|^{\frac{1}{2}} \sim 0.045$; or
- *inverted hierarchical (IH)*: $m_3 \ll m_1 < m_2$,
  with $m_{1,2} \cong |\Delta m^2_{\rm A}|^{\frac{1}{2}} \sim 0.045$ eV; or
- *quasi-degenerate (QD)*: $m_1 \cong m_2 \cong m_3 \cong m_0$, $m_j^2 \gg |\Delta m^2_{\rm A}|$. In this case one has $m_0 \gtrsim 0.20$ eV.

After the spectacular experimental progress made in the studies of neutrino oscillations, further understanding of the structure of neutrino masses and neutrino mixing, of their origins and of the status of CP-symmetry in the lepton sector requires an extensive and challenging program of research to be pursued in neutrino physics. The main goals of this research program should include [66]:

• High precision measurement of the solar and atmospheric neutrino oscillations parameters, Δm^2_{21}, θ_{21}, and Δm^2_{31}, θ_{23}.

• Measurement of, or improving by at least a factor of (5 - 10) the existing upper limit on, θ_{13} - the only small mixing angle in $U_{\rm PMNS}$.

• Determination of the $sign(\Delta m^2_{31})$ and of the type of ν-mass spectrum (NH, IH, QD, etc.).

• Determining or obtaining significant constraints on the absolute scale of ν-masses, or on $min(m_j)$.

• Determining the nature–Dirac or Majorana, of massive neutrinos ν_j.

• Establishing whether the CP-symmetry is violated in the lepton sector a) due to the Dirac phase δ, and/or b) due to the Majorana phases α and β if ν_j are Majorana particles.

• Searching with increased sensitivity for possible manifestations, other than flavour neutrino oscillations, of the non-conservation of the individual lepton charges L_l, $l = e, \mu, \tau$, such as $\mu \rightarrow e + \gamma$, $\tau \rightarrow \mu + \gamma$, etc. decays.

• Understanding at fundamental level the mechanism giving rise to neutrino masses and mixing and to L_l-non-conservation, i.e., finding the *Theory of neutrino mixing*. This includes understanding the origin of the patterns of ν-mixing and ν-masses suggested by the data. Are the observed patterns of ν-mixing and of $\Delta m^2_{21,31}$ related to the existence of new fundamental symmetry of particle interactions? Is there any relations between quark mixing and neutrino mixing, e.g., does the relation $\theta_{12} + \theta_c = \pi/4$, where θ_c is the Cabibbo angle, hold? Is $\theta_{23} = \pi/4$, or $\theta_{23} > \pi/4$ or else $\theta_{23} < \pi/4$? What is the physical origin of CPV phases in $U_{\rm PMNS}$? Is there any relation (correlation) between the (values of) CPV phases and mixing angles in $U_{\rm PMNS}$? Progress in the theory of ν-mixing might also lead, in particular, to a better understanding of the mechanism of generation of baryon asymmetry of the Universe [67].

Obviously, the successful realization of the experimental part of this research program would be a formidable task and would require many years.

The mixing angles, θ_{21}, θ_{23} and θ_{13}, Dirac CPV phase δ and Δm^2_{21} and Δm^2_{31} can, in principle, be measured with a sufficiently high precision in a variety of ν-oscillation experiments (see further). These experiments, however, cannot provide information on the absolute scale of ν- masses and on the nature of massive neutrinos ν_j. The flavour neutrino oscillations are insensitive to the Majorana CPV phases α and β [60,27]. Establishing whether ν_j have distinct antiparticles (Dirac fermions) or not (Majorana fermions) is of fundamental importance for understanding the underlying symmetries of particle interactions [18] and the origin of ν-masses. If ν_j are Majorana fermions, getting experimental information about the Majorana CPVphases in $U_{\rm PMNS}$ is a remarkably challenging problem. [68,69,70]. The phases α and β can affect significantly the predictions for the rates of the (LFV) decays $\mu \to e + \gamma$, $\tau \to \mu + \gamma$, etc. in a large class of supersymmetric theories with see-saw mechanism of neutrino mass generation (see, e.g., [71]). Majorana CPV phases might be at the origin of the baryon asymmetry of the Universe [l] [67].

6. Instead of Conclusions

We are at the beginning of the "road" leading to a comprehensive understanding of the patterns of neutrino masses and mixing and of their origin. There are no doubts that progress in the studies of neutrino mixing will lead to more profound understanding of the fundamental forces governing particle interactions and of the Universe we are living in.

References

1. B. Pontecorvo, Zh. Eksp. Teor. Fiz. (JETP) **33** (1957) 549 and **34** (1958) 247.
2. B. Pontecorvo, JETP **53** (1967) 1717.
3. B.T. Cleveland *et al.*, Astrophys. J. **496** (1998) 505; J.N. Abdurashitov *et al.*, astro-ph/0204245; C. Cattadori, Talk given at ν'04 Int. Conference, June 14-19, 2004, Paris, France.
4. Y. Fukuda *et al.*, Phys. Rev. Lett. **77** (1996) 1683.
5. S. Fukuda *et al.*, Phys. Rev. Lett. **86** (2001) 5651 and 5656; M. Nakahata, Talk given at ν'04 Int. Conference, June 14-19, 2004, Paris, France.
6. J. Wilkerson, Talk given at ν'04 Int. Conference, June 14-19, 2004, Paris, France; Q. R. Ahmad *et al.*, Phys. Rev. Lett. **87** (2001) 071301; *ibid.* **89** (2002) 011301 and 011302; S. N. Ahmed *et al.*, Phys. Rev. Lett. **92** (2004) 181301.
7. K. Eguchi *et al.*, Phys.Rev.Lett.**90** (2003) 021802.
8. L. Wolfenstein, Phys. Rev. **D17** (1978) 2369.
9. S.P. Mikheyev and A.Yu. Smirnov, Sov. J. Nucl. Phys. **42** (1985) 913.
10. Y. Fukuda *et al.*, Phys. Rev. Lett. **81** (1998) 1562; E. Kearns, Talk given at ν'04 Int. Conference, June 14-19, 2004, Paris, France;
11. Y. Ashie *et al.*, Phys. Rev. Lett. **93** (2004) 101801.
12. M.H. Ahn *et al.*, Phys. Rev. Lett. **90** (2003) 041801.
13. C. Athanassopoulos *et al.*, Phys. Rev. Lett. **81** (1998) 1774.

---

[l]For further discussion see, e.g., [66] and the references quoted therein.

14. T. Araki *et al.* [KamLAND Collaboration], hep-ex/0406035.
15. E. Aliu *et al.* [K2K Collaboration], hep-ex/0411038.
16. V. Gribov and B. Pontecorvo, Phys. Lett. **B28** (1969) 493.
17. S.M. Bilenky and B. Pontecorvo, Phys. Rep. **41** (1978) 225.
18. S.M. Bilenky and S.T. Petcov, Rev. Mod. Phys. **59** (1987) 671.
19. S.M. Bilenky, W. Grimus and C. Giunti, Prog. Part. Nucl. Phys. **43** (1999) 1.
20. S. Nussinov, Phys. Lett. B**63** (1976) 201; B. Kayser, Phys. Rev. **D24** (1981) 110; J. Rich, Phys. Rev. **D48** (1993) 4318.
21. C. Giunti, Found. Phys.Lett. **17** (2004) 103 and hep-ph/0402217; A.D. Dolgov *et al.*, hep-ph/0407189, and articles quoted therein; see also J.H. Field, hep-ph/0410051.
22. A. Bandyopadhyay *et al.*, hep-ph/0409224.
23. G. Raffelt, hep-ph/0208024 and articles quoted therein.
24. M.M. Guzzo, A. Masiero and S.T. Petcov, Phys. Lett. B **260** (1991) 154; V. Barger, R.J.N. Phillips and K. Whisnant, Phys. Rev. D **44** (1991) 1629; E. Roulet, Phys. Rev. D **44** (1991) R935; S. Bergmann, Nucl. Phys. B **515** (1998) 363; M.M. Guzzo, P.C. de Holanda and O.L.G. Peres, Phys. Lett. B **591** (2004) 1.
25. V. Barger et al., Phys. Rev. **D22** (1980) 2718.
26. P. Langacker, J.P. Leveille and J. Sheiman, Phys. Rev. **D27** (1983) 1228.
27. P. Langacker et al., Nucl. Phys. **B282** (1987) 289.
28. P.I. Krastev and S.T. Petcov, Phys. Lett. B**205** (1988) 84; T.K. Kuo and J. Pantaleone, Phys. Lett. B**198** (1987) 406.
29. A. D. Dziewonski and D. L. Anderson, Physics of the Earth and Planetary Interiors **25**, 297 (1981).
30. S. T. Petcov, Phys. Lett. B **434** (1998) 321, (E) *ibid.* B **444** (1998) 584.
31. M. V. Chizhov and S.T. Petcov, Phys. Rev. Lett. **83** (1999) 1096, and Phys. Rev. Lett. **85** (2000) 3979; Phys. Rev. D **63** (2001) 073003.
32. J. Bernabéu, S. Palomares-Ruiz and S. T. Petcov, Nucl. Phys. B **669** (2003) 255; S. Palomares-Ruiz and S. T. Petcov, hep-ph/0406096.
33. J.N. Bahcall, *Neutrino Astrophysics*, Cambridge University Press, Cambridge, 1989.
34. S.T. Petcov, Lecture Notes in Physics, v. 512 (eds. H. Gausterer and C.B. Lang, Springer, 1998), p. 281, hep-ph/9806466.
35. D. Cowen, these Proceedings.
36. J.N. Bahcall, M. Pinsonneault and S. Basu, Astrophys. J. **555** (2001) 990.
37. S.T. Petcov, Phys. Lett. B**191** (1987) 299.
38. W. Haxton, Phys. Rev. D **35**, 2352 (1987); A. Dar et al., Phys. Rev. D **35**, 3607 (1987); Notzold, Phys. Rev. D **36**, 1625 (1987).
39. S.T. Petcov, Phys. Lett. **200B** (1988) 373.
40. T. Kaneko, Prog. Theor. Phys. **78** (1987) 532; M. Ito, T. Kaneko and M. Nakagawa, *ibid.* **79** (1988) 13.
41. S.T. Petcov, Phys. Lett. B**406** (1997) 355.
42. H. Bateman and A. Erdelyi, *Higher Transcendental Functions* (McGraw-Hill, New York, 1953).

43. C. Cohen-Tannoudji, B. Diu and F. Laloe, *Quantum Mechanics*, vol. 1 (Hermann, Paris, and John Wiley & Sons, New York, 1977).
44. S.T. Petcov, Phys. Lett. B**214** (1988) 139.
45. S.T. Petcov and J. Rich, Phys. Lett. B**224** (1989) 401.
46. P.I. Krastev and S.T. Petcov, Phys. Lett. B**207** (1988) 64; Proc. of the Moriond Workshop on Neutrinos and Exotic Phenomena, Les Arcs, France, January 1988 (ed. J. Tran Thanh Van, Ed. Frontières, Gif-sur-Yvette), p. 173.
47. M. Bruggen, W.C. Haxton and Y.-Z. Quian, Phys. Rev. D **51** (1995) 4028.
48. E. Lisi et al., Phys. Rev. D **63** (2000) 093002.
49. L.D. Landau, Phys. Z. USSR **1** (1932) 426;
C. Zener, Proc. R. Soc. A **137** (1932) 696.
50. W. Haxton, Phys. Rev. Lett. **57** (1987) 1271; S. Parke, *ibid.* **57** (1987) 1275.
51. A. Messiah, Proc. of the VIth Moriond Workshop on Massive Neutrinos in Astrophysics and in Particle Physics (eds. O. Fackler and J. Tran Thanh Van, Ed. Frontières, Gif-sur-Yvette, 1986), p. 373.
52. G. Giacomelli, these Proceedings.
53. D. Autiero, these Proceedings.
54. A. Bandyopadhyay *et al.*, hep-ph/0406328.
55. A. Bandyopadhyay *et al.*, Phys. Lett. **B583** (2004) 134.
56. Z. Maki, M. Nakagawa and S. Sakata, Prog. Theor. Phys. **28** (1962) 870.
57. S. Brice *et al.*, Talk given at ν'04 Int. Conference, June 14-19, 2004, Paris.
58. M. Maltoni et al., hep-ph/0405172.
59. M. Sorel, J. Conrad and M. Shaevitz, hep-ph/0305255.
60. S.M. Bilenky, J. Hosek and S.T. Petcov, Phys. Lett. **B94** (1980) 495.
61. J. Schechter and J.W.F. Valle, Phys. Rev. D **23** (1980) 2227;
M. Doi *et al.*, Phys. Lett. **B102** (1981) 323.
62. S.M. Bilenky, S. Pascoli and S.T. Petcov, Phys. Rev. D **64** (2001) 053010.
63. M. Apollonio *et al.*, Phys. Lett. **B466** (1999) 415;
F. Boehm *et al.*, Phys. Rev. Lett. **84** (2000) 3764.
64. S.T. Petcov, Phys. Lett. **B214** (1988) 259.
65. J.N. Bahcall, M.C. Gonzalez-Garcia and C. Peña-Garay, hep-ph/0406294.
66. S.T. Petcov, hep-ph/0412410.
67. M. Fukugita and T. Yanagida, Phys. Lett. B **174** (1986) 45; see also W. Buchmuller *et al.*, Nucl. Phys. B **665** (2003) 445, G. F. Giudice, *et al.*, Nucl. Phys. B **685** (2004) 89, and references therein.
68. S.M. Bilenky *et al.*, Phys. Rev. D **56** (1996) 4432.
69. V. Barger *et al.*, Phys. Lett. **B540** (2002) 247; A. de Gouvea, B. Kayser and R. Mohapatra, Phys. Rev. D **67** (2003) 053004.
70. S. Pascoli, S.T. Petcov and W. Rodejohann, Phys. Lett. **B549** (2002) 177.
71. S. Pascoli, S.T. Petcov and C.E. Yaguna, Phys. Lett. **B564** (2003) 241.

SHORT AND LONG BASELINE NEUTRINO EXPERIMENTS

DARIO AUTIERO

IN2P3, IPN Lyon, Rue Enrico Fermi,4
69400 Villeurbanne Cedex, France

These two lectures discuss the past and current neutrino oscillation experiments performed with man-made neutrino sources, like accelerators and nuclear reactors. The search for neutrino oscillations is a remarkable effort, which has been performed over three decades. It is therefore interesting to discuss the short and long baseline neutrino experiments in their historical context and to see how this line of research evolved up to the present generation of experiments, looking at what was learnt from past experiments and how this experience is used in the current ones. The first lecture focuses on the past generation of short baseline experiments (NOMAD and CHORUS) performed at CERN and ends with LSND and MINIBOONE. The second lecture discusses how after the CHOOZ and the atmospheric neutrino results the line of the long baseline experiments developed and presents in details the K2K and MINOS experiments and the CNGS program.

1. Introduction to neutrino oscillations

Neutrino mixing was first hypothesized by Pontecorvo in 1958 and then by Maki, Nakagawa and Sakata in 1962. Neutrinos are massive particles and they mix similarly to quarks: in the today's favorite 3 neutrinos framework the flavor eigenstates ν_e, ν_μ, ν_τ are not mass eigenstates but linear superpositions of the mass eigenstates ν_1, ν_2, ν_3. with eigenvalues m_1, m_2, m_3. A unitary mixing matrix U describes this superposition. The U matrix is usually parameterized in terms of three mixing angles (θ_{12}, θ_{23}, θ_{13}) and one Dirac-like CP phase δ (two extra phases should be included in case of Majorana neutrinos) [1].

Let us consider the time evolution of a flavor eigenstate ν_a, produced at t=0. It is evident that different masses imply different time evolution of the phases of the corresponding mass eigenstates, which are composing the initial state. Projecting back the initial state after its time evolution on the flavor basis one can obtain the probability of finding a flavor not present at t=0. The possibility of detecting a new flavor not present at the production is called appearance. It is instructive to look at the simplified case of two-neutrinos mixing. Given the presence of only the flavor ν_a at t=0 the probability of detecting the flavor ν_b at the instant t has an oscillatory behavior determined by two parameters: the mixing angle θ which is related to the amplitude of the oscillation and the squared masses difference Δm^2, which is related to the wavelength of the oscillation:

$$P_{ab}(t)=\sin^2(2\theta)\ \sin^2(\Delta m^2\ t/4E)$$

For very large Δm^2 the oscillation becomes very fast and averages over the dimensions of the source and of the detector. The Baseline of the oscillation is given by the L/E ratio of the experimental setup: short baseline experiments are for historical reasons the ones sensitive to large Δm^2 (>1 eV^2) while long baseline experiments are sensitive to Δm^2 of interest for the atmospheric neutrino anomaly ($<10^{-2}$ eV^2).

Appearance experiments (ν_e, ν_τ appearance at accelerators) will put in evidence the presence of a new flavor not present in the original neutrino beam and which can be explained only through the oscillation mechanism. This kind of experiments relies on the control of the purity of the initial beam and on the control of the background processes in the detector, which could mimic the appearance of a new flavor. Neutrino beams at accelerators are almost pure ν_μ beams: ν_τ are practically absent, ν_e are present at the level of 1%.

Disappearance experiments (nuclear reactors, accelerators with low energy beams) instead measure the survival probability at a certain distance from the source of the neutrino flavor produced at the source. These experiments rely on the knowledge of the initial neutrino flux at the source to which they have to compare in order to claim for an effect. This knowledge is the main systematic limitation by measuring the un-oscillated flux with a near detector.

2. The short baseline experiments at CERN

We will now discuss the situation for neutrino oscillation searches at the beginning of the 90s. It does not look a very far past in everyday life but it was a completely different epoch in neutrino physics. At that time, people were concerned by the long standing (since 1968) problem of the solar neutrino deficit opened by the Homestake measurements and confirmed by Kamiokande since 1986. In 1992, the first Gallex results confirmed this deficit also for neutrinos directly produced in the pp cycle [2]. The atmospheric neutrino anomaly was still quite weak. At that time, the controlled observation of neutrino oscillations with an accelerator neutrino beam would have been a great discovery. Prejudices were orienting the searches towards small mixing angles and large Δm^2. By taking the MSW solution of the solar neutrino deficit and considering the see-saw mechanism and a strong neutrino masses hierarchy one was lead to think that the mass of the state ν_3, almost coinciding with the ν_τ due to the small mixing, was about 30 eV [3]. This value was of cosmological relevance, implying that neutrinos were an important component of the dark matter.

CERN started an experimental program focused on the short-baseline search for ν_μ -ν_τ oscillations by looking for ν_τ appearance in the West Area Neutrino Facility beam. This beam had a sufficiently large average energy of 24 GeV, allowing for ν_τ appearance while the average distance in between the experiments and the neutrino source was 600 m. The WANF was an almost pure ν_μ beam with 6% contamination of anti-ν_μ and about 1% contamination of electron neutrinos. The prompt contamination of ν_τ was negligible.

The CHORUS and NOMAD experiments were looking for ν_τ through their charged current interactions followed by the tau decay [4]. The decay can be identified by using two different techniques: the first one is based on the measurement of the tau decay topology, seen as a sharp change of trajectory (the kink) in the single prong decay channels; the second one on the measurement of the kinematic of the tau decay, which is characterized by the presence of one or two neutrinos in the final state, seen as missing transverse momentum, and by the visible tau decay daughters.

The CHORUS experiment was based on the detection of the decay kink. For this task a high space resolution detector is required like nuclear emulsions, the main decay channel exploited is the muonic decay channel of the tau. The NOMAD experiment was instead pursuing the kinematical method, which needs a high-resolution spectrometer and good calorimetry.

The most promising decay channel for NOMAD was the electronic decay channel, exploiting the fact that the beam background from charged current events is about 100 times smaller for the ν_e than for the ν_μ. Both experiments keep an actual interest since they pioneered the tau appearance techniques which are used in the present long baseline experiments, furthermore they collected important samples of neutrino interactions finely reconstructed which greatly improved the knowledge of "standard" neutrino physics in the multi-GeV energy range. The sensitivity of this generation of experiments was covering the region down to a Δm^2 of about 1 eV^2 and of mixing angles as small as a few 10^{-4}. The use of kinematics to extract a ν_τ signal was first proposed in 1979 in ref. [5] and it became finally applied with the NOMAD experiment [4]. This was a magnetic spectrometer where drift chambers were acting at the same time as neutrino target (with a fiducial mass of 2.7 tons) and as tracking detector with an average density of 0.1 g/cm^3 and a typical momentum resolution of 3.5% for momenta smaller than 10 GeV/c. NOMAD had excellent electron identification capabilities, thanks to a transition radiation detector, a pre-shower and a lead-glass electromagnetic calorimer. The overall pion/electron rejection power was about 10^6.

NOMAD collected in 4 years a statistics of 1.7 millions of neutrino interactions. Closing the kinematics on the transverse plane and measuring the missing momentum and the angular correlations between this vector and the one of the hadronic system and of the visible tau decay daughter achieved the separation of the signal from the background. There are no single cuts allowing to achieve in one shot a 10^5 rejection factor like the one needed to kill the ν_e charged currents (CC) background in the electronic decay channel. This rejection power was obtained by combining in likelihood functions many variables describing how much the event was unbalanced on the transverse plane and how much the tau decay daughter candidate was isolated with respect to the hadronic system. Corrections of the discrepancies between data and Monte Carlo were estimated with the data themselves by comparing the sample of ν_μ CC events in the data and MC. This method is not applicable for the search in the muonic decay channel, where the signal would directly interfere with the corrections. Therefore, NOMAD sacrificed this particular channel and exploited the ν_μ CC events to correct the simulation for all the other channels. NOMAD performed a completely blind analysis in which the events falling in the region where the signal was expected were looked only after the agreement with data of the background prediction was proved in all the other regions with negligible presence of the signal. This was done as well on the sample of events with a positively charged tau decay daughter, which is completely background dominated.

The CHORUS experiment had a target of 800 Kg of nuclear emulsions [4]. A scintillating fibres detector with high space resolution was used for the localization of neutrino interactions in the emulsions. A magnetic spectrometer was devoted to the identification of the muons and the measurement of their charge. This is essential in order to fight against the background coming from the decays of charmed particles, which mainly affects the tau searches based on the kink topology. The CHORUS collaboration developed the technique of automatic scanning microscopes for the analysis of the data recorded in the nuclear emulsions.

3. The LSND and Karmen experiments (1993-2001)

While the short-baseline experiments at CERN were starting, an evidence for neutrino oscillations was claimed by the LSND experiment at Los Alamos. LSND was an accelerator experiment performed on a high intensity, low energy neutrino beam. Protons with 800 MeV kinetic energy were stopped in a dump and neutrinos were coming from pions and muons decaying at rest. Oscillations were searched by looking at a distance of about 30 m from the source for the appearance of anti-ν_e oscillated from anti-ν_μ (originating from the decays of positive muons) with a detector of 167 tons of liquid scintillator. The final state of the anti-ν_e charged current interactions includes a positron and a neutron. The detection pattern implies a prompt signal from the positron plus a delayed signal from the neutron capture. The intrinsic anti-ν_e contamination of the beam was strongly reduced by: a) the smaller production of negative pions; b) the fact that these are mostly absorbed in the dump; c) the negative muons originating from the few decays in flight of the negative pions are mostly captured [6].

The Karmen experiment, performed in parallel at the Rutherford Laboratories did not observe any evidence of oscillations [6]. However, the Karmen experiment was penalized by beam intensity about a factor 5 smaller than LNSD and a target mass three times smaller, moreover, also the distance source-detector was almost a factor two smaller. On the other hand, an advantage of Karmen was the time structure of the beam with a repetition rate of 50 Hz. The oscillations signal, coming from the anti-ν_μ from the decay of positive muons is expected to be within 10 microseconds from the beam extraction. LNSD observed an excess of 88 events with respect to the background estimation, equivalent to a 3.8 sigma fluctuation of the background and corresponding to an oscillation probability of 2.6 10^{-3}. Karmen observed instead 15 events, consistent with the background prediction. The observations of the two experiments were still compatible in a region of smaller mixing angles and of smaller Δm^2.

A search for ν_μ-ν_e oscillations was performed in NOMAD in the years 1995-2000 in order to check the LNSD results, exploiting the excellent electron identification capabilities of the detector and the low ν_e contamination in the beam. A ν_μ-ν_e oscillation probability at the level of 10^{-3} would have resulted in a 10% increase in the ν_e flux as measured by NOMAD, furthermore the oscillated ν_e would also have a lower energy and narrower radial distribution of the prompt ν_e. In order to perform this search, a careful simulation of the beam line, with a systematic uncertainty on the ν_e /ν_μ ratio smaller than 5%, was needed. The NOMAD data were in agreement with the expectations without oscillations and this result excluded the LSND result above 10 eV^2. The anti-ν_μ-anti-ν_e oscillation claim of LSND complicated the global scenario of neutrino oscillations. With three neutrinos, it is possible to build only two independent Δm^2. The Δm^2 indicated by LSND (around 1 eV^2) is not compatible with the mass differences coming from the oscillations of solar neutrinos and atmospheric neutrinos. At least four neutrinos are needed in order to reconcile all these results but it is known from LEP that he number of active light neutrinos is three, so the fourth neutrino must be sterile. Even under this assumption, the global fit of oscillations signal is poor (oscillations involving sterile neutrinos are disfavored for the atmospheric and Solar neutrinos, more sophisticated mechanisms like CPT violation must be advocated).

4. MiniBooNE

The MiniBoone experiment was proposed is order to provide a definitive confirmation with different systematic, energy and statistics of the LSND claim. 8 GeV protons accelerated by the Fermilab Booster are extracted on a beryllium target and produce a wide band neutrino beam with a focalization system and a 50m long decay tube. The average energy of the ν_μ is 500 MeV, about 10 times larger than in LSND and the source-detector distance is increase accordingly to 540 m. The detector is a sphere containing a fiducial mass of 445 tons of mineral oil watched by photomultipliers. Particle identification in the final state (electrons, muons, π^0) is possible by looking at the pattern of Cerenkov rings. Correspondingly, to the LSND oscillation probability, 1000 signal events are expected with two years of data taking, to be compared to a background of about 2500 events (not taking yet into account the different energy distributions of the signal and background events). First results are expected during 2005 [7].

5. The CHOOZ and Palo Verde experiments

The Kamiokande results on the atmospheric neutrino anomaly (1994-1997) were showing that the double ratio of the ν_μ/ν_e fluxes (measured/expected) was around 0.6 and this was interpretable both in terms of ν_μ-ν_τ or ν_μ-ν_e oscillations with a Δm^2 around 10^{-2} eV^2. The zenith angle dependence was still quite weak. In 1995, Perkins et al. published a paper where they proposed the interpretation of the solar and the atmospheric data just in terms of one ν_μ-ν_e oscillation at 10^{-2} eV^2 [8]. This simple and fascinating interpretation, which was only discouraged by the energy dependency of the solar neutrino deficit, was complemented by Acker-Pakvasa in 1996 including also LSND in a three-neutrino framework. A series of medium-baseline experiments were discussed at that time to check this hypothesis.

The clarification of this scenario came from the CHOOZ and Palo Verde experiments, each an anti-ν_e disappearance experiment at a nuclear reactor [9]. The distance source-detector was around 1 km and the corresponding L/E was about

300 km/GeV. The CHOOZ detector was a target of 5 tons of liquid scintillator doped with 0.09% Gadolinium and it was shielded with about 300 m of water equivalent. The anti-ν_e detection was performed as usual via the inverse beta decay by looking for the prompt signal coming from the positron annihilation and the delayed signal of the neutron capture. About 25 events per day were expected in absence of oscillations with a reactor-off background of 1.2 events/day. The reactor flux was known at 2.7%. No significant deficit of anti-ν_e was observed and this result was published in advance with respect to the results on atmospheric neutrinos, which were showing for the first time a strong zenith angle dependence of the ν_μ disappearance and a ν_e flux compatible with expectations. The favorite interpretation of the Super-Kamiokande and MACRO data was in terms of ν_μ-ν_τ oscillations with a Δm^2 of a few 10^{-3} eV^2 [10,11]. CHOOZ, which has been the first long-baseline experiment had already killed the interpretation of the atmospheric neutrino anomaly in terms of ν_μ-ν_e oscillations and put a limit which is still the most stringent one on the θ_{13} parameter (θ_{13} less than 11 degrees).

6. K2K

K2K, started in 1999, was the first long-baseline experiment performed at accelerators to check the atmospheric neutrino oscillations [12]. It is based on a ν_μ beam with average energy of 1.3 GeV sent from the KEK accelerator to the Super-Kamiokande detector (250 Km away). K2K looks for ν_μ disappearance and the corresponding energy spectral distortion induced by the oscillations. In order to measure precisely the disappearance effect, a near detector is placed at about 250 m from the target at KEK. This detector allows measuring the ν_μ flux in absence of oscillations as well as to check the beam direction. The near detector is made of 3 different detectors: a small replica of Super-Kamiokande with a fiducial mass of 25 tons, a scintillating fibres detector in water (a fine grained water target with a fiducial mass of 6 tons) followed by a lead glass calorimeter; a muon range detector including 330 tons of fiducial mass. Since October 2003 the lead glass calorimeter has been replaced by scintillating bars detectors of 11 tons fiducial mass designed to study accurately the topology of low energy neutrino interactions. K2K has recorded 108 events in Super-Kamiokande in coincidence with the beam extraction. The number of events expected in absence of oscillations is 151. For one ring events in Super-Kamiokande the neutrino energy can be reconstructed starting from the muon energy under the hypothesis of a quasi-elastic neutrino interaction. The distortion of the energy spectrum strongly reinforces the evidence for neutrino oscillations and the best fit extracts a Δm^2 very close to the one from the atmospheric neutrino data.

7. MINOS

The MINOS experiment is designed in order to confirm the interpretation of the atmospheric neutrino anomaly in terms of neutrino oscillations and measure the oscillation parameters [13]. MINOS is a two-detector setup on the NUMI beam going from Fermilab to the Soudan mine in Minnesota (735 Km distance). The MINOS beam is obtained from 120 GeV protons extracted from the main injector. The optics of the beam is adjustable by moving the horns and the target in order to get different energy spectra and better tune the beam energy (from a few GeV average energy to about 15 GeV) to the baseline to be explored.

The MINOS far detector is a steel and scintillator tracking calorimeter with a total mass of 5400 tons. The iron is magnetized at 1.5 T. The iron plates of 2.54 cm thickness are sandwiched with scintillator strips of 1cm thickness and 4 cm width. The near detector has the same design as the far detector but a mass of just one Kton and faster electronics. The search for oscillations is performed by looking at the disappearance of ν_μ and its relative spectral distortion. The experiment can also look for ν_e appearance and discriminate between the ν_μ-ν_τ or ν_μ-$\nu_{sterile}$ hypothesis by measuring the neutral currents over charged currents ratio in the far and near detector. This will be the same in the two detectors in case of oscillations in the sterile neutrino while it would be larger in the far detector in case on ν_μ-ν_τ oscillations which just imply a deficit of the charged current events. The NUMI beam is actually under commissioning. Start of data taking is foreseen by the spring 2005.

8. The CNGS program

The CNGS program was proposed in order to provide an unambiguous evidence for ν_μ-ν_τ oscillations in the region of atmospheric neutrinos by looking for ν_τ appearance and to search for the sub-leading ν_μ-ν_e oscillations (measurement of θ_{13}). The program consists in the beam (CNGS approved in 1999) going from CERN to the Gran Sasso underground laboratory in Italy (732 Km distance) and by the two ν_τ appearance experiments: OPERA approved in the year 2000 and ICARUS approved in the year 2002. It is not foreseen to have near detectors since the intrinsic ν_τ contamination of the beam is negligible and the knowledge of the beam is less important like in the case of disappearance experiments where beam systematic is important and the spectral distortion has to be measured. The number of ν_τ events detected by the experiments is a convolution of the ν_μ flux, the oscillation probability and the ν_τ charged current cross-section. It can be easily demonstrated that for a distance small compared to the oscillations length (the CNGS has an average L/E of 43 Km/GeV compared to the 515 Km/GeV corresponding to the first oscillation peak of the atmospheric neutrinos) the rate dependency on Δm^2 factors out with respect to the dependency on the energy. Therefore, the rate of oscillated events will depend on $(\Delta m^2)^2$ and will be practically constant with the distance from the source. The energy dependency is completely defined by the convolution of the flux with the cross section divide by the energy squared. This integral can be optimized by matching the ν_μ fluence to the product of the cross-section for ν_τ CC by the oscillation probability. The resulting CNGS beam is therefore optimized in order to produce the maximum number of ν_τ charged current interactions in the detectors at Gran Sasso [14].

The detection of a ν_τ signal is a very challenging task, as seen from the past generation of short-baseline experiments at CERN NOMAD and CHORUS. There are two conflicting requirements: the high granularity (energy resolution/space resolution) needed by the ν_τ detection techniques and the large detector masses needed because of the distance from the neutrino source (the flux is 5 orders of magnitude smaller than in the case of short-baseline experiments) and because of the oscillation probability. By just rescaling these factors to the conventional techniques of the CERN short baseline experiments one would need to build detectors about three orders of magnitude larger than NOMAD and/or CHORUS. OPERA looks for the tau decay topology like CHORUS, needs high space resolution, and uses nuclear emulsions. ICARUS performs a kinematical search "a la NOMAD" and needs very good energy and angular resolution. Given the impossibility of a bare scaling of the conventional techniques this problem is solved by OPERA with the concept of the ECC (Emulsion Cloud Chamber) which is not anymore a target made of bulk emulsions but a sandwich of lead (which provides the mass for neutrino interactions) and emulsions. ICARUS is based on liquid argon calorimetry over large volumes and performing the readout from the surface solves the conflict between the large mass and the high granularity. Both experiments benefit of the experience gained by the past generation (NOMAD and CHORUS) of tau appearance experiments but the scaling to target masses three orders of magnitude larger was made possible only thanks to many years of R&D on the automatic fast scanning of large surfaces of emulsions (OPERA) and on the realization of the liquid Argon TPC and the liquid Argon purification

(ICARUS). Both detectors have excellent electron identification capabilities, which also allow studying the ν_e appearance.

9. ICARUS

ICARUS is a liquid Argon Time Projection Chamber. It detects the primary ionization in Argon (a minimum ionizing particle produces about 20000 electrons in 3 mm). The detector provides 3D event reconstruction with 1 mm space resolution and high-resolution calorimetric measurement of the electromagnetic and hadronic showers. Photomultipliers provide the detection of the UV scintillation light in Argon in order to define the time at which the event occurred. The drift field is about 1KV/cm and the drift length is typically 1.5 m, which requires keeping the level of oxygen contamination below 0.1 ppb.

A first 600 tons module has been built by the ICARUS collaboration. The complete project foresees the construction of a 3000 ton detector, in two modules of 1.2 kton and one of 600 tons. The ICARUS detector has also important applications in the non-accelerator physics like the search for proton decay, neutrinos from supernovae and atmospheric and solar neutrinos [14].

The ν_τ appearance search proceeds quite similarly as in the NOMAD analysis for the electronic decay channel of the tau by building a likelihood function with the visible energy of the electron, the missing transverse momentum and the angular correlations on the transverse plane. In 5 years of data taking with the full 3000 tons detector about 12 events of signal are expected (given the present oscillations parameters of the atmospheric neutrinos) with a background of 0.7 events

10. OPERA

The basic unit of the OPERA experiment is the "brick". The brick is based on the concept of the Emulsion Cloud Chambers [15] and it is a sandwich of 56 sheets of lead (1 mm thick) and 56 emulsion layers. By integrating the lead with the emulsions the brick solves the problem of the need for a large target mass and a high space resolution in a completely modular way. The bricks are indeed standalone particle detectors: they allow for the reconstruction of the neutrino interaction vertex and the tau kink decay topology and for the measurement of the momenta of the secondary charged particles by multiple Coulomb scattering; they allow for pion/muon separation at low energy by measuring the dE/dx. The bricks also provide electron identification and the measurement of the energies of electrons and photons by micro-calorimetry. The technique of the bricks has already been validated for the tau search by the DONUT experiment in the year 2000 with the first direct observation of the ν_τ.

The bricks are passive objects and do not give any information concerning the events trigger and the localization of the neutrino interactions. Muon identification and the muon momentum/charge measurement cannot be performed with the bricks. Electronic detectors integrated with the bricks in a hybrid experiment provide this kind of information. The electronic detectors localize in which brick the neutrino interacts and perform the muon identification. The localized brick can then be analyzed at microscopic level. Finally, by merging the information of the brick with the ones on the muon identification in the electronic detectors it is possible to define a tau candidate.

The OPERA experiment is organized in two super modules: each super module includes a target section and a magnetic spectrometer. The total mass of the bricks is about 1.8 Ktons. The target section is made of walls of bricks interleaved with planes of scintillator strips. The scintillator strips are used for the trigger, brick localization and to start the muon tracking and identification, which is finally completed in the spectrometers. Each magnetic spectrometer is made of iron slabs (5 cm thick) interleaved with planes of RPC (22 gaps/spectrometer) allowing for a fine measurement of the muon range. The iron is magnetized at 1.55 T and stations of drift tubes with 300 microns resolution measure the curvature of the trajectory of the muon. Automatic microscopes will scan the emulsion sheets at speeds larger than 20 cm^2/hour [14].

OPERA is sensitive to the electronic, muonic and single charged hadron decay channels of the $\tau^{\pm}$. The analysis includes both the decays where the kink topology can be completely measured (long decays, 40%) and the decays occurring in the same lead plates were the neutrino interacted (short decays, 60%). In this last case, the background rejection is based on impact parameter cuts. The main background comes from the decays of charmed particles. In case of the muonic channel this background is reduced by about a factor 18 by the measurement of the charge of the muon, the major background is represented for this channel by the Coulomb scattering of muons from ν_μ CC in the lead sheets. In 5 years, OPERA expects about 12 signal events with the present values of the atmospheric neutrino parameters and a background of about 0.7 events. OPERA can look for ν_e appearance with also the advantage of being capable of removing the "background" of electrons coming from the tau decays on a single event basis. The major background is the intrinsic ν_e contamination in the beam. The θ_{13} sensitivity of OPERA can be enhanced by performing a simultaneous fit of the visible energy, the electron energy and the missing transverse momentum. OPERA is sensitive in 5 years to values of θ_{13} down to 7°.

11. Conclusions

The past generation of short baseline experiments has been important to setup the techniques now used for tau appearance in the long baseline experiments. They allowed collecting very large samples of neutrino interactions allowing studying in details charm physics and the details of the hadronic system. This is the richest sample of neutrino interactions in the multi-GeV range. NOMAD also acquired very important experience in the control of beam systematic and the ν_e contamination. These experiments unfortunately did not see any oscillation signal.

In the actual scenario, it is very important to clarify the pending situation of LSND and its revolutionary implications on the neutrino oscillation framework. MiniBooNE will come out with first results probably in 2005. The current generation of long-baseline experiments represents a huge effort towards the final assessment of neutrino oscillations in the region of atmospheric neutrinos, a more precise measurement of the oscillation parameters and a first attempt to measure θ_{13}. The use of accelerators for the study of neutrino oscillations will have a long future with the era of precision measurements and the search for leptonic CP violation (superbeams, beta-beams, neutrino factories) [16]. As usual, the techniques currently studied for ν_τ appearance will be very useful in the future: like for instance the study of the ν_e-ν_τ oscillations to disentangle parameter ambiguities at neutrino factories.

Acknowledgments

I would like to thank the members of the various neutrino oscillation experiments for their cooperation.

References

1. S. Petcov, Neutrino Masses, Mixing and Oscillations, This School.
2. D.F. Cowen, Solar and Reactor Neutrinos, This School.
3. J. Ellis et al., Phys. Lett. **B292**(1992)189.
4. E. Eskut et al., CHORUS, Nucl. Instrum. Meth. A401(1997)7.
   J. Altegoer et al., NOMAD, Nucl. Instrum. Meth. A404(1998)96.
5. C.H. Albrigth et al., Phys. Rev. **D**20(1979)2177.
6. C. Athanassopoulos et al., LSND, Phys. Rev. Lett. 81(1998)1774.
   B. Armbruster et al., Karmen, Phys. Rev. **D**65(2002)112001.
7. H.L. Ray, Current status of the MiniBooNE experiment, hep-ex/0411022.

8. P.F. Harrison et al., Phys. Lett. **B349**(1995)137.
9. M. Apollonio et al., Palo Verde, CHOOZ, Phys. Lett. **B**466(1999)415.
   F. Boehm et al., Phys. Rev. Lett. 84 (2000) 3764.
10. G. Giacomelli and M. Giorgini, Atmospheric Neutrino Oscillations, This School.
11. Y. Fukuda et al., SuperKamiokande, Phys. Rev. Lett. 81(1998)1562.
    M. Ambrosio et al., MACRO, Phys. Lett. **B**434(1998)451.
12. E. Aliu et al., K2K, Evidence for muon neutrino oscillations in an accelerator-based experiment, hep-ex/0411038.
13. R. Saakian, Status and prospects of the MINOS experiment, Phys. Atom. Nucl. 67(2004)1084.
14. D. Duchesneau, The CERN - Gran Sasso neutrino program, Nucl. Phys. Proc. Suppl. 123(2003)279.
    S. Amerio et al., ICARUS, Nucl. Instrum. Meth. A527(2004)329.
    M. Dracos, The OPERA experiment, Phys. Atom. Nucl. 67(2004)1092.
15. K Kodama et al., DONUT, Phys. Lett. B504(2001)218.
16. R.A. Carrigan Jr., Future accelerators, neutrino factories and muon colliders, This School.

ATMOSPHERIC NEUTRINO OSCILLATIONS

G. GIACOMELLI AND M. GIORGINI
Physics Dept, University of Bologna, and INFN Sezione di Bologna
V.le Berti Pichat 6/2, I-40127 Bologna, Italy
E-mail: giacomelli@bo.infn.it, miriam.giorgini@bo.infn.it

The latest results from the Soudan 2, MACRO and SuperKamiokande experiments on atmospheric neutrino oscillations are summarised and discussed. In particular a discussion is made on the Monte Carlo simulations used for the atmospheric neutrino flux.

1. Introduction

Atmospheric neutrinos are generated in the decays of hadrons produced in high energy cosmic ray (CR) interactions. A high energy primary cosmic ray, proton or nucleus, interacts in the upper atmosphere producing a large number of pions and kaons, which decay yielding muons and muon neutrinos; also the muons decay yielding muon and electron neutrinos. The ratio of the numbers of muon to electron neutrinos is about 2 and $N_\nu/N_{\overline{\nu}} \simeq 1$. Atmospheric neutrinos are produced in a spherical surface at about 10-20 km above ground and they proceed towards the earth.

If neutrinos have non-zero masses, one has to consider the *weak flavour eigenstates* ν_e, ν_μ, ν_τ and the *mass eigenstates* ν_1, ν_2, ν_3. The weak flavour eigenstates ν_l are linear combinations of the mass eigenstates ν_m through the elements of the mixing matrix U_{lm}:

$$\nu_l = \sum_{m=1}^{3} U_{lm}\, \nu_m \tag{1}$$

For 2 flavour eigenstates (ν_μ, ν_τ) and 2 mass eigenstates (ν_2, ν_3)

$$\begin{cases} \nu_\mu = \nu_2 \cos\,\theta_{23} + \nu_3 \sin\,\theta_{23} \\ \nu_\tau = -\nu_2 \sin\,\theta_{23} + \nu_3 \cos\,\theta_{23} \end{cases} \tag{2}$$

where θ_{23} is the mixing angle. The survival probability of a ν_μ beam is

$$P(\nu_\mu \rightarrow \nu_\mu) = 1 - \sin^2 2\theta_{23}\ \sin^2\left(\frac{1.27\Delta m^2 \cdot L}{E_\nu}\right) \tag{3}$$

where $\Delta m^2 = m_3^2 - m_2^2$, L is the distance travelled by the ν from production to detection. The probability for the initial ν_μ to oscillate into a ν_τ is

$$P(\nu_\mu \rightarrow \nu_\tau) = 1 - P(\nu_\mu \rightarrow \nu_\mu) = \sin^2 2\theta_{23}\ \sin^2\left(\frac{1.27\Delta m^2 \cdot L}{E_\nu}\right) \tag{4}$$

Atmospheric neutrinos are well suited for the study of neutrino oscillations, since they have energies from a fraction of GeV up to more than 100 GeV and they travel distances L from few tens of km up to 13000 km; thus L/E_ν ranges from ~ 1 km/GeV to 10^5 km/GeV. Moreover one may consider that there are two identical sources for a single detector: a near one (downgoing neutrinos) and a far one (upgoing neutrinos). Atmospheric neutrinos are particularly useful to study oscillations for small Δm^2, and matter effects can be studied with their high energy component.

The early water Cherenkov detectors and the tracking calorimeters measured ν_μ and ν_e charged current interactions. The results were expressed in terms of the double ratio $R' = R_{obs}/R_{MC}$, where $R_{obs} = (N_{\nu_\mu}/N_{\nu_e})_{obs}$ is the ratio of observed μ and e events and $R_{MC} = (N_{\nu_\mu}/N_{\nu_e})_{MC}$ is the same ratio for Monte Carlo (MC) events. The R' double ratios from IMB[1] and Kamiokande[2] were smaller than expectations, while the NUSEX[3] and Frejus[4] R' agreed with expectations. The Baksan[5] scintillation telescope detected upthroughgoing muons at the expected rate but gave indications of an anomalous angular distribution. Later, the Soudan 2 tracking and shower calorimeter detector confirmed the anomaly in the ν_μ/ν_e double ratio for contained events[6]. MACRO reported in 1995 a measurement of upthroughgoing muons coming from ν_μ of $\langle E_\nu \rangle \sim 50$ GeV, in which there was an anomalous zenith distribution and a deficit in the total number of observed upgoing muons[7]. SuperKamiokande (SK) confirmed the anomalous double ratio and provided a wealth of informations for sub-GeV and multi-GeV ν_μ, ν_e and for higher energy upthroughgoing muons and stopping muons. In 1998 Soudan 2, MACRO and SK provided strong indications in favour of $\nu_\mu \longleftrightarrow \nu_\tau$ oscillations[8–11]. After 1998 new results were presented by the 3 experiments[12–14]. Here we shall review their results.

2. Atmospheric neutrino flux calculations

In the past use was made of unidimensional Monte Carlo codes, Bartol96[15] and HKKM95[16]. Recently new improved MC predictions for neutrino fluxes were made available by the HKKM01[17] and FLUKA[18] groups. They include three dimensional calculations of hadron production and decays and of neutrino interactions, improved hadronic model and new fits of the primary CR flux. The two MCs yield predictions for the non oscillated and oscillated ν_μ fluxes equal to within few %. The shapes of the angular distributions for oscillated and non oscillated Bartol96, HKKM95, new FLUKA and new HKKM01 fluxes are the same to within few %. The absolute values of the MACRO upthroughgoing muon data are about 25% higher than those predicted by the new FLUKA and HKKM01 MC[13], while the shapes of the oscillated and non oscillated angular distributions differ by no more than 5%, see Sec. 4.

A similar situation is found in the SK data[14], see Sec. 5. The electron-like events were in agreement with the HKKM95[16] MC predictions in absence of oscillations, while they are higher than the HKKM01[17] non oscillated MC, Fig. 7. For the muon-like events, the new MC predictions are low for the SK data, especially for the high energy upthroughgoing events[14]. Previous comparisons between the SK muon data and the HKKM95[16] predictions showed a global deficit of events and a zenith distribution in agreement with $\nu_\mu \longleftrightarrow \nu_\tau$ oscillations[14].

The difference between the new and old MC predictions is very probably due to the use of a new fit of the cosmic ray data[19].

Recent results by the L3C and BESS experiments[20] on the primary cosmic ray fit show good agreement with the Bartol96 and HKKM95 predictions and a disagreement with the new fit of the cosmic ray data[19].

The calculations of HKKM01 and FLUKA are in good agreement when using the old fit or the new fit to the primary CR flux; this confirms the improvement in the hadronic model.

In SK the evidence for neutrino oscillations lies in the shapes of the angular distribution and in the ratio of μ/e data. In MACRO the evidence is due mainly to the shape of the high energy angular distribution and the prediction is the same in all simulations.

The MACRO data suggest that the FLUKA normalisation should be raised by $\sim 25\%$ at $E_\nu \sim 50$ GeV and by 12% at $E_\nu \sim 2-3$ GeV [13]. Similar conclusions are reached by SK[14].

3. Results from the Soudan 2 experiment

The Soudan 2 experiment used a modular fine grained tracking and showering calorimeter of 963 t, located 2100 m.w.e. underground in the Soudan

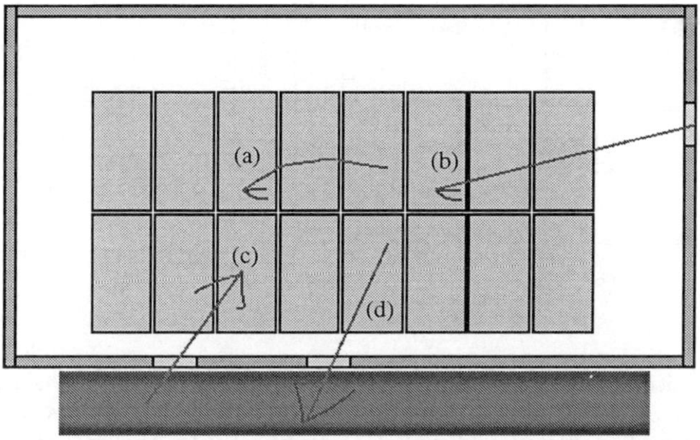

Figure 1. Longitudinal cross section of the Soudan 2 detector and observed event topologies: (a) Fully Contained Events $\langle E_\nu \rangle \sim 1$ GeV, (b) Partially Contained Events $\langle E_\nu \rangle \sim 6$ GeV, (c) In-down muons $\langle E_\nu \rangle \sim 2.4$ GeV, (d) Up-stopping muons $\langle E_\nu \rangle \sim 6.2$ GeV.

Gold mine in Minnesota. The bulk of the mass consisted of 1.6 mm thick corrugated steel sheets interleaved with drift tubes. The detector was surrounded by an anticoincidence shield. Figure 1 shows a longitudinal cross section of the apparatus and the topologies of the events observed. The final analysis used the Fully Contained High-Resolution events[12], Table 1.

Table 1. Soudan 2 Hi-Res data.

	Data	$MC_{\text{no osc}}$[15]
Track	101.9 ± 12.7	193.1
Showers	146.7 ± 12.5	179.0

The fully contained events consist mostly of quasi-elastic neutrino reactions, but include a background of photons and neutrons from cosmic ray muon interactions in the surrounding rock. The track and shower events for a 5.9 kt·yr exposure are summarised in Table 1, where they are compared with MC predictions based on the Bartol96 neutrino flux[15].

After corrections for background and selecting a high resolution (Hi-Res) sample of events, the Soudan 2 double ratio for the whole zenith angle range ($-1 \leq \cos\Theta \leq 1$) is $R' = (N_\mu/N_e)_{DATA}/(N_\mu/N_e)_{MC} = 0.69 \pm 0.12$, consistent with muon neutrino oscillations.

The ν_e data agree with the no oscillation MC predictions, while the ν_μ data are lower, except in the forward zenith bin. The double peak structure

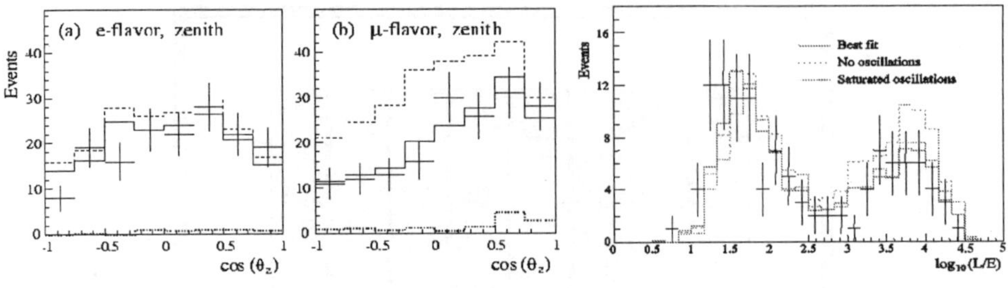

(a) (b) (c)

Figure 2. Zenith angle distribution for Hi-Res $e-$flavour (a) and $\mu-$flavour (b) events. The points with error bars are the data, the dashed and solid histograms are the predicted non oscillated and oscillated ν distributions, respectively, the dotted histograms are the contribution of the rock background. (c) Distribution in $\log L/E_\nu$ for the Hi-Res Fully Contained events (black crosses) compared with the MC predictions for oscillations (solid histogram), no oscillations (dashed histogram) and saturated oscillations (dotted histogram).

arises from the acceptance of the apparatus. The roughly interpolated 90% C.L. allowed region in the $\sin^2 2\theta - \Delta m^2$ plane, computed using the Feldman-Cousins method[21] is shown in Fig. 8, where it is compared with the allowed regions obtained by the SK and MACRO experiments.

4. Results from the MACRO experiment

MACRO was a large area multipurpose underground detector designed to search for rare events and rare phenomena in the penetrating cosmic radiation. It was located in Hall B of the Gran Sasso Lab at an average rock overburden of 3700 m.w.e.; it started data taking with part of the apparatus in 1989; it was completed in 1995 and was running in its final configuration until the end of 2000. The detector had global dimensions of $76.6 \times 12 \times 9.3$ m^3; vertically it was divided into a lower part, which contained 10 horizontal layers of streamer tubes, 7 of rock absorbers and 2 layers of liquid scintillators, and an upper part which contained the electronics and was covered by 1 scintillator layer and 4 layers of streamer tubes. The sides were covered with 1 vertical scintillator layer and 6 of limited streamer tubes[22].

MACRO detected upgoing ν_μ's via charged current interactions $\nu_\mu \to \mu$; upgoing muons were identified with the streamer tube system (for tracking) and the scintillator system (for time-of-flight measurement). The events measured and expected for the 3 measured topologies, Table 2, and the

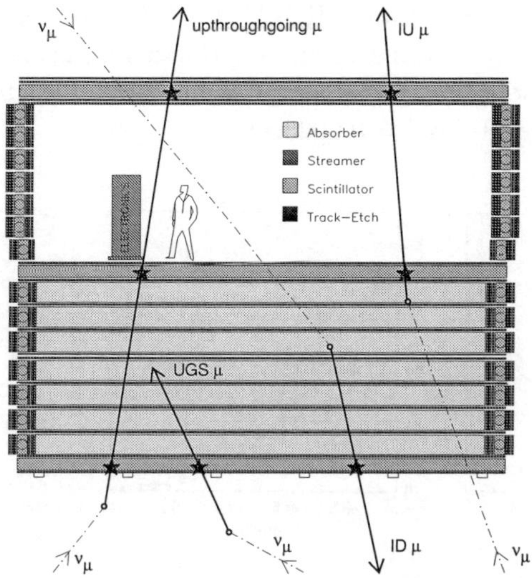

Figure 3. Cross section of the MACRO detector. Event topologies induced by ν_μ interactions in or around the detector. IU_μ = semicontained Internal Upgoing μ; ID_μ = Internal Downgoing μ; UGS_μ = Upgoing Stopping μ; Upthroughgoing = upward throughgoing μ.

L/E_ν distribution, Fig. 5a, deviate from MC expectations without oscillations; the deviations point to the same $\nu_\mu \longleftrightarrow \nu_\tau$ oscillation scenario[13].

Table 2. MACRO events.

	Events	MCno osc[15]	R = Data/MCno osc
Upthr.	857	1169	0.73
IU	157	285	0.55
ID+UGS	262	375	0.70

Upthroughgoing muons ($E_\mu > 1$ GeV) come from interactions in the rock below the detector of ν_μ with $\langle E_\nu \rangle \sim 50$ GeV. The MC uncertainties arising from the neutrino flux, cross section and muon propagation on the expected flux of muons are estimated to be $\sim 17\%$; this systematic error on the upthroughgoing muons flux is mainly a scale uncertainty.

In order to verify that different flux simulations affect the zenith distribution at the level of only a few percent (while there is an effect of the order of $\sim 25\%$ on the event rates) MACRO compared the predictions of the Bartol96[15], FLUKA[18] and HKKM01[17] MCs. In Fig. 4a the MACRO data are compared with the oscillated Bartol96, the new HKKM01 and FLUKA calculations using the new CR fit. All predicted curves are for maximal mixing and $\Delta m^2 = 0.0023$ eV^2.

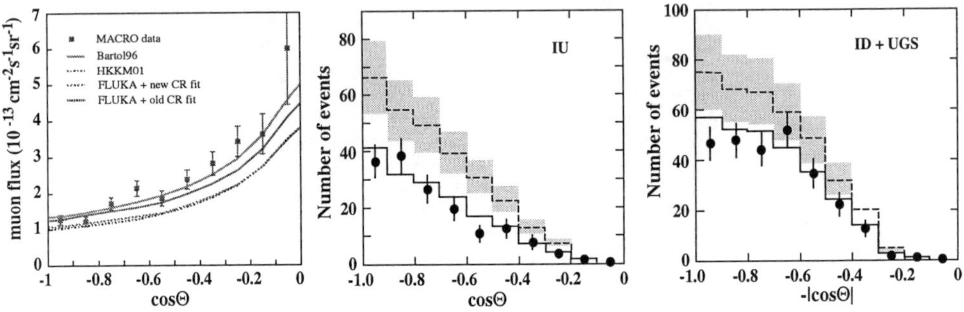

(a) (b) (c)

Figure 4. (a) Comparison between the zenith distribution of the MACRO upthroughgoing muons and the oscillated MC predictions given by Bartol96 (solid curve), HKKM01 (dash-dotted line), FLUKA fitted to the new CR measurements (dashed curve) and FLUKA with the old CR fit (dotted curve). Zenith distributions for (b) IU and for (c) ID+UGS MACRO events (black points) compared with the no oscillation Bartol96 MC (dashed line with a scale error band) and with the $\nu_\mu \longleftrightarrow \nu_\tau$ predictions with $\Delta m^2 = 2.3 \cdot 10^{-3}$ eV^2 and maximal mixing.

Low energy events. *Semicontained upgoing muons* (IU) come from ν_μ interactions inside the lower apparatus. The average parent neutrino energy for these events is $\sim 2-3$ GeV. *Up stopping muons* (UGS) are due to external ν_μ interactions yielding upgoing muons stopping in the detector; the *semicontained downgoing muons* (ID) are due to downgoing ν_μ's with interaction vertices in the lower detector. The lack of time information prevents to distinguish between the two subsamples. An almost equal number of UGS and ID events is expected. The average parent neutrino energy for these events is $\sim 2-3$ GeV. The number of events and the angular distributions are compared with the MC predictions without oscillations in Table 2 and Fig. 4b,c. The low energy data show a uniform deficit of the measured number of events over the whole angular distribution with respect to the Bartol96 predictions.

$\nu_\mu \longleftrightarrow \nu_\tau$ against $\nu_\mu \longleftrightarrow \nu_{sterile}$. Matter effects due to the difference between the weak interaction effective potential for muon neutrinos with respect to sterile neutrinos, which have null potential, yield different total number and different zenith distributions of upthroughgoing muons. The measured ratio between the events with $-1 < \cos\Theta < -0.7$ and with $-0.4 < \cos\Theta < 0$ was used[13]. In this ratio most of the theoretical uncertainties on neutrino flux and cross sections cancel. Combining the experimental and theoretical errors in quadrature, a global uncertainty of 6% is obtained. The measured ratio is $R_{meas} = 1.38$, to be compared with $R_\tau = 1.61$ and $R_{sterile} = 2.03$. One concludes that $\nu_\mu \longleftrightarrow \nu_{sterile}$ oscillations (with any mixing) are excluded at the 99.8% C.L. compared to the $\nu_\mu \longleftrightarrow \nu_\tau$ channel with maximal mixing and $\Delta m^2 = 2.3 \cdot 10^{-3}$ eV^2.

ν_μ energy estimate by Multiple Coulomb Scattering of muons. Since MACRO was not equipped with a magnet, the only way to estimate the muon energy is through their Multiple Coulomb Scattering (MCS) in the absorbers. Two analyses were performed[23]. The first was made studying the deflection of upthroughgoing muons using the streamer tubes in digital mode. This method had a spatial resolution of ~ 1 cm. The second analysis was performed using the streamer tubes in "drift mode"[23]. The space resolution was $\simeq 3$ mm. For each muon, 7 MCS variables were defined and given in input to a Neural Network, previously trained with MC events of known energy crossing the detector at different zenith angles. The output of this program gave the muon energy estimate event by event. The sample of upthroughgoing muons was separated in 4 subsamples with average energies E_μ of 12, 20, 50 and 100 GeV. The ratios Data/MC$_{\text{no osc}}$ as a function of $\log_{10}(L/E_\nu)$ obtained from upthroughgoing muons are plotted in Fig. 5a; they are in agreement with the $\nu_\mu \longleftrightarrow \nu_\tau$ oscillation hypothesis[13].

New determination of the oscillation parameters. In previous analyses MACRO fitted the shape of the upthroughgoing muon zenith distribution and the absolute flux compared to Bartol96. This yielded $\Delta m^2 = 2.5 \cdot 10^{-3}$ eV^2 and maximal mixing[13]. Later, in order to reduce the effects of systematic uncertainties in the MC simulations, MACRO used the following three independent ratios. It was checked that FLUKA, HKKM01 and Bartol96 Monte Carlo simulations yield the same predictions to within $\sim 5\%$.

(i) High Energy Data: zenith distribution ratio: $R_1 = N_{vert}/N_{hor}$

(ii) High Energy Data, ν energy measurement ratio: $R_2 = N_{low}/N_{high}$

(iii) Low Energy Data: $R_3 = (Data/MC)_{IU}/(Data/MC)_{ID+UGS}$.

The no oscillation hypothesis has a probability $P \sim 3 \cdot 10^{-7}$ and is thus ruled out by $\sim 5\sigma$. By fitting the 3 ratios to the $\nu_\mu \longleftrightarrow \nu_\tau$

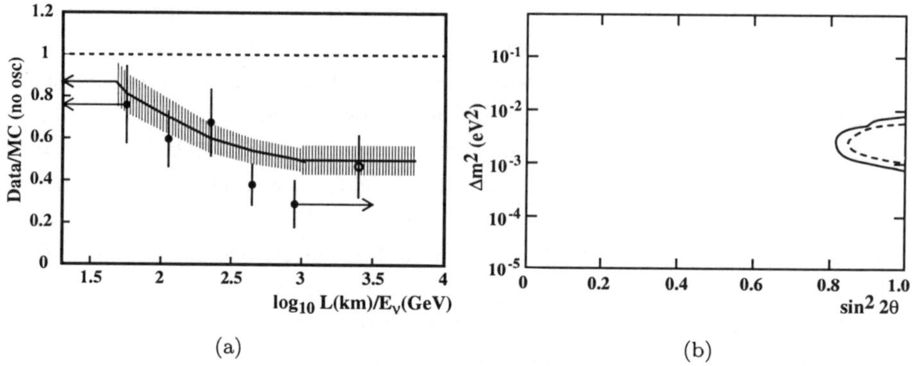

Figure 5. (a) Ratio Data/MCno osc as a function of the estimated L/E_ν for the upthroughgoing muon sample (black points). The solid line is the MC expectation assuming $\Delta m^2 = 2.3 \cdot 10^{-3}$ eV^2 and $\sin^2 2\theta = 1$. The last point (empty circle) is obtained from the IU sample. (b) Interpolated 90% C.L. contour plots of the allowed regions in the $\sin^2 2\theta - \Delta m^2$ plane for the MACRO data using only the ratios R_1, R_2, R_3 (continuous line) and adding also the information on the absolute values R_4, R_5 (dotted line).

oscillation formulae, MACRO obtained $\sin^2 2\theta = 1$, $\Delta m^2 = 2.3 \cdot 10^{-3}$ eV^2 and the allowed region indicated by the solid line in Fig. 5b. There is a good consistency between the old and new methods.

Using the Bartol96 flux, it is possible to add the information on the absolute flux values of the

(iv) High energy data (systematic error $\simeq 17\%$) $R_4 = N_{meas}/N_{MC}$.

(v) Low energy semicontained muons (scale error 21%) $R_5 = N_{meas}/N_{MC}$.

These informations reduce the area of the allowed region in the $\sin^2 2\theta - \Delta m^2$ plane, as indicated by the dashed line in Fig. 5b. The final MACRO best fit is $\Delta m^2 = 2.3 \cdot 10^{-3}$ eV^2 and $\sin^2 2\theta = 1$ (6σ significance).

5. Results from the SuperKamiokande experiment

SuperKamiokande[14] is a large cylindrical water Cherenkov detector of 39 m diameter and 41 m height containing 50 kt of water (the fiducial mass of the detector for atmospheric neutrino analyses is 22.5 kt); it was seen by 11146, 50-cm-diameter inner-facing phototubes. The 2 m thick outer layer of water, acting as an anticoincidence, was seen by 1885 smaller outward-facing photomultipliers. The ultra pure water has a light attenuation of almost 100 m. The detector is located in the Kamioka mine, Japan, under

2700 m.w.e. SK took data in its full configuration from April 1996 till November 2001, when an accident happened. It resumed operation with about half of PMTs in October 2002.

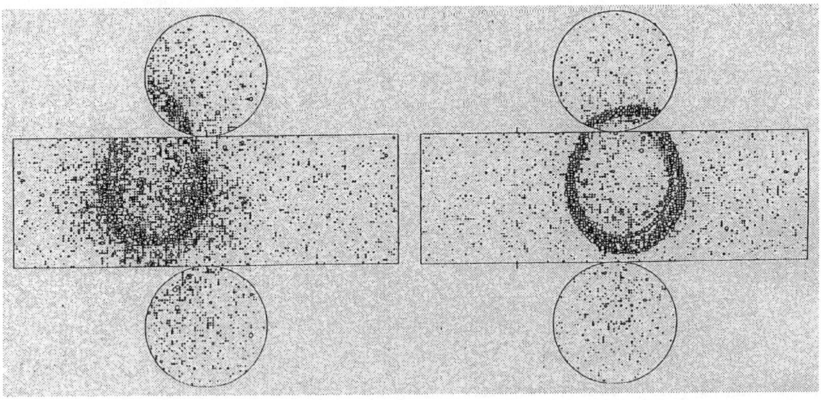

(a) (b)

Figure 6. Sharpness of Cherenkov rings produced in the SK detector by (a) an electron and (b) a muon.

Atmospheric neutrinos are detected in SK by measuring the Cherenkov light generated by the charged particles produced in the neutrino CC interactions with the protons and oxygen nuclei. The large detector mass and the possibility of clearly defining a large inner volume allow to collect a high statistics sample of *fully contained* events (FC) up to relatively high energies (up to $\sim$ 5 GeV). The FC events have both the neutrino vertex and the resulting particle tracks entirely within the fiducial volume; they yield rings of Cherenkov light on the PMTs. Fully contained events can be further subdivided into two subsets, the so-called *sub-GeV* and *multi-GeV* events, with energies below and above 1.33 GeV, respectively.

Another sub-sample, defined as the *partially contained* events (PC), is represented by those CC interactions where the vertex is still within the fiducial volume, but at least a primary charged particle, typically the muon, exits the detector without releasing all of its energy. In this case the light pattern is a filled circle. For these events the energy resolution is worse than for FC interactions. *Upward-going muons* (UPMU), produced by neutrinos coming from below and interacting in the rock, are further subdivided into *stopping muons* ($\langle E_\nu \rangle \sim 7$ GeV) and *upthroughgoing muons* ($\langle E_\nu \rangle \sim 70 \div 80$ GeV), according to whether or not they stop in the detector.

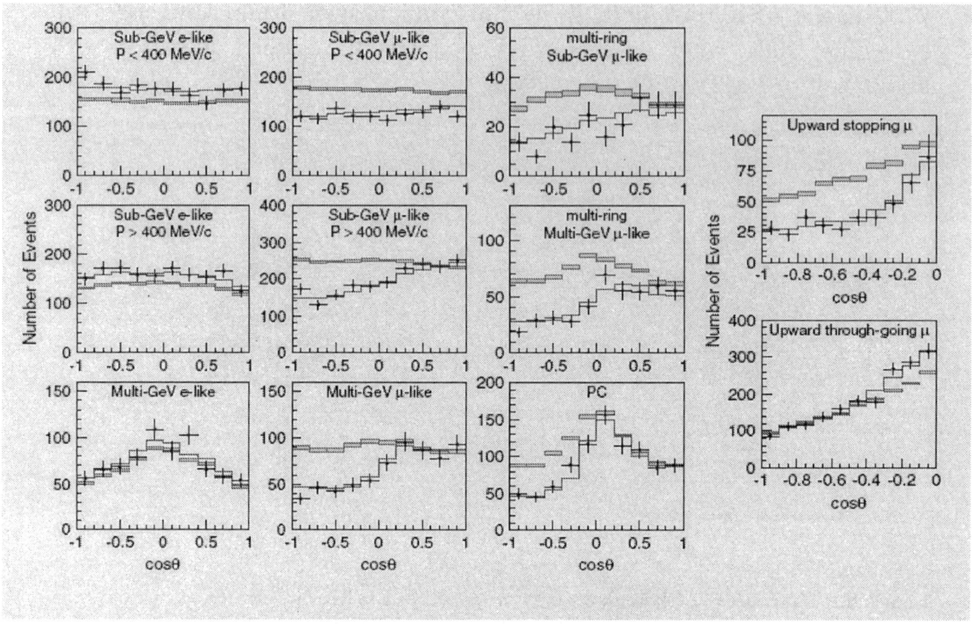

Figure 7. SK data taken from 1996 till 2001 with the detector in full configuration (1489 days for FC+PC events and 1646 days for upgoing μ). Zenith distributions for e-like and μ-like sub-GeV and multi-GeV events, for partially contained events and for upthroughgoing and stopping muons (black points). The boxes are the no oscillation HKKM01 predictions, the solid lines refer to $\nu_\mu \longleftrightarrow \nu_\tau$ oscillations with maximal mixing and $\Delta m^2 = 2.4 \cdot 10^{-3}$ eV^2.

Particle identification in SuperKamiokande is performed using likelihood functions to parametrise the sharpness of the Cherenkov rings, which are more diffuse for electrons than for muons, Fig. 6. The algorithms are able to discriminate the two flavours with high purity (of the order of 98% for single track events). The zenith angle distributions for e-like and μ-like sub-GeV and multi-GeV events, for PC events and for upward throughgoing or stopping muons are shown in Fig. 7. These data were taken from 1996 till 2001 with the detector in full configuration (1489 days for FC+PC events and 1646 days for upgoing μ). The data and MC behaviour shows the problem with the new HKKM01 MC discussed in Sec. 2. The number of measured and expected μ-like events are summarised in Table 3. The new data, taken in 2003 with about half of the PMTs and referring to 311 days for FC+PC events and 243 days for upgoing muons, show the same behaviour[14] as the older data.

Table 3. SK μ data taken from 1996 to 2001.

	Data	$MC_{\text{no osc}}$[17]
Sub-GeV 1-ring	3227	4213
Sub-GeV multi-ring	208	323
Multi-GeV 1-ring	651	900
Multi-GeV multi-ring	439	712
PC μ	647	1034
Up-stop	418	721
Upthr.	1842	1684

New analyses have been performed leaving free the normalisation. The last value for the double ratio R' reported by SK is $0.658 \pm 0.016_{stat} \pm 0.032_{sys}$ for the sub-GeV sample and $0.702 \pm 0.031_{stat} \pm 0.099_{sys}$ for the multi-GeV sample (both FC and PC).

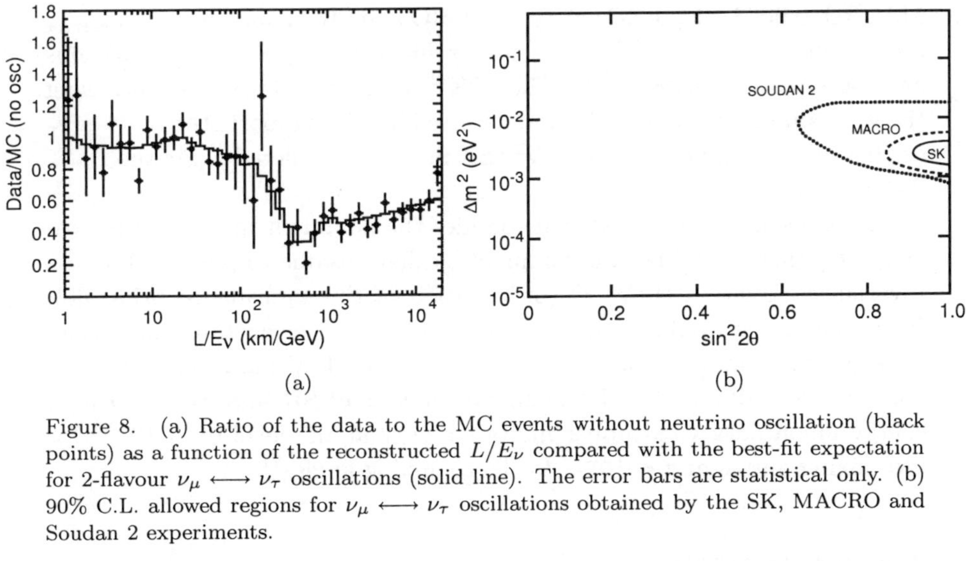

(a) (b)

Figure 8. (a) Ratio of the data to the MC events without neutrino oscillation (black points) as a function of the reconstructed L/E_ν compared with the best-fit expectation for 2-flavour $\nu_\mu \longleftrightarrow \nu_\tau$ oscillations (solid line). The error bars are statistical only. (b) 90% C.L. allowed regions for $\nu_\mu \longleftrightarrow \nu_\tau$ oscillations obtained by the SK, MACRO and Soudan 2 experiments.

SK used also a selected sample of events with good resolution in L/E_ν, to search for the dip in the oscillation probability expected when the argument of the second sine-squared term in Eq. (4) is $\pi/2$. A dip in the

L/E_ν distribution is observed at $L/E_\nu \simeq 500$ km/GeV, see Fig. 8a. This is another proof in favour of ν oscillations and a further constraint on Δm^2. Alternative models that could explain the zenith angle and energy dependent deficit of the atmospheric muon neutrinos are disfavoured, since they do not predict any dip in the L/E_ν distribution[14].

Interpreting the μ-like event deficit as the result of $\nu_\mu \longleftrightarrow \nu_\tau$ oscillations in the two-flavour mixing scheme, SK computed an allowed domain for the oscillation parameters[14], see Fig. 8b. The events were binned in a multidimensional space defined by particle type, energy and zenith angle, plus a set of parameters to account for systematic uncertainties. The best fit using FC, PC, UPMU and MRING events[14] corresponds to maximal mixing and $\Delta m^2 = 2.4 \cdot 10^{-3}$ eV^2. In Fig. 8b, the 90% SK allowed region in the $\sin^2 2\theta - \Delta m^2$ plane is compared with the MACRO and Soudan 2 ones. The limit lines represent smoothed interpolations and are qualitative.

6. Conclusions

The atmospheric neutrino data strongly favour $\nu_\mu \longleftrightarrow \nu_\tau$ oscillations with maximal mixing and $\Delta m^2 = 0.0023 - 0.0052$ eV^2 (Soudan 2: 0.0052, MACRO: 0.0023, SK: 0.0024 eV^2). MACRO and SK exclude $\nu_\mu \longleftrightarrow \nu_{sterile}$ oscillations at the level of 99.9%; SK excludes also $\nu_\mu \longleftrightarrow \nu_e$ oscillations and some exotic processes[14,24]. The K2K [25] long baseline experiment, using the ν_μ beam produced at KEK and detected by SK, supports the $\nu_\mu \longleftrightarrow \nu_\tau$ oscillations with $\Delta m^2 = 0.0027$ eV^2. SK finds the predicted dip in the L/E_ν distribution.

It has been hypothesized that, besides the dominant mass neutrino oscillations, there could be sub-dominant oscillations due to possible Lorentz Invariance Violation (LIV) [26] (or violation of the equivalence principle). In this case, one would consider a mixing of flavour eigenstates and velocity eigenstates and estimate upper limits on the LIV parameters $\delta v/2 = (v_3 - v_2)/2$ and $\sin^2 2\theta_v$. Preliminary analyses of SK and MACRO high energy muon data yield upper limits for $\delta v/2$ at the level of $\sim 10^{-23}$ for low mixing angles and $\sim 10^{-26}$ for large mixing angles[27].

Acknowledgements

We would like to acknowledge the cooperation of many experimental and theoretical colleagues from the MACRO, Soudan 2 and SK experiments. Particular thanks to the colleagues of the Bologna group.

References

1. IMB Coll., R. Becker-Szendy et al., *Phys. Rev.* **D46**, 372 (1992).
2. Kamiokande Coll., Y. Fukuda et al., *Phys. Lett.* **B335**, 237 (1994).
3. NUSEX Coll., M. Aglietta et al., 23rd ICRC Proc., Vol. 4 (1993) 446.
4. Frejus Coll., K. Daum et al., *Z. Phys.* **C66**, 417 (1995).
5. Baksan Coll., S. Mikheyev, 5th TAUP Workshop, Gran Sasso, Italy, 1997.
6. Soudan 2 Coll., W.W.M. Allison et al., *Phys. Lett.* **B391**, 491 (1997).
7. MACRO Coll., S. Ahlen et al., *Phys. Lett.* **B357**, 481 (1995).
8. Soudan 2 Coll., W.W.M. Allison et al., *Phys. Lett.* **B449**, 137 (1999).
9. MACRO Coll., M. Ambrosio et al., *Phys. Lett.* **B434**, 451 (1998); F. Ronga et al., *Nucl. Phys.* B *Proc. Suppl.* **77**, 117 (1999).
10. SuperKamiokande Coll., Y. Fukuda et al., *Phys. Rev. Lett.* **81**, 1562 (1998); *Phys. Lett.* **B433**, 9 (1998).
11. G. Giacomelli et al., hep-ph/9901355 (1999).
12. Soudan 2 Coll., M. Sanchez et al., *Phys. Rev.* **D68**, 113004 (2003).
13. MACRO Coll., M. Ambrosio et al., *Phys. Lett.* **B478**, 5 (2000); *Phys. Lett.* **B517**, 59 (2001); *Eur. Phys. J.* **C36**, 323 (2004). G. Giacomelli et al., *Phys. Atom. Nucl.* **67**, 1139 (2004); hep-ex/0110021; hep-ex/0201032; *Braz. J. Phys.* **33**, 211 (2003); *Eur. Phys. J.* **C33**, 5826 (2004). M. Giorgini et al., hep-ex/0210008.
14. SuperKamiokande Coll., Y. Fukuda et al., *Phys. Rev. Lett.* **85**, 3999 (2000); *Nucl. Phys.* B *Proc. Suppl.* **91**, 127 (2001). Y. Ashie et al., *Phys. Rev. Lett.* **93**, 101801 (2004).
15. V. Agrawal et al., *Phys. Rev.* **D53**, 1314 (1996).
16. M. Honda et al., *Phys. Rev.* **D52**, 4985 (1995).
17. M. Honda et al., *Phys. Rev.* **D64**, 053011 (2001); *Phys. Rev.* **D70**, 043008 (2004).
18. G. Battistoni et al., *Astrop. Phys.* **19**, 269 (2003).
19. T.K. Gaisser et al., *Nucl. Phys.* B *Proc. Suppl.* **118**, 109 (2003).
20. L3C Coll., P. Le Coultre et al.; BESS Coll., T. Sanuki et al., Proc. of the Neutrino Oscillation Workshop NOW2004, Otranto, Italy, 2004.
21. G.J. Feldman and R.D. Cousins, *Phys. Rev.* **D57**, 3873 (1998).
22. MACRO Coll., S. Ahlen et al., *Nucl. Instr. Meth.* **A324**, 337 (1993). M. Ambrosio et al., *Nucl. Instr. Meth.* **A486**, 663 (2002).
23. MACRO Coll., M. Ambrosio et al., *Nucl. Instr. Meth.* **A492**, 376 (2002); *Phys. Lett.* **B566**, 35 (2003).
24. A. Habig et al., Proc. of the 28th ICRC, Tsukuba, Japan (2003), Academic Press, pag. 1255.
25. D. Kielczewska et al., *Acta Phys. Polon.* **B35**, 1933 (2004). T. Ishii et al., hep-ex/0406055.
26. S. Coleman and S.L. Glashow, *Phys. Lett.* **B405**, 249 (1997). G. Lambiase, *Phys. Lett.* **B560**, 1 (2003). S.L. Glashow, hep-ph/0407087 (2004).
27. G. Fogli et al., *Phys. Rev.* **D60**, 053006 (1999). G. Giacomelli et al., Proc. of the Neutrino Oscillation Workshop NOW2004, Otranto, Italy, 2004.

SOLAR AND REACTOR NEUTRINOS

D.F. COWEN
Pennsylvania State University

An overview of the current state of solar and reactor neutrino measurements is presented, with an emphasis on the history of the field and the experiments that made and are making relevant measurements. The distinction between radiochemical and real-time Cherenkov experiments is described, and the results from the SNO and KAMLAND experiments and future plans are covered in some detail.

1. The Solar Neutrino Problem

In the early 1960s, Bahcall and Davis proposed building a radiochemical detector in the Homestake mine [1] to confirm, through detection of solar neutrinos, that the sun was powered by fusion reactions in the solar core. The Homestake experiment proceeded to measure a persistent *deficit* of solar neutrinos, and this deficit was later confirmed some decades later by other radiochemical experiments, GALLEX and SAGE [2], and by the water Cherenkov experiments, Kamiokande and SuperKamiokande [3].

These experiments were sensitive to different regions of the solar neutrino energy spectrum, as shown in the figure below. The Homestake experiment consistently reported a flux deficit for nearly three decades, and once it became clear that all solar neutrino experiments, using a variety of techniques and sensitive to different regions of the solar neutrino energy spectrum, also measured a deficit, there was impetus to construct experiments that would specifically address the issue. These dedicated experiments are the heavy water Cherenkov experiment, SNO, and the reactor-neutrino experiment, KamLAND [4]. The solar neutrino "pp" chain, the spectral sensitivities of many of these experiments, and a summary of their results, are given in Figure 1, Figure 2, and Figure 3, respectively.

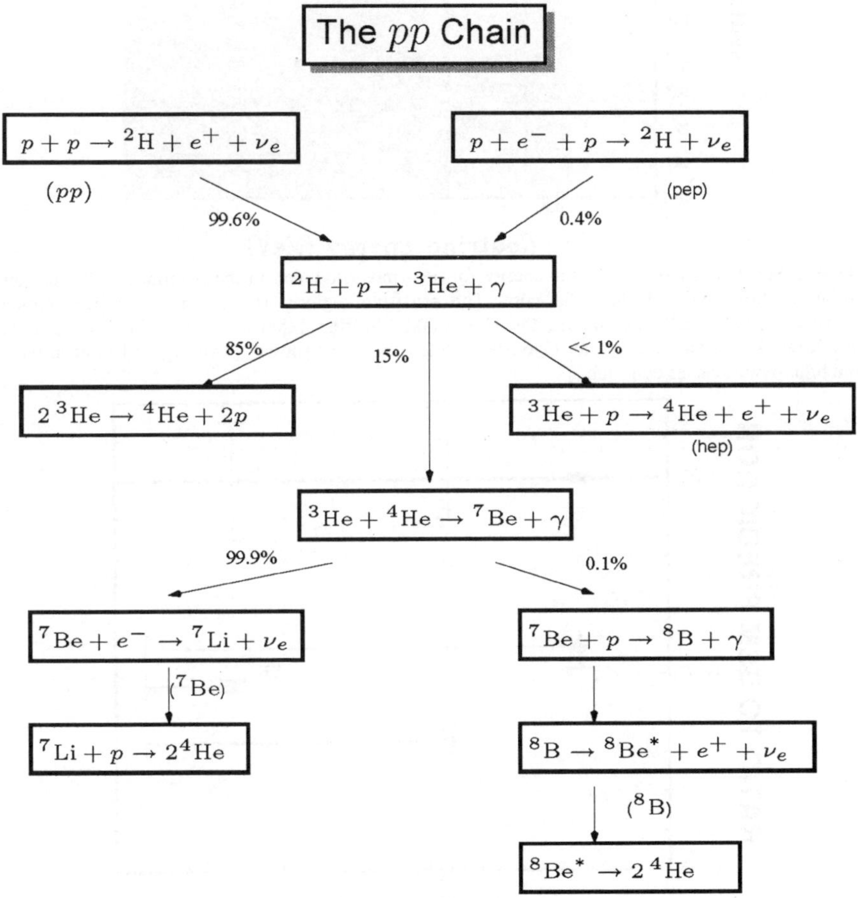

Figure 1: Solar neutrino reaction chain [5].

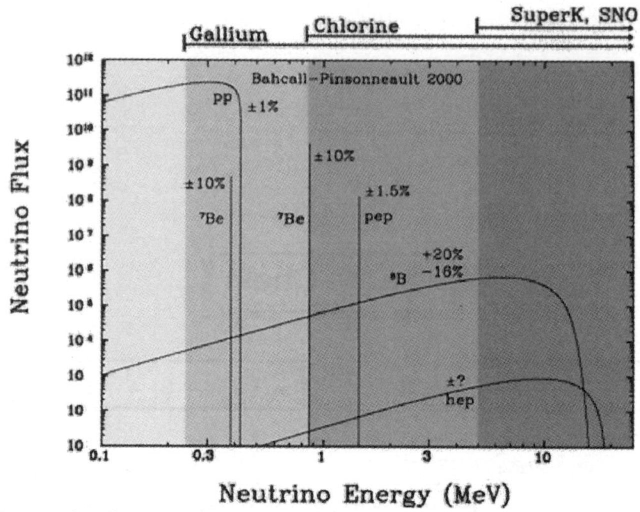

Figure 2: Solar neutrino flux versus energy for neutrinos produced in the reactions of the proton-proton reaction chain in the solar core. The sensitive regions of the Gallium radiochemical experiments, GALLEX and SAGE, the Homestake chlorine experiment, and the Kamiokande, SuperKamiokande and SNO water Cherenkov detectors are as shown. [This figure has been taken from http://www.sns.ias.edu/~jnb/.]

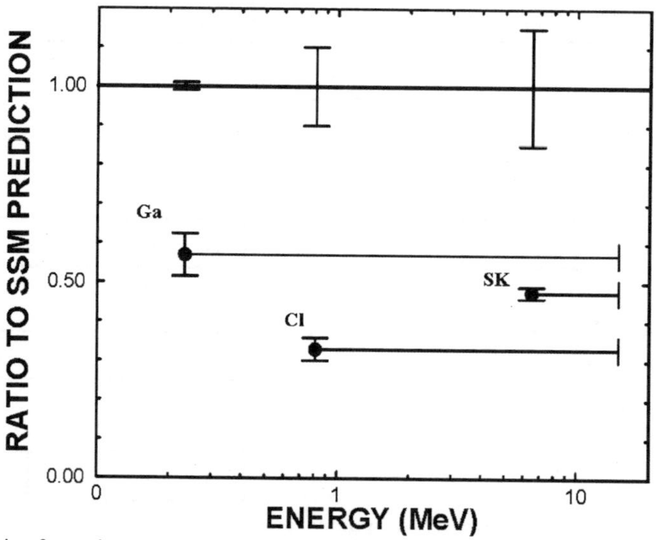

Figure 3: Ratio of experimental solar neutrino measurements to the Standard Solar Model (SSM) [6], indicating the deficit relative to predicted neutrino flux seen by all types of experiments. (Figure taken from astro-ph/0204245.)

2. The Early Detectors

To understand why the early solar neutrino detectors measured a neutrino flux deficit, we must first understand how these detectors were built.

2.1. *Homestake*

The Homestake experiment consisted of a tank 20 feet in diameter and 48 feet long. The tank contained 615 tons of perchloroethylene (cleaning fluid) and was located 4,900 feet below the ground in a mine in Lead, South Dakota, USA.

The Homestake experiment measured neutrinos through the reaction

$$\nu_e + {}^{37}\mathrm{Cl} \text{ ⑧ } e^- + {}^{37}\mathrm{Ar}.$$

This reaction has an energy threshold of 0.814 MeV. The ^{37}Ar atoms were detected in a cold trap when they decayed, at the rate of roughly one per day. Note that Homestake could only detect electron-flavor neutrinos and that the experiment was insensitive to other flavors.

2.2. *GALLEX and SAGE*

The GALLEX and SAGE experiments were also radiochemical experiments, but they used the reaction

$$\nu_e + {}^{71}\mathrm{Ga} \text{ ⑧ } e^- + {}^{71}\mathrm{Ge}.$$

This reaction has an energy threshold of 0.233 MeV, low enough to be sensitive to the low energy "pp" neutrinos produced in the solar core. To make the flux measurement, Germanium atoms are extracted and measured at the rate of roughly one per day. As with Homestake, the GALLEX and SAGE experiments could only detect electron-flavor neutrinos.

2.3. *Kamiokande and SuperKamiokande*

Kamiokande and its successor, SuperKamiokande, were the first large-scale water Cherenkov detectors that were sensitive to low energy solar electron neutrinos (and, to a lesser extent, muon and tau neutrinos, but without the ability to distinguish them from one another). The SuperKamiokande detector consists of approximately 11,000 photomultiplier tubes (PMTs) arranged in a cylindrical 40m x 40m geometry surrounding a 50 kton purified water volume, of which about 22 kton is used as the fiducial volume. The device is located about 1 km underground in Japan, and detected roughly 150 neutrinos daily.

The chief advantages of this type of detector over the radiochemical variety are that it can detect many more neutrinos, detect them in real time and measure the travel direction of the detected neutrino (allowing them to identify the sun as the source of the neutrinos). These advantages greatly reduce systematic problems and background contamination, increase the types of measurements that can be made, and outweigh the chief disadvantage—a higher energy threshold (see Figure 2).

Water Cherenkov devices detect neutrino interactions by sensing the Cherenkov light cone emitted by the daughter products of the neutrino interaction. In the Kamiokande and SuperKamiokande experiments, the daughter product is a relativistic electron, which produces a "fuzzy" cone of Cherenkov light that gets projected onto a cylindrical array of PMTs. The cone is fuzzy due to the multiple scattering suffered by the electron. (Sharper patterns are produced by the muon daughters of muon neutrino interactions since muons do not suffer multiple scattering to the degree that electrons do. Note, however, that at solar neutrino energies neutrinos cannot produce muons.)

3. The Solution to the Solar Neutrino Problem

Until relatively recently, the variety of conclusions one could draw from this situation were that the experiments were wrong, the solar models were wrong, that neutrinos had mass and were oscillating, or some combination of the above. As we now know, the solution is that neutrinos have mass and oscillate, which rendered a significant fraction of the solar electron neutrinos invisible to the radiochemical and water Cherenkov detectors described above. It was not until the advent of the Sudbury Neutrino Observatory (SNO), a *heavy* water Cherenkov detector sensitive to all neutrino flavors and capable in some ways of distinguishing one flavor from the other, that the oscillation picture was validated. This result was subsequently confirmed and refined by the reactor neutrino measurement in Japan, the Kamioka Liquid-scintillator Anti-Neutrino Detector (KamLAND).

3.1. *The Sudbury Neutrino Observatory (SNO)*

SNO was able to solve the solar neutrino problem because it was sensitive to solar neutrinos in three ways: 1) via elastic scattering of primarily electron neutrinos off of electrons; 2) via charged-current scattering of exclusively electron neutrinos off of the deuteron; and 3) via neutral-current scattering of all neutrino flavors off of the deuteron. Reaction 1 is the same interaction as measured by light water Cherenkov detectors. Reaction 2 is unique to SNO and provided it with a way to make an exclusive measurement of solar electron neutrinos, essentially undiluted with neutrinos of other flavors. Reaction 3 is also unique to SNO and enabled it to measure the total solar neutrino flux, independent of the neutrino flavor.

SNO could demonstrate the presence of solar electron neutrino oscillations by showing that the ratio of the fluxes measured in Reactions 2 and 3 was significantly less than one. However, in the first year or so of data taking, to benefit from superior statistics SNO initially used the SuperKamiokande measurement as a normalization [7]. Then, once SNO had accumulated enough statistics on its own, it used all three Reactions to show that the solar neutrino flux had a significant component of non-electron flavor neutrinos contained within [8]. This result is summarized in Figure 4 below.

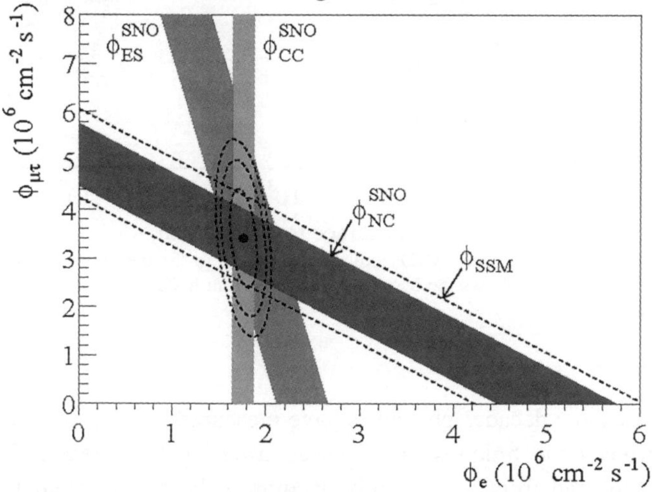

Figure 4: The result from SNO using all three Reactions (see text for details), indicating the presence of a non-electron neutrino component in the solar neutrino flux. The result shows that the non-ν_e flux is statistically significant at the level of 5.3σ. Figure taken from nucl-ex/0204008.

3.2. *The Kamioka Liquid-scintillator Anti-Neutrino Detector (KamLAND)*

The KamLAND detector confirmed the results of SNO using anti-neutrinos from reactors in Japan and South Korea. KamLAND uses a kton of liquid scintillator, surrounded by nearly 2,000 PMTs, to measure the signals produced by the reaction of anti-electron-neutrinos on protons in the liquid. These reactions create a nearly-background-free coincident signal: a positron which emits Cherenkov light, and a free neutron, which bounces around for a while and then emits Cherenkov light when it gets captured.

From the SNO results, KamLAND should have seen a deficit of anti-electron neutrinos, which is indeed what was observed. This result is summarized in Figure 5, where it is clear that one of the chief advantages of

KamLAND over previous reactor experiment was the distance between the neutrino source(s) and the detector.

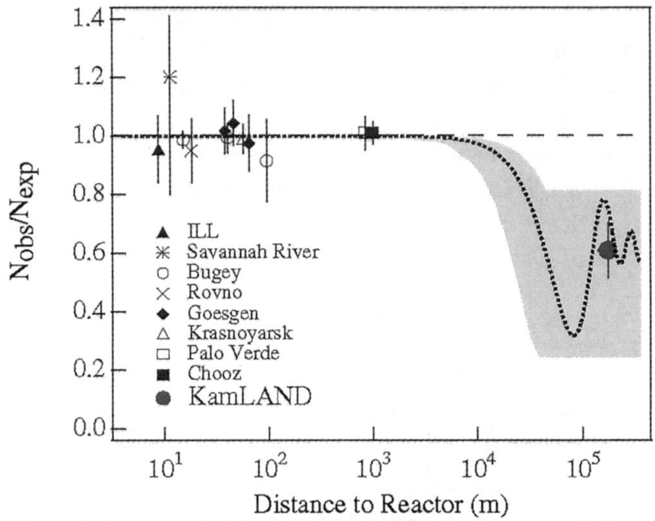

Figure 5: Initial results from KamLAND, indicating a deficit in the measured flux of reactor anti-neutrinos. The dotted curve shows the best-fit predictions from SNO and other experiments. Figure taken from hep-ex/0212021.

4. Conclusion

Over the past three decades physicists have measured the flux of solar neutrinos using a variety of techniques. In all cases, a deficit was measured relative to what was expected from increasingly accurate solar model predictions. In the past five years, it has been shown that the source of this discrepancy is due to the intrinsic properties of the neutrinos themselves. On the way from the solar core to the earth, these neutrinos oscillate, resulting in a significant flux of non-electron neutrinos in earth-bound detectors. Until the advent of SNO, however, these detectors were essentially insensitive to non-electron neutrinos. By using heavy water, SNO was able to demonstrate convincingly that a large fraction of the solar neutrinos arriving at earth were either muon or tau flavor, and thus that neutrinos underwent oscillations. Finally, the KamLAND experiment confirmed the SNO result using anti-neutrinos produced in reactors in Japan and South Korea, a result which eluded previous reactor experiments due to their overly short baselines.

References

1. For information about the Homestake experiment and Ray Davis, please see http://www.bnl.gov/bnlweb/raydavis.
2. GALLEX Collaboration, Phys. Lett. B490(1-2) (2000) 16-26; SAGE Collaboration, Phys. Rev. Lett. 83 (1999) 4686-4689
3. See, for example, Phys. Rev. Lett. 86 (2001) 5651-5655.
4. KamLAND Collaboration, Phys. Rev. Lett. 90 (2003) 021802
5. John N. Bahcall and Carlos Peña-Garay, New J.Phys. 6 (2004) 63
6. Ibid.
7. SNO Collaboration, Phys. Rev. Lett. 87 (2001) 071301.
8. SNO Collaboration, Phys. Rev. Lett. 89 (2002) 011301

NEUTRINO ASTRONOMY

JOHN CARR
Centre de Physique des Particules de Marseille
IN2P3/CNRS France

Neutrino Astronomy opens a new window for observations of our Universe. These lectures describe the detectors in operation and under construction to perform this new field of research as well as the scientific objectives of the domain.

1. Introduction

Neutrino Astronomy has developed in the past few decades and was one of the reasons for the attribution of the 2002 Nobel Prize in physics to R. Davis and M. Koshiba: "for pioneering contributions to astrophysics, in particular for the detection of cosmic neutrinos". The present lecture notes cover the motivations for this new astronomy and the development of the current generation of neutrino telescopes which pursue it.

One of the main objectives of neutrino astronomy is the discovery and understanding of the sites of acceleration of high energy particles in the universe. Since their original discovery one hundred years ago the origin of the high flux of charged cosmic ray arriving at the Earth is unknown. Linked to this objective is the study of the sources discovered and measured in multi-wavelength astronomy such as: Supernova Remnants (SNR); Active Galactic Nuclei (AGN); Microquasars (MQ) and Gamma Ray Bursts (GRB). An important further objective of neutrino telescopes is the search for dark matter in the form of neutralinos. In supersymmetric theories with R-parity conservation, the relic neutralinos from the Big-Bang would concentrate in massive bodies at sites such as the centres of the Earth, Sun and Galaxy. In these sites neutralino annihilations and the subsequent decays of the resulting particles would yield neutrinos detectable in neutrino telescopes of the scale currently in operation and being constructed.

In the past decades several neutrino telescope projects have been launched. At the present time there are two operating neutrino telescopes (AMANDA and Baikal) and three projects in the Mediterranean Sea building current detectors and developing technology for future detectors (ANTARES, NEMO and NESTOR). In these lecture notes the techniques of neutrino telescopes are presented and illustrated using mainly examples from the ANTARES telescope which is currently under construction in the deep sea.

2. Composition of the Universe

In our current understanding, the universe originated 1.4×10^{10} years ago with the Big-Bang. At the present time the average density of mass/energy in the universe is $\sim 10^{-29}$ gms/cm^3 $\approx 5\times10^3$ eV/c^2/cm^3 and measurements indicate that only a small fraction of this density is in the well known form of luminous stars and atoms. These measurements give a proportion of 73% dark energy, 23% cold dark matter, ~3.5% dark baryonic matter and ~0.5% luminous stars. In addition there is ~0.02% of radiation in various forms.

The stars are distributed in galaxies and have life cycles which depend strongly on their masses. Small stars, like the sun, eventually burn all available hydrogen and end up as burnt out black dwarfs; while more massive stars, of several times the solar mass, are not stable under gravity after burn-out and end in supernova explosions. After the supernova, a fraction of the original mass remains as a compact object (either a neutron star or a black hole) and a fraction of the mass expands continually as a supernova remnant. The radiation emitted from a star consists of electromagnetic radiation, neutrinos and charged cosmic rays all of low energy (on the scale relevant to the discussion in these lectures). The electromagnetic radiation peaks at the wavelengths of visible light, with the spectrum extending to higher energies for the higher surface temperatures corresponding to larger star masses, but never extending to the high energy range of X-rays and gamma rays. The neutrinos emitted are in the MeV energy range from the nuclear fusion reactions and the charged cosmic rays, of energies up to 1 GeV, originate near the star surface.

Stars and the sun have been extensively studied since antiquity with observations made by the naked human eye and the human eye aided by optical telescopes. This classical astronomy is biased to observe objects only dominant in the visible energy range and higher energy objects escaped observation until the developments of multi-wavelength astronomy in the second half of the twentieth century. Extensions at both ends of the spectrum of electro-magnetic radiation from radio to gamma rays have added a wealth of information to our knowledge of the universe. Among the objects discovered by this new astronomy are: quasars or active galactic nuclei at the centres of distant galaxies, microquasars in the local galaxy and gamma ray bursts. These later objects have been studied extensively during the past 20 years and their nature is just recently becoming understood. Extension of this new astronomy to include new forms of radiation (e. g. neutrinos, charged cosmic rays and gravitational waves) to become "multi-messenger astronomy" is the next development. Neutrino astronomy in this context has unique advantages which will be detailed later.

Dark energy and dark matter make up the majority of the matter/energy density in the Universe. These subjects have been covered in other courses in this school and while the search for dark matter in the form of Weakly Interacting Massive Particles (WIMPS) is one of the major scientific objectives of neutrino telescopes, this will not be dealt with in detail here.

In proportion a minor element but in importance a major element, the radiation in the universe has several components. The Cosmic Microwave Background Radiation (CMBR) consists of the relic photons left after the "epoch of last scattering" when the formation of stable atoms took place after the Big-Bang. This component constitutes only 0.005% of the total energy density but carries vital information on the conditions of the Big-Bang which lead to many of our conclusions of the present composition of the universe. Other electromagnetic radiation constitutes ~0.01% of the energy density and with it multi-messenger has given us our knowledge of the majority of objects and structures in the universe. Charged cosmic rays have a similar energy density as the electromagnetic radiation. Since the discovery nearly one hundred years ago the origin of the bulk of cosmic rays arriving on Earth has been a mystery. Neutrino astronomy offers a new method to resolve this enigma.

3. Astronomy with Neutrinos

Neutrinos provide an entirely new way to observe astronomical objects. Like photons but unlike charged cosmic rays, they propagate without deviation in electromagnetic fields; however their property of weak interaction with matter gives neutrinos unique features compared to photons of all energies. Neutrinos pass through large amounts of matter without interaction enabling probes, for example: through dust clouds in the galactic plane; through dense accretion disks of matter around massive central sources such as black holes and to the centres of stars and planets including the Sun and Earth.

Neutrinos are necessarily produced in any regions of space where high energy charged particles or gamma rays interact with matter. In extremely energetic astronomical sources, high energy neutrinos are emitted as secondary products produced in interactions of charged cosmic rays; the charged cosmic rays being accelerated in shock processes in the sources. Typically the interactions are of high energy protons with nucleons in the interstellar matter or with photons from the local radiation field, e.g.: $p + p \rightarrow \pi^0 + \pi^\pm + \ldots$. Neutrinos are produced in decays of charged pions: $\pi^\pm \rightarrow \nu_\mu\, \mu$, $\mu \rightarrow \nu_e\, \nu_\mu\, e$ and high energy gamma rays are produced in the same reactions from the decay of neutral pions: $\pi^0 \rightarrow \gamma\, \gamma$. The resulting fluxes of neutrinos and gammas from pp interactions is roughly the same while for pγ interactions it is expected that the flux of gammas is roughly four times that of neutrinos due to the dominance of the Δ resonance in the pγ mode. Due to the multiple secondary particles, the neutrino only carries a fraction of the primary proton energy, typically 10% or less. These reactions can occur close to the acceleration source where the matter density is likely to be high or in interstellar space or molecular clouds in the Galaxy.

The relative fractions of neutrino flavours produced at the source is approximately $\nu_e : \nu_\mu : \nu_\tau \cong 1 : 2 : 10^{-5}$. After propagation over large astronomical distances of the order of kpc to Earth, neutrino oscillations lead to similar numbers of each neutrino flavour, $\nu_e : \nu_\mu : \nu_\tau \cong 1 : 1 : 1$.

At present the only observations of extra-terrestrial neutrinos have been from the sun and from SN1987a in the Large Magellanic Cloud. These neutrinos originate in nuclear reactions and have energies in the MeV range. The neutrinos which are searched for by the present generation of neutrino telescopes from the pp and pγ reactions are of much higher energy in the TeV to PeV range.

The penetrating nature of the neutrino is the feature which makes neutrino astronomy unique; however it is also the feature which leads to the necessity for massive detectors, even for very intense sources. The neutrino interaction cross-section with matter is linear with energy up to a few TeV becoming $\propto E_\nu^{0.4}$ at higher energies with $\sigma \approx 10^{-34}$ cm^2 at 100 TeV. With this cross-section a neutrino of energy 100 TeV has a probability of 63% to interact in crossing the diameter of the Earth. The present generation of neutrino telescopes have effective masses ~1 Gigatonne of target material. To set the scale of the neutrino fluxes necessary for detection, as an example in the ANTARES detector detection of 1 event per year requires an integral flux above energy 1 TeV of $\sim 2\times10^{-11}$ neutrinos/cm^2/sec corresponding to a typical differential flux of $\sim 2\times10^{-8}$ E_ν^2 GeV/cm^2/sec with the assumption that a typical neutrino source will have a power law energy spectrum $\propto E_\nu^{-2}$.

4. Extreme Cosmic Sources of High Energy Radiation

The most intense sources of continual high energy radiation currently known, are active galactic nuclei (AGN) with luminosities of $\sim 10^{46}$ erg/sec ($\approx 6\times10^{45}$ TeV/sec); to set the scale of these sources it should be recalled that the photon luminosity of the sun is 4×10^{33} erg/sec. Taking a typical cosmological distance of 100 Mpc ($\approx 3\times10^{26}$ cm), the energy flux arriving at Earth would be $\sim 5\times10^{-9}$ TeV/cm^2/sec if the energy emission of a typical individual AGN were isotropic.

Hence if a significant fraction of this energy flux were carried by TeV neutrinos, detection would be possible in present generation neutrino telescopes. In fact, a significant part of the energy from AGNs is not isotropic but collimated in jets, and for those AGNs (blazars) where the jet is pointed towards the Earth the flux could be higher. In the galaxy there exist miniature versions of AGNs : microquasars with luminosities $\sim 10^{38}$ erg/sec at distances of ~ 10 kpc and compared to AGNs the reduced luminosity is roughly compensated by the smaller distance scale giving expected fluxes on Earth of the same order of magnitude as blazars.

Quasars and microquasars while not constant in flux continue their emission due to the influx of the accreted matter from the surrounding environment. There exist other intense sources which convert their progenitor mass to radiated energy in a very short period of time. The best understood sources of this type are supernovae, the last events in the life cycles of typical stars. Typical supernova explosions emit $\sim 10^{49}$ ergs in light, $\sim 10^{51}$ ergs in kinetic energy of the debris and $\sim 10^{53}$ in neutrinos of tens of MeV energy. These low energy neutrinos are emitted in the first few seconds after the event, the light in a few months and the kinetic energy in the debris is converted to charged cosmic rays in the supernova remnant for ~1000 years. Gamma ray bursts, discovered in the 1960's, are even more cataclysmic events of still debatable nature. Here energies of $\sim 10^{52}$ ergs are emitted in pulses of gamma rays of duration 0.1-10 seconds.

The resolution of the enigma of the source of cosmic rays may lie in supernova remnants (SNR). An argument exists, (see for instance Gaisser [1]) that the power carried in SNR in the galaxy, taking a rate of one SN every 30 years, is $\sim 10^{42}$ ergs/sec and is of the same order of magnitude of the total power in cosmic rays in the galaxy which is $\sim 5\times 10^{40}$ ergs/sec such that a few per cent of efficiency of converse of the kinetic energy of SNR to cosmic rays is sufficient to explain their origin.

5. Techniques of Neutrino Telescopes

Neutrino telescopes are sensitive to all three flavours of neutrinos but the detection efficiency of each mode can be very different depending on the detection technique of the telescope. In all techniques the neutrinos are detected via the secondary particles produced in interactions with matter, either inside or around the detector. For charged current interactions of neutrinos with nucleons the lepton produced corresponds to the flavour of the neutrino: $\nu_e N \rightarrow eX$, $\nu_\mu N \rightarrow \mu X$, $\nu_\tau N \rightarrow \tau X$, where X represents the hadrons resulting from the nucleon recoil. In neutral current reactions the neutrino scatters inelastically: $\nu N \rightarrow \nu X$, and the event topology is similar for all flavours. In neutrino telescopes the lepton plays the main role in the detection efficiency and while the hadrons, X, are detected they usually have little effect on the efficiency. Due to this, the charged current modes generally dominate the efficiency and the ν_μ mode dominates over the other flavours due to the long range of muons in matter.

While neutrino data exist from underground detectors in caverns, this write-up only discusses neutrino telescopes using large volumes of water or ice in the deep sea, deep lakes or deep glacier. These telescopes are based on the detection of Cherenkov from the secondary particles produced in the neutrino-matter interaction. As mentioned above the detection mode with the highest sensitivity is that of ν_μ and figure 1 illustrates the principle of a deep sea neutrino telescope using this channel. A matrix of light detectors, in the form of photomultipliers in glass spheres, "optical modules", is deployed near the sea bed.

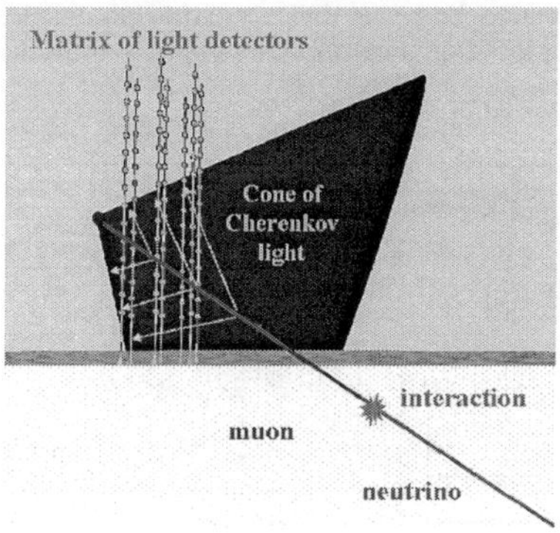

Figure 1. Principle of detection of high energy neutrinos in an underwater neutrino telescope

This matrix of light detectors enables the direction of the muon track to be measured with a precision of a few tenths of a degree and at high energies the muon track direction is closely aligned with that of the neutrino such that the neutrino direction is measured with similar precision. As an example figure 2a shows the simulated angular resolution for the ANTARES neutrino telescope. This figure shows the angular resolution for muons and for neutrinos, the difference being due to the deep inelastic scattering interaction where at higher and higher energies the neutrinos follow more and more closely the muon direction. Above 10 TeV the angular resolution becomes ~0.2°. By using the total light collected in the detector it is possible to obtain a measurement of the energy of the muon track and hence that of the neutrino. The accuracy of this measurement is limited by the fluctuations in the energy loss measurements of the muon. The resolution possible is shown in figure 2b where the error on the logarithm of the muon energy is plotted. The achievable energy resolution is ~0.4 in log(E_μ) at 10 TeV, decreasing to ~0.3 in log(E_μ) at 100 TeV which corresponds to a factor 2 in E_μ.

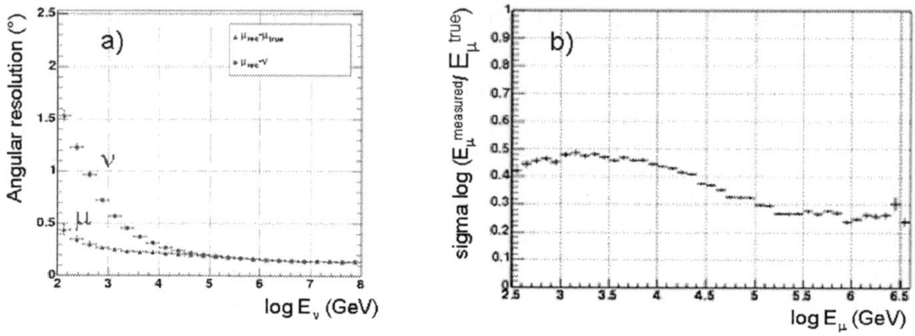

Figure 2. Simulated resolutions for ANTARES underwater neutrino telescope: a) angular resolution showing the difference between the reconstructed muon direction and the true muon direction and the muon and the true neutrino direction as indicated on the figure; b) energy resolution.

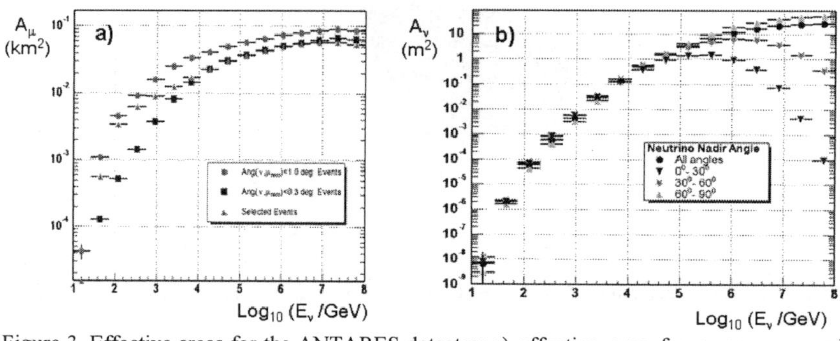

Figure 3. Effective areas for the ANTARES detector: a) effective area for muons averaged over all neutrino incident directions, the different symbols are for selection cuts on the quality of the events, the triangles are the standard cuts, squares are events with angular resolution better than 0.3° and the circles for resolution better than 1°; b) effective area for neutrinos, the different symbols indicate different incidence angles as given in the figure.

The rate of neutrino detection in a neutrino telescope is the product of several factors characterizing the neutrino interactions with matter and the detector properties. For the ν_μ mode the long range of the muon contributes to increase the event rate. The observed event rate is given by: $N_\nu = \Phi_\nu \times A_\nu$ where Φ_ν is the flux of neutrinos arriving at the Earth and A_ν is the effective area of the detector for neutrinos. For ν_μ events: $A_\nu = P_{Earth} \times \sigma_\nu \times \rho N_{A\nu} \times R_\mu \times A_\mu$, where P_{Earth} is the survival probability that the neutrino crosses the Earth to the detector; σ_ν is the neutrino interaction cross-section; $\rho N_{A\nu}$ is the number density of target nucleon in the rock or water; R_μ is the range of the muon and A_μ is the effective area of the detector for muons. The geometry and properties of the neutrino telescope enter only in A_μ and this quantity is obtained from detailed simulations of the detector. Figure 3a and 3b show the effective muon area and the corresponding neutrino effective area for ANTARES.

6. Neutrino Telescope Projects

There are currently two neutrino telescopes in operation in the world: Baikal [2] at a depth of 1200 m in the water of Lake Baikal in Siberia and AMANDA [3] at a depth of 2000 m in the ice at the South Pole in Antarctica. In addition there are a number of groups developing neutrino telescopes in the deep sea: ANTARES, NEMO and NESTOR. A deep sea-water telescope has significant advantages over ice and lake-water experiments due to the better optical properties of the medium. There are however, serious technological challenges to overcome to deploy and operate a detector in the sea. The pioneer sea-water project, DUMAND which worked from 1980 to 1995 to build a detector off the coast of Hawaii, did not overcome these challenges and the project was cancelled. In contrast the projects AMANDA and Baikal which deploy from the solid glacial ice and the frozen ice surface of the lake, respectively, have developed workable deployment systems. The advantages of sea-water neutrino telescopes are significantly better angular resolution e.g. ~0.2°, as shown earlier, for ANTARES compared to ~3° for AMANDA, as well as more uniform efficiency due to the homogeneous medium. A disadvantage of a sea-water detector is the higher optical background due to radioactive decay of ^{40}K and light emission from living organisms: bioluminescence. These backgrounds can be overcome in the design of the detector by having a higher density of optical modules and high bandwidth data readout.

AMANDA was installed in stages in holes in the glacial ice made with a hot water drilling technique. The first detector elements were deployed in 1993 at depths of 810 to 1000m; however, measurements of the ice transparency at those depths showed that the light scattering was unacceptable for operation of a detector. Subsequent strings were deployed at depths of 1500 to 2000m where the ice properties are better. In 1997 the AMANDA B10 detector had 300 optical modules on 10 strings and the data so far published from AMANDA comes from this detector. Since then extra strings have been added with improved signal readout technology. The present AMANDA II detector has 19 strings and about 700 optical modules. A much larger neutrino telescope, ICECUBE, with 4800 optical modules will begin installation at the South Pole site at the beginning of 2005.

In the Mediterranean Sea there are three sites under evaluation for Neutrino Telescopes. The most advanced project is that of the ANTARES collaboration which is building a detector with initially 900 optical modules at a site off the south coast of France near Toulon. The NEMO collaboration is exploring a site off Sicily and developing technology for a future large detector. Since 1990 the ANTARES and NEMO collaborations have been working together on the detector at the Toulon site with the intention to choose the best site for a future larger telescope. The NESTOR collaboration intends to build a detector with 168 optical modules at a site near Pylos off the coast of Greece.

The ANTARES collaboration started in 1996 to develop and construct a detector at a site off the French coast with a depth of 2400m. The first phase of the ANTARES project was to fully evaluate this site in terms of water quality, sedimentation rate and geological stability. The absorption length light at the site was measured to be 45-60 m in the blue and 25-30 m in the ultra-violet, the scattering length for large angle scatters is greater than 100m and the loss of light transmission through the glass housings of the optical modules has been evaluated in measurements lasting 8 months to be less than 2 % / year. Extensive studies of the bioluminescence rate at the site have been carried out and lead to the conclusion that this background will give an acceptable dead time in the photo-multipliers given the electronics design of the detector.

The design of the ANTARES detector array is to have optical modules suspended on individual mooring lines, with readout via cables connected to the bottom of the lines. This technology is similar to the solution originally chosen by the DUMAND collaboration. As with DUMAND, the ANTARES detector requires connections made on the seabed by underwater vehicles but since in the last 10 years the relevant underwater technology has advanced dramatically due to the needs of the offshore-oil industry, the ANTARES instrumentation is based on industrial products and more reliable. Currently a wide range of suitable deep-sea connectors is available and extensively used in industry, including electro-optical connectors wet mateable on the site. Many commercial underwater vehicles exist capable of making these connections. The ANTARES readout design maximizes the reliability of the detector by dividing the system into independent sections such that there is no single active component in the sea whose failure causes the loss of the whole detector. The detector signals are digitized in local electronics in the sea and than transmitted to the shore on high bandwidth optical links. On the shore, a computer farm makes the trigger decisions to decide which data is recorded to tape. A major aspect of the

ANTARES approach is the possibility to recover and repair all elements of the detector deployed in the sea.

7. Potential Sources for Neutrino Telescopes

As stated earlier, one of the major objectives of neutrino astronomy is the search for the origin of cosmic rays. In 2002 a paper [4] appeared from the CANGAROO collaboration using a 3.8m Gamma Ray Telescope located in Australia making the claim of the observation of the acceleration of cosmic ray protons in the supernova remnant RX J1713.7-3946. The observation was of gamma rays up to 10 TeV with an energy spectrum $E^{-2.8}$ where the observed energy spectrum could not be explained only with π^0 decay from hadronic interactions. The positive observation of TeV gamma rays from this source has recently been confirmed by the HESS collaboration; however with a different energy spectrum [5]. A subsequent paper by J. Alvarez-Muñiz and F. Halzen [6] calculated the expected neutrino flux from this source assuming the integral energy carried in neutrinos was the same as that in gamma rays with a neutrino flux spectrum E^{-2}. The conclusion of this later paper is that a northern hemisphere detector such as ANTARES would observe a few events per year from this source and in a few years of operation would have a clear signal with a probability less than 1% that the events seen were background from atmospheric neutrinos. It seems likely that other similar supernova remnants are present in the Galaxy hence providing more possibilities for clear observations.

Predictions from supernova remnants with central pulsars: plerions, such are the Crab Nebula give rather low events rates in neutrino telescopes. Rates have been calculated by Bednarek [7] with a model for pulsar wind nebulae in which most of the observed gamma ray emission comes from leptonic processes and hadronic processes only contribute to the high energy part of the spectrum. This model gives around 1 event/year/1km^2 for the Crab Nebula and for the Vela Nebula: negligible rates for detection in ANTARES. Another model by Guetta and Amato [8] assumes that the observed gamma rays with energies above 2 TeV originate from pion decay implying hadronic processes which also give neutrinos. This assumption gives higher event rates for neutrino detection up to 10 event/year/1km^2 for some plerions and so a possibility for detection in ANTARES.

While in the Galaxy there are 220 supernova remnants in the catalogue of Green with about 10% having central pulsars, the rate of supernova explosions in the Galaxy is believed to be 1 to 3 per 100 years. If chance were to give a galactic supernova explosion during the lifetime of a neutrino telescope the observable signals would be very large. In the initial stages of the supernova the neutrinos are in the MeV energy range giving high rates of uncorrelated counts in the optical modules. These MeV neutrinos will be detectable in ice neutrino telescopes but not in sea water detectors because of the higher optical noise backgrounds. At later stages of the supernova explosion higher energy neutrinos are emitted which could be detectable in all types of neutrino telescopes. Waxman and Loeb [9] predict the rate of TeV neutrinos originating when the shock of a type II supernova breaks out of the envelope of the massive star. This break out occurs about 10 hours after the original explosion and the burst of neutrinos lasts around 1 hour giving ~100 events/km^2, a signal easily detectable in any neutrino telescope because of the short time window which would make the background negligible. An energetic pulsar in young supernova remnants can power high energy acceleration of particles and several predictions exist for such processes. For instance, a model by Protheroe et al. [10] where a young pulsar accelerates iron nuclei which photo-disintegrate to produce hadrons and so neutrinos. This model gives high rates of neutrinos but only for a few months after the supernova explosion.

Among other galactic sources with predictions of high observable neutrino rates are microquasars. The rate calculations presented by Distefano et al. [11] are based on model predictions by Levinson and Waxman [12]. This model assumes that inhomogeneities in the microquasar jet can cause internal shocks which accelerate protons and electrons. The maximum proton energy attainable is of the order of 10 PeV in the jet frame and the protons interact with a photon field around the object to produce secondary neutrinos. Using measured properties of several known microquasars and a number of assumptions for the parameters in the model, including a fraction of 10% of the jet energy carried by protons, the rates of neutrinos from two microquasars (GX339-4 and SS433) are high (respectively 183 and 252 events/km^2/year) leading to a positive detection in ANTARES in around one year.

A review of neutrino rate predictions from galactic sources has been made by W. Bednarek et al. [13] explaining in more detail the prediction for sources described above together with others. Table 1 gives a summary of the neutrino rate predictions from this paper. The range of numbers for the event rate illustrates the uncertainties and the large variation of the predictions from different models.

Beyond the local galaxy there are also point sources where the flux of neutrinos could be large enough to be detected by the neutrino telescopes in operation and under construction. The known extragalactic sources of high energy cosmic radiation are AGN's and GRB's and in neutrino telescopes individual sources could be observed as discrete point sources. In addition, the integral over all such sources in the universe could give an observable diffuse flux of neutrinos. This later diffuse flux must be distinguished from the background of atmospheric neutrinos by a harder energy spectrum, necessitating a good knowledge of the high energy contributions to the background flux. Predictions of fluxes from individual GRB sources are generally lower than for AGN sources, however coincidences with observations from gamma ray detectors in space can be used to vastly reduce the atmospheric neutrino background by searching for events within a short (~second) time window relative to the gamma ray signal.

Table 1. Summary of neutrino event rate predictions (after W. Bednarek et al. [13])

Source Type	Neutrino events/ km2/yr
Supernovae	50-1000
Plerions	1-10
Shell SNR	40-100
Pulsars+Clouds	1-30
Binary Systems	a few
Microquasars	1-300

In addition to observations on known sources, it is well possible that neutrino telescopes could discover hitherto unknown sources. The unique penetrating properties of neutrinos allow many speculations; only neutrinos can exit from regions of high matter density which might completely obscure certain objects from observations with other cosmic messengers. Sources observable with TeV gamma rays are limited to distances of tens of kilo parsecs while charged cosmic rays in this energy range do not point to the source. Further, the large acceptance in solid angle of neutrino telescopes opens the possibility to discover close sources, in principle observable with TeV gamma ray telescopes but as yet unobserved due to the narrow angular acceptance and necessary pointing strategy of most gamma ray detectors. The new chance discovery of a TeV gamma source HESS J1303-63 reported recently [14] emphasises these possibilities for neutrino telescopes.

8. Existing Results

At the present time all observations of high energy neutrinos observed in the past and present generation of neutrino telescopes are consistent with the flux of neutrinos from the atmosphere of the Earth; there is as yet no evidence for high energy neutrinos from extra-terrestrial sources. Figures 4 and 5, from T. Montaruli [15], summarize existing and future limits for neutrino point sources and diffuse fluxes respectively. The measured diffuse fluxes are consistent with the expectations of atmospheric neutrinos based on models using measurement of cosmic ray fluxes.

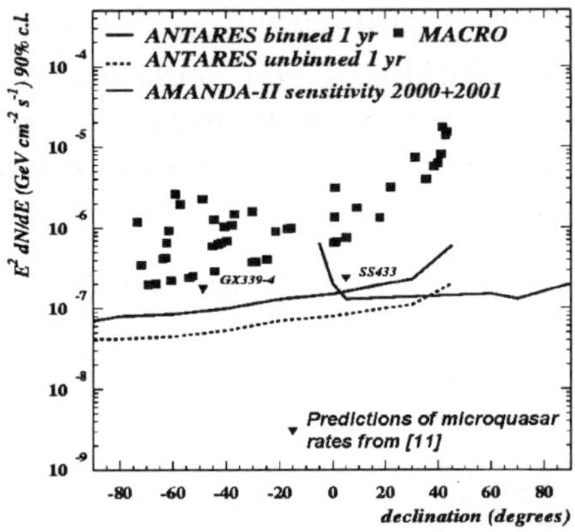

Figure 4. Existing limits, expected limits and some predictions of rates for discrete sources of neutrinos assuming an E^{-2} neutrino flux. The squares are published limits from the MACRO experiment [16]. The lines indicate expected limits from AMANDA II for the 2000-2001 data and for ANTARES after one year of date taking. The inverted triangles are rate predictions for microquasars from Distefano et al. [11].

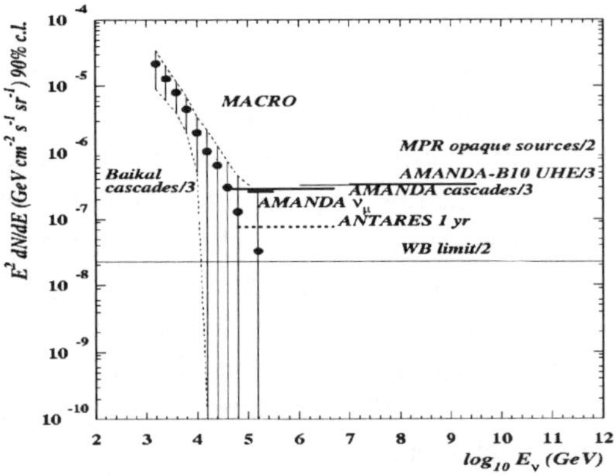

Figure 5. Existing limits, expected limits and some upper limit predictions of rates for a diffuse flux of neutrinos with an E^{-2} spectrum. The points are measured fluxes of atmospheric neutrinos from the AMANDA experiment. The solid lines are measured limits from MACRO, Baikal and AMANDA as indicated with the expectation from ANTARES as a dotted line. The grey lines are predicted limits from Mannheim et al. [17] and Waxman and Bahcall [18].

The existing limits on both point sources and diffuse fluxes start to test interesting model predictions. The forthcoming neutrino telescopes in the Northern Hemisphere such as ANTARES will be able to probe for source at and close to the centre of the galaxy, a region rich in sources of high energy gamma rays. Complementary searches with neutrinos will give conclusive evidence on the nature of the acceleration mechanisms for the gamma ray observations, being able to distinguish between hadronic and leptonic processes.

9. Conclusions

The field of neutrino astronomy has made rapid advances in the last decade. Two large neutrino telescopes, AMANDA and Baikal, are operating and publishing results. Both projects are in the process of expanding their detectors, AMANDA with IceCube and Baikal with extra outlier lines. In the deep sea, the three Mediterranean projects have all made very significant progress in recent years and ANTARES and NESTOR will have completed detectors in the next few years. Together with the NEMO group they have combined in the "KM3NET" design project for the next stage towards a km3 Mediterranean Neutrino Telescope.

References

1. T. K. Gaisser, *Cosmic Rays and Particle Physics,* Cambridge University Press (1990).
2. I.A. Belolaptikov et al., (Baikal Coll.), Astropart. Phys. 7 (1997) 263.
3. E. Andres et al., (AMANDA Collaboration), Nature 410 (2001) 441.
4. R. Enomoto et al. (CANGAROO Collaboration) Nature 416 (2002) 823.
5. D. Berge, in proceedings of Gamma Ray Astronomy Symposium, Heidelberg (2004).
6. J. Alvarez-Muñiz and F. Halzen, Astrophys. J. 576 (2002) L33.
7. W. Bednarek, A&A 407 (2002) 1.
8. D. Guetta and E. Amato, APh 19 (2003) 403.
9. E. Waxman and A. Loeb, Phys. Rev. Lett. 87 (2001) 071101.
10. R.J. Protheroe, W. Bednarek and Q. Luo, APh. 9 (1998) 1.
11. C. Distefano et al., ApJ. 575 (2002) 378.
12. A. Levinson and E. Waxman, Phys. Rev. Lett. 87 (2001) 17.
13. W. Bednarek et al., astro-ph/0404534.
14. M. Beilicke, in proceedings of Gamma Ray Astronomy Symposium, Heidelberg (2004).
15. T. Montaruli, proceedings of ISMD2004 symposium, Somona, CA, 2004.
16. M. Ambrosio et al., (MACRO Collaboration) ApJ 546 (2001) 1038.
17. K. Mannheim et al., Phys. Rev. D63 (2001) 023003.
18. E. Waxman and J.N. Bahcall, Phys. Rev. D59 (1999) 023002.

DARK MATTER AND DARK ENERGY

P. ULLIO

Scuola Internazionale Superiore di Studi Avanzati,
Via Beirut 2-4,
I-34014 Trieste, Italy
E-mail: ullio@sissa.it

With the recent remarkable progresses in our understanding of the dynamics of the Universe and of structure formation, the evidence for non-baryonic dark matter and dark energy has become stronger and stronger. We summarize here the steps that have led to the current picture and the expectations for the future. On the other hand, we are still far from a comprehensive understanding of the dark side of the Universe from a particle physics point of view. We present here some of the most popular frameworks to embed dark energy in a high energy setup, putting emphasis on the fundamental problems which still have to be addressed. For what concerns dark matter, several interesting ideas have been proposed and are being tested, possibly already with some hints that we might be on the right track in the quest for dark matter.

1. The dark side of the Universe

The latest years will be remembered in Science as those that marked our entrance in the era of precision cosmology, with dramatic improvements in our understanding of the dynamics of the Universe and of the theory of structure formation: it is by now well established that the Universe has a flat geometry (or close to flat), with the largest contribution to its mean energy density today provided by a term with negative pressure, usually dubbed "dark energy", and the next-to-largest in the form of non-baryonic cold dark matter, the building block of all structure we see in the Universe.

On the other hand, little progress has been made in identifying the nature of the dark matter and dark energy in terms of elementary components: the Standard Model (SM) of elementary particles has been shown to describe very successfully some aspects of the hot Big Bang model for the Universe, especially regarding its thermal evolution in the radiation dominated phase; it is a very pressing, but at the same time challenging, task to single out the extension to the SM in which the dark components can be embedded.

We will briefly review here how our view of the Universe has changed in the recent years, and present some of the ideas that have been put forward to formulate a unifying scenario and to test it.

2. Towards the era of precision cosmology

2.1. *The long-standing issue of dark matter*

The observation that most of the matter in the Universe is not in the form of luminous stars or hot (X-ray emitting) gas is definitely not recent. Back in 1933, Zwicky found evidence for the presence of dark matter in the nearby Coma cluster. A galaxy cluster is a large, gravitationally bound group of galaxies; assuming that the system has relaxed to dynamical equilibrium, the virial theorem, $K + U/2 = 0$, can be applied. Zwicky found that the kinetic energy term K, estimated by measuring proper velocities of individual galaxies in Coma, was much larger than the gravitational potential energy term U, computed assuming that the mass in the cluster was the sum of the mass of galaxies; the mass to light ratio obtained in this way was about $M/L \simeq 300\, M_\odot/L_\odot$. Actually, including the contribution from gas as inferred from X-ray maps, one finds that the ratio between the total mass in Coma (and analogously in other clusters) and the "visible" mass (i.e. gas plus stars) is about: $M/M_{\rm vis} \simeq 20$[1].

In the seventies, systematic studies of rotation curves of galaxies, showed on the other hand that large amounts of non-visible matter were present in galaxies as well. Here the evidence for dark matter follows simply from the application of Kepler's third law, $v_{\rm circ}(r) = \sqrt{\frac{G_N M(r)}{r}}$, which links the circular velocity at a given galactocentric distance r to the total mass M within that same radius (for simplicity we are discussing the case for a spherically symmetric body; in case of axial symmetry the picture is analogous). Assuming that $M(r)$ is dominated by stars and gas, at radii much larger than the scale within which these are observed, one expects a Keplerian fall-off of the circular velocity, i.e. $v_{\rm circ}(r) \propto r^{-1/2}$, while observationally one finds that rotation curves tend to remain flat. One needs again to invoke an extra dark component, extending to very large radii: one finds that in spiral and elliptical galaxies the mass to light ratio is about $M/L \sim 10-30\, M_\odot/L_\odot$[2], while in smaller objects such as low-surface-brightness or dwarf galaxies it can be as large as $M/L \sim 200-600\, M_\odot/L_\odot$[3].

A method to estimate the total amount of ordinary (baryonic) matter in the Universe was understood and developed in the late sixties. Soon after the formulation of the hot Big Bang model, it was realized when the temperature of the Universe is $T \sim 1$ MeV, at an age of the Universe of about 1 s, neutrons go out of thermal equilibrium and their thermal leftovers get trapped in light nuclei states before having time to decay (for a review, see, e.g., [4]). Applying the knowledge on conversion rates derived from laboratory experiments to the environment of an expanding and cooling Universe, the primordial abundance of ^4He, D, ^3He and ^7Li can be computed (the process of synthesis of light elements in the early Universe is usually referred to as Big Bang Nucleosynthesis (BBN)). Assuming the SM as particle physics framework, there is only one free parameter in BBN, the total number density of baryons, which one could hope to determine comparing predictions to measurements of primordial light elements abundances. This kind of measurements are experimentally challenging, as environments with no significant star formation have to be identified: a very important step forward in the field came in 1998 with the high precision measurement of the D abundance, looking at absorption lines in very distant quasars[5].

Cosmologists usually measure abundances in terms of the the critical density $\rho_c \equiv 3[H_0]^2/(8\pi G_N)$ (with H_0 the expansion rate of the Universe today, usually written as $H_0 = h \cdot 100\,\mathrm{km/sec/Mpc}$; observationally, $h \simeq 0.7$), i.e. the mean density, at the present time, of a Universe with flat geometry. The contribution to the energy density today of a given species i is written as $\Omega_i \equiv \rho_i/\rho_c$. From BBN, the baryon contribution is determined to be $\Omega_\mathrm{b} h^2 \simeq 0.02 \pm 0.002$, or $\Omega_\mathrm{b} \simeq 0.04$, much larger than the contribution of stars only, which from photometric maps is found to be about $\Omega_\star \simeq 0.005$. At the same time, Ω_b is much smaller than the lower limit on the total matter term as estimated from galaxy clusters (from X-ray maps and hydrodynamics) $\Omega_\mathrm{M} \sim 5 - 10\ \Omega_\mathrm{b} \sim 0.2 - 0.4$. It follows that most of the dark matter in the Universe has to be non-baryonic!

2.2. *The surprise of dark energy*

In 1998, another breakthrough wiped out the cosmological framework which had been regarded as theoretically favored up to that time, i.e. the $\Omega_\mathrm{M} = 1$ Einstein – de Sitter model: Two independent teams[6,7] reported the discovery that Universe is presently accelerating, rather than decelerating, as predicted for a matter dominated Universe. This was inferred by mapping the Hubble's diagram of distant supernovae, as obtained by measuring apparent magnitudes (i.e. luminosity distances) and redshifts in a sample of

Type Ia supernova standard candles. The evolution in time (redshift) of the scale factor of the Universe $a(t)$ is consistent with a dynamics dominated at recent times by a fluid component, dubbed "dark energy", defined by an equation of state:

$$p = w\rho \tag{1}$$

with w negative and smaller than about -0.6[8] (this upper limit depends on the value of $\Omega_{\rm M}$ and is valid in case of a constant w; there is also the possibility that w varies in time, and in this case the upper limit refers approximately to the value of w today). The simplest setup of this kind (although, possibly, not the most natural) is the cosmological constant scenario, that in which a constant term Λ is included in Einstein's equations, giving rise to an effective component with $p_\Lambda = -\rho_\Lambda$, i.e. $w = -1$.

This picture has been confirmed and reinforced in the latest few years by the first accurate maps of temperature anisotropies in the cosmic microwave background (CMB) radiation: the Boomerang[9] and Maxima[10] instruments on balloons produced the first precision measurements on a small patch of the sky in 2000, the WMAP instrument[11] on a satellite mission caught the first high precision full map of the sky in 2003 (progress has been made also thanks to several other recent experiments on different angular scales). Anisotropy maps measure the acoustic peaks of the baryon-photon fluid at the surface of last scattering, when matter becomes neutral and photon started to propagate essentially freely throughout the Universe. A very large amount of information is imprinted on the surface of last scattering. First of all, the position of the first acoustic peak marks the angular size of the sound horizon on the surface of last scattering, a physical quantity which is easy to estimate and can then be used as the reference size we need to measure the geometry of the Universe: the data give $\Omega_{\rm tot} = 1.056 \pm 0.045$[12], i.e., to a good approximation the Universe is flat (a flat Universe is one of the key prediction of inflation). At the same time, acoustic peaks contains other informations including the total amount of matter interacting with photons (baryons) and the total amount of matter contributing to gravitational potential wells (cold dark matter CDM): interpreting the data one finds, in particular, a value of $\Omega_{\rm b}$ in good agreement with estimates from BBN, and that $\Omega_{\rm M}$ is certainly inconsistent with being 1, while its favoured value is about 0.3. It follows that, to match the condition of flat Universe there is again the need for an additional term, the dark energy component, which does not cluster or form structure, but provides about 70% of the energy density of the Universe today.

2.3. *The structure formation picture*

Another recent ambitious goal in cosmology has been to map the large scale structure of the Universe. The 2dF[13] and the SDSS[14] experiments have recently produced the first extended three-dimensional maps of the distribution of galaxies in the Universe, to be compared with predictions of the theory of structure formation. This kind of information is very powerful in discriminating between two feasible scenarios: the case of hot dark matter (HDM) versus the case of cold dark matter (CDM). Hot candidates, such as, e.g., massive but light neutrinos, are particles which are relativistic at the collapse epoch in the radiation dominated phase and free-stream out of galaxy-sized overdense regions after matter-radiation equality: this give rise to a top-down structure formation scenario, with very large structures forming early and then fragmenting into smaller ones. Such scenario is excluded by current data. The CDM picture holds instead for particles that are massive (masses in the GeV range or above), and hence non-relativistic: in this case structures form in a hierarchical, bottom-up scenario, with small structures forming first and merging in larger and larger bodies (for a more detailed presentation, see, e.g.,[15]).

Maps of distribution of galaxies point again to a energy density budget for the Universe with $\Omega_{\rm M} \sim 0.3$ and $\Omega_\Lambda \sim 0.7$, but also put tight upper bounds on HDM terms. Since neither baryons nor neutrinos can account for dark matter, it follows that the SM of elementary particles does not embed any viable dark matter candidate.

Other techniques have been developed to measure cosmological parameters; those that give information on dark matter and dark energy include, e.g., gravitational weak or strong lensing, Lyman-α forest data, the Sunyayev-Zel'dovich effect. Global best fit values of energy density terms [12] give: $\Omega_{\rm tot} = 1.056 \pm 0.045$, with $\Omega_\Lambda \simeq 0.73$ and $\Omega_{\rm M} \simeq 0.27$, with the latter being the sum of a baryonic term $\Omega_{\rm b} \simeq 0.049$ and a CDM term $\Omega_{\rm CDM} \simeq 0.22$.

3. Dark energy in a particle physics context

Today's pressing target is to use the feedback from the astrophysical and cosmological observations to formulate a new consistent particle physics model, embedding dark matter and dark energy. For the former, several ideas have been put forward, with detection strategies being developed and already operative. Regarding dark energy, there is no detailed model fully working, since there are several puzzling elements making the construction of such a model very hard. We discuss these issues here, and sketch the dark matter issue in the next Section.

To see how a negative pressure can appear in a particle physics context, consider the lagrangian density of a scalar particle (the case for fermions and vectors is perfectly analogous):

$$L = \frac{1}{2}\partial_\mu \phi \partial^\mu \phi - V(\phi) \,. \tag{2}$$

Under the assumption of isotropy and homogeneity of Universe, the stress energy tensor T_μ^ν, which acts as the source term in Einstein's equation, has only diagonal entries, i.e. density and pressure; the contribution to them from the field we have considered are in the form:

$$\begin{aligned} \rho \equiv T_0^0 &= \frac{1}{2}\dot{\phi}^2 + V(\phi) \\ p \equiv -T_i^i &= \frac{1}{2}\dot{\phi}^2 - V(\phi) \,. \end{aligned} \tag{3}$$

Assuming that the field is in a configuration close to the minimum of the potential V_0, and that V_0 is large, one finds $p = -\rho = -V_0$, i.e a cosmological constant behavior.

The first problem one has to face is the fact that the measured cosmological constant is very small compared to energy scales we are familiar with in particle physics: $\rho_\Lambda = \frac{\Lambda}{8\pi G} = \Omega_\Lambda \cdot \rho_c(t_0) \simeq 2.5 \cdot 10^{-47}$ GeV^4. It follows that, when considering an explicit model, a huge fine-tuning is usually required: we mention here two examples, one in a classical field theory model, one at a quantum level.

Consider a case of spontaneous symmetry breaking, induced by the transition from a potential with minimum in $\phi = 0$ into the "Mexican hat" potential: $V(\phi) = V_0 - \frac{1}{2}\mu^2\phi^2 + \frac{1}{4}\lambda\phi^4$, with ground states in $\phi = +\sigma$ or $\phi = -\sigma$, being $\sigma = \sqrt{\mu^2/\lambda}$. If you start with $V_0 = 0$ you get a large negative cosmological constant, $V(\phi = \sigma) = -\mu^4/4\lambda$. You are then forced to choose V_0 such that $V_0 - \mu^4/4\lambda \simeq 10^{-47}$ GeV^4. Applying this idea to the Higgs mechanism, you expect $V_0 \sim (100 \text{ GeV})^4$ and you are forced to introduce a fine-tuning of 1 part in 10^{55}!

In a quantum field theory framework, zero-point vacuum fluctuations generate a V_0 term even starting from a zero value at the classical level. For both bosons and fermions the generated effective cosmological constant is divergent. In the example above:

$$\rho_{\text{vac}} = \frac{\Lambda}{8\pi G} = \langle T_{00} \rangle_{\text{vac}} \propto \int_0^\infty \sqrt{k^2 + m^2} k^2 dk \,. \tag{4}$$

To face this problem you need to introduce an ultraviolet cutoff, $\rho_{\rm vac} \simeq k_c^4/(16\,\pi^2)$; the natural cutoff would be the Planck mass, but then Λ would be 120 orders of magnitude too large! Since fermions and bosons give contributions with opposite signs, an option may be to consider a supersymmetric (SUSY) theory, i.e. a theory with the same number of fermionic and bosonic degrees of freedom, and such that these are associated to particles with equal masses (unbroken SUSY). However, laboratory experiments show that SUSY is not one of the ingredients of the low energy particle model, and that, if indeed it enters in the definition of model at high energy, SUSY needs to be broken at a scale of about 1 TeV or above, and the fine-tuning problem on the generated cosmological constant comes back in. More recently, attempts to explain the smallness of the cosmological constants have been mainly in the direction of linking this problem to the one of possible large extra dimensions, roughly speaking "diluting" a large value of the cosmological constant in our low energy 4-dimensional world into a higher dimensional space which gravity only would sense.

Another option to address the fine-tuning problem is consider a setup with a time varying equation of state $p(t) = w(t)\,\rho(t)$. This is possible even within the Lagrangian introduced above in Eq. 2, with still ρ and p as given in Eq. 3, but supposing now that ϕ is far from the potential minimum $V_0 = 0$. Depending on the value of kinetic term with respect to the potential term, w can vary between -1 and 1: we have now a dynamical configuration with ϕ that can start at a large value and roll down the potential (with a effective "friction" term induced by the expansion of the Universe). For potentials that are sufficiently steep, $\Gamma \equiv \frac{V''V}{(V')^2} \geq 1$, a common evolutionary path is reached from a wide range of initial conditions[16]: this is the tracker trajectory, i.e. an evolution path which traces the main background component (radiation or matter) and hence explaining the reason why the dark energy term is small today. For this class of models dark energy is usually dubbed as "quintessence"[17]; one example is., e.g., a scalar field theory with $V(\phi) = V_0/\phi^\alpha$. Dark energy properties will be measured with much higher accuracy in the future, hopefully disentangling whether w is evolving in time or not.

Even quintessence models, on the other hand, do not address the so called "cosmological coincidence problem", i.e. the problem related to the fact that in our Universe setup, with matter and radiation rapidly scaling with the Universe scale factor, respectively as a^{-3} and a^{-4}, it happens that we live in the very special epoch when the Universe has just started accelerating. To obtain this feature in a real model fine-tuning is needed again and light mass parameters have to be brought back in.

4. Relic particles as dark matter candidates

A popular framework to introduce non-baryonic dark matter candidates is to assume that these may appear as thermal relics from the early Universe, analogously to the radiation and the baryonic components (see, e.g., Ref.[18] for a more extended discussion). Any new particle of given mass M_χ has a finite relic abundance if such particle is stable and it has some non-zero coupling to SM particles. Such coupling guaranties that, in the early Universe, extrapolating to sufficiently high temperatures, χ would be in thermal equilibrium, as enforced by pair annihilation/production processes:

$$\chi\bar{\chi} \leftrightarrow l\bar{l}\,, \tag{5}$$

with l some lighter SM particle. Regardless of the specific mechanism at work, the number density of χ in thermal equilibrium is just related to statistical properties, i.e. it is in the form:

$$n_\chi^{eq} = \frac{g_\chi}{(2\pi)^3} \int d^3p \, \frac{1}{\exp(E/T) \pm 1}\,, \tag{6}$$

where the phase space distribution function depends only on the energy $E = \sqrt{p^2 + M_\chi^2}$, and the $+$ sign applies for fermions, while the $-$ sign for bosons.

As the Universe cools down while expanding, χ remains in thermal equilibrium up to the freeze-out temperature T_f at which pair annihilation (as well as pair production) in the thermal bath becomes inefficient. As a rule of thumb, one can show that T_f corresponds to the time when the Universe expansion rate gets equal to the annihilation rate, i.e.[18]

$$H(T_f) \simeq \Gamma(T_f) = n_\chi^{eq}(T_f) \langle \sigma_A v \rangle_{T=T_f}\,, \tag{7}$$

with the second term on the right-hand-side being the thermally-averaged annihilation cross section. After freeze out, when $\Gamma \ll H$, the number density of χ is not any more depleted by pair annihilations (nor replenished by pair production) and is just diluted in volume by the Universe expansion; normalizing it to to the entropy density $s(T)$, one can easily extract its contribution to the energy density today:

$$\left(\frac{n_\chi}{s}\right)_{T=T_0} = \left(\frac{n_\chi}{s}\right)_{T=T_f}\,, \tag{8}$$

with the entropy density today being $s_0 \simeq 3000\ \mathrm{cm}^{-3}$ and and thermal scaling $s(T)$ extrapolated within the SM.

There are two opposite regimes: the case for particles that are relativistic at T_f and that for non-relativistic ones. For the former, both n_χ and s scale like T^3 and the dependence on T_f factorizes out (actually some mild dependence remains, related to the number of effective relativistic degrees of freedom from SM particles at a given temperature). Assuming $M_\chi > T_0$, the energy density today scales linearly with the mass of χ:

$$\rho_\chi(T_0) = M_\chi \cdot s_0 \left(\frac{n_\chi}{s}\right)_{T=T_f} = \text{const} \cdot M_\chi \,. \tag{9}$$

Light neutrinos belong to this class of candidates; in terms of the critical density, one finds:

$$\Omega_\nu h^2 = \frac{\rho_\nu}{\rho_c/h^2} = \frac{\sum_i M_{\nu_i}}{94.4\,\text{eV}} \,. \tag{10}$$

As already mentioned, structure formation disfavors large contributions from HDM components. Upper limits on Ω_ν can be immediately rephrased as upper limits on the sum of the light neutrino masses. Limits quoted in the literature depend quite sensitively on the underlying assumptions; in the analysis of Ref. [12] a seven-parameter model is fitted against the SDSS galaxy survey and the WMAP data on CMB temperature anisotropies, giving the upper limits $\Omega_\nu h^2 < 0.12 \cdot \Omega_{CDM} h^2$ or $\sum_i M_{\nu_i} < 1.7\,\text{eV}$.

The decoupling in the non-relativistic regime involves instead a massive particle with equilibrium number density exponentially decreasing with T_f (Maxwell-Boltzmann tail). To find an indication on the relic abundance scalings, the approximate relation in Eq. 7 suffices. The equilibrium number density at freeze-out can be written as:

$$n_\chi^{eq}(T_f) \simeq \frac{H(T_f)}{\langle \sigma_A v \rangle_{T=T_f}} \,, \tag{11}$$

with the scaling of H as appropriate in the radiation dominated era, and $T_f \simeq M_\chi/20$ (as can be found, a posteriori, from the exact solution of the number density evolution equation, the so-called "Boltzmann equation"). One finds that[19]:

$$\Omega_\chi h^2 = \frac{M_\chi \, n_{\chi,0}}{\rho_c} \simeq \frac{3 \cdot 10^{-27} \text{cm}^{-3} \text{s}^{-1}}{\langle \sigma_A v \rangle_{T=T_f}} \,; \tag{12}$$

on the right-hand-side, there is no explicit dependence on the particle mass, but just on the inverse of the annihilation rate.

From this expression is clear that the stronger the coupling of the particle, the smaller the relic density: this is intuitively easy to understand, since a larger annihilations rate implies that decoupling takes place at a later time, with the freeze-in of the number density at a smaller equilibrium value. The numerical value in the numerator on the right-hand-side of Eq. 12 gives explicitly which annihilation strength is needed to generate viable dark matter candidates. In fact, if the coupling of χ to lighter particles is of weak interaction type, then:

$$\langle \sigma_A v \rangle \sim \frac{\alpha^2}{(100\,\mathrm{GeV})^2} \sim 10^{-25}\mathrm{cm}^{-3}\mathrm{s}^{-1} \tag{13}$$

with $\alpha \sim 10^{-2}$, i.e. a Weakly Interacting Massive Particle (WIMP) is naturally a good dark matter candidate.

Reversing the argument, requiring that thermal relic density of the particle χ accounts for a CDM density of about $0.081 < \Omega_{CDM}h^2 < 0.125$ (value obtained from SDSS and WMAP data in the nine-parameter model fit of Ref. [12]), then a weak interaction rate is required for $\langle \sigma_A v \rangle_{T=T_f}$.

The leading WIMP dark matter candidate is the lightest supersymmetric particle (LSP), most likely the lightest neutralino χ_1^0. In the minimal supersymmetric extension to the standard model (MSSM), neutralinos are mass eigenstates obtained from the mixture of the supersymmetric partners of the photon, the Z boson and neutral parts of two Higgs doublets. The lightest of these states is massive (with mass in the range between a few GeVs to few TeVs), weakly interacting (since it has zero electric and color charges) and stable (in case of R-parity conserving SUSY models; R-parity is a symmetry which was introduced as simple recipe to prevent dangerous flavor-changing-neutral-current terms, but implies also that SUSY particles enter reactions in even numbers only). In the MSSM it is then quite natural to find models with cosmologically relevant relic abundance for χ_1^0, i.e. this happens in relatively large portions of the huge MSSM parameter space, see, e.g.,[20]. Neutralinos are obviously just one possibility; among other WIMP DM candidates that have recently been studied, there are, e.g., the lightest Kaluza-Klein particle in models with universal large extra dimensions[23] and LIMPs, i.e WIMPs which are just coupled to leptons[24].

At the same time, although in many respects very attractive, the idea to introduce CDM as a thermal relic may not be the right approach. The lightest neutralino itself may appear as the leftover in some non-thermal contest[21]. Several alternative mechanisms have been proposed for non-thermal CDM candidates, stemming from very different contexts: e.g., the axion was introduced to solve the problem of weakness of CP violation in strong interactions[22]; in some scenarios, the gravitino, the SUSY partner of the graviton is the LSP and may play the role of CDM candidate (as well as of the warm dark matter candidate, at the border between HDM and CDM, if its mass is in the MeV range). Another possibility is that the DM puzzle may be related to inflation, giving rise to super-heavy relics such as wimpzillas or "Q-balls", i.e. extended topological objects, eventually supersymmetric.

As opposed to the case of WIMP detection, which will be discusses in the next Section, since non-thermal candidates do not fall any more in a definite class of models, detection techniques have to tuned in each scenario and for some of them, such as for particles which are just gravitationally coupled, detection prospects are rather discouraging.

5. WIMP dark matter identification

By definition WIMPs have a small but finite coupling with SM particles: such coupling makes feasible, although difficult, their detection if indeed they are the building block of dark matter halos, including the halo of our own Galaxy, the Milky Way. Several techniques have been studied to search for dark matter WIMPs (for thorough reviews and comprehensive lists of references, see, e.g., Refs.[19,25]); they may divided into two main groups. On one side, there is direct detection[26], i.e. the attempt to measure recoil energies of WIMPs while they cross a well-shield low-background detector, and indirect detection through the search for energetic neutrinos produced by the annihilation of WIMPs that have accumulated at the center of massive bodies, such as the Sun and/or the Earth[27]. Both techniques rely on the properties of scattering on ordinary matter, the first being more sensitive to the coherent or scalar (spin-independent) WIMP-nucleon coupling, the second scaling with the WIMP capture rate, which, in the Sun, is mainly related to the axial-vector (spin-dependent) coupling,,,,, while, in the Earth, is linked again the scalar interactions. The second possibility is to exploit the property of WIMPs to annihilate in pairs: their mean density in the Milky Way or external galaxies is much smaller than typical densities in the early Universe or in the center of massive bodies, but still there is a finite probability to generate SM particles. These particles hadronize and/or decay into stable species and the strategy is then to search for those species with small backgrounds, as it is the case with gamma-rays and antimatter cosmic-rays[28].

Recently, the most notable progresses have been made in the field of direct detection, with several competing experimental groups implementing different detection strategies and materials. Present detectors start to have sensitivities at the level of scattering rates predicted for WIMP candidates, such as SUSY models. Some configurations have been excluded, as well as an effects consistent with detection has been reported[30]: the method to discriminate a signal versus eventual background components is to single out features expected for the local WIMP distribution in momentum space[29]. One possibility is to build detectors which are sensitive to the direction of the incident WIMP; there are some attempts in this directions, but at very early stages[31]. The second possibilty is to search for a modulation in the total event rate, i.e signal plus background, since the signal is expected to have a (rather small) daily modulation and (a slightly larger, at about the 5% level) annual modulation because of the earth rotation on its axis and of its motion along the orbit around the Sun.

An effect compatible with being an annual modulation signal has been reported by the DAMA Collaboration[30], by now on a seven year period, with a very large exposure and high statistics. The most likely explanation, in case SUSY dark matter is advocated, is to suppose that the effect is driven by spin-independent couplings, however this interpretation has not been confirmed by competing experiments which, so far, did not find any evidence for a signal and hence just produced exclusion plots. Claims of incompatibility have been put forward by the EDELWEISS Collaboration[32] and by the CDMS Collaboration[33]; these are probably inconclusive statements, since there are several possible caveats in such comparisons[30], mainly stemming from the fact that these experiments use different materials and experimental techniques, and that the explanation in terms of spin-independent coupling is just one of the viable hypothesis to interpret the result of the DAMA Collaboration.

Dark matter searches with other detection methods are reaching as well the precision level needed for an eventual discovery. Some (very) weak hints of possible signals are already present in current cosmic-ray and γ-ray data: An excess in the galactic positron flux has been reported[34] and it has been shown that this may be compatible with a signal from neutralino annihilations[35]. Data on the γ-ray flux in the Galactic Center direction are hardly consistent with standard emission models, both in the few-GeV energy range [36] and in the region between a few hundreds GeV and a few TeV [37]; this could be a hint for the presence of a component from the production (and subsequent decay) of neutral pions due to WIMP annihilations[38], since CDM particles are expected to have an enhanced density towards the Galactic Center. It will be however rather hard to single out unambiguously such signal, since its signature is rather weak. A spectacular confirmation of this hypothesis could come instead from the detection of monochromatic γ-rays generated in prompt two-body final states from WIMP annihilations at one-loop level[39]: there is no plausible astrophysical background for such high energy photons (i.e. energies matching the WIMP mass). Detection prospects in this channel, as in any other channel, vary with the definition of the WIMP model: most notably, different detection techniques are complementary, as well as dark matter searches and tests of the model at accelerators may integrate each other.

6. Conclusions

With the recent remarkable progresses in our understanding of the dynamics of the Universe and of structure formation, the evidence for non-baryonic dark matter and dark energy has become stronger and stronger. We are however still far from having a comprehensive understanding of the dark side of the Universe from a particle physics point of view. For what concerns dark energy, some approaches have been explored, but there are still fundamental problems which are very hard to address; some help may come from future, more refined cosmological observations. For what concerns dark matter, several interesting ideas have been put forward and are being tested; there may be even some indications that the quest for dark matter may be on the right track.

References

1. V. Hradecky et al., *Astrophys. J.* **543**, 521 (2000).
2. D. Zarinsky and S.D.M. White, *Astrophys. J.* **435**, 599 (1994).
3. M.L. Mateo, *Ann. Rev. Astron. and Astrophys.* **36**, 435 (1998).
4. K.A. Olive, G. Steigman and T.P. Walker, *Phys. Rept.* **333**, 389 (2000).
5. David Tytler et al., *Astron. J.* **117**, 63 (1999).
6. S. Perlmutter et al., *Astrophys. J.* **517**, 565 (1999).
7. A.G. Riess et al., *Astron. J.* **116**,1009 (1998).
8. B.J. Barris et al., *Astrophys. J.* **602**, 571 (2004).
9. P. de Bernardis et al., *Nature* **404**, 955 (2000).
10. S. Hanany et al., *Astrophys. J.* **545**, L5 (2000).
11. C.L. Bennett et al., *Astrophys. J. Suppl.* **148**, 1 (2003).
12. M. Tegmark et al., *Phys. Rev.* **D69**, 103501 (2004).
13. M. Colless et al., *Mon. Not. Roy. Astron. Soc.* **328**, 1039 (2001).
14. A. Connolly et al., *Astrophys. J.* **579**, 42 (2002).
15. J.A. Peacock, *Cosmological Physics*, Cambridge University Press, (1999).
16. P.J. Steinhardt, L.M. Wang and I. Zlatev, *Phys. Rev.* **D59**, 123504 (1999).

17. P.J.E. Peebles and B. Ratra, *Astrophys. J.* **325**, L17 (1988); M.S. Turner and M. White, *Phys. Rev.* **D56**, 4439 (1997); R.R. Caldwell, R. Dave and P.J. Steinhardt, *Phys. Rev. Lett.* **80**, 1582 (1998); E.J. Copeland, A.R. Liddle and D. Wands, *Phys. Rev.* **D57**, 4686 (1998).
18. E.K. Kolb and M.S. Turner, *The Early Universe*, Westview Press, (1994).
19. G. Jungman, M. Kamionkowski and K. Griest, *Phys. Rept.* **267**, 195 (1996).
20. A. Masiero, S. Profumo and P. Ullio, hep-ph/0412058; P. Gondolo, J. Edsjo, P. Ullio, L. Bergstrom, M. Schelke and E.A. Baltz, *JCAP* **0407**, 008 (2004).
21. M. Fujii and K. Hamaguchi, *Phys. Lett.* **B525**, 143 (2002); M. Fujii and K. Hamaguchi, *Phys. Rev.* **D66**, 083501 (2002); R. Jeannerot, X. Zhang and R.H. Brandenberger, *JHEP* **9912**, 003 (1999).
22. J.P. Preskill, M.B. Wise and F. Wilczek, *Phys. Lett.* **B120**, 127 (1983); M. Dine and W. Fischler, *Phys. Lett.* **B120**, 137 (1983).
23. G. Servant and T.M.P. Tait, *Nucl. Phys.* **B650**, 391 (2003).
24. L.M. Krauss, S. Nasri and M. Trodden, *Phys. Rev.* **D67**, 085002 (2003).
25. L. Bergström, *Rept. Prog. Phys.* **63**, 793 (2000).
26. M. W. Goodman and E. Witten, *Phys. Rev.* **D31**, 3059 (1986); I. Wasserman, *Phys. Rev.* **D33**, 2071 (1986).
27. W.H. Press and D.N Spergel, *Astrophys. J.* **296**, 679 (1985); J. Silk, K. Olive and M. Srednicki, *Phys. Rev. Lett.* **55**, 257 (1985); L. Krauss, M. Srednicki and F. Wilczek, *Phys. Rev.* **D33**, 2079 (1986).
28. J. Silk and M. Srednicki, *Phys. Rev. Lett.* **53**, 624 (1984); J. Silk and H. Bloemen, *Astrophys. J.* **313**, L47 (1987); S. Rudaz and F.W. Stecker, *Astrophys. J.* **325**, 16 (1988).
29. A. Drukier, K. Freese and D. Spergel, *Phys. Rev.* **D30**, 3495 (1986); K. Freese, J. A. Frieman and A. Gould, *Phys. Rev.* **D37**, 3388 (1988).
30. R. Bernabei et al., *Riv. N. Cim.* **26**, 1 (2003).
31. M. J. Lehner et al., astro-ph/9905074.
32. A. Benoit et al., *Phys. Lett.* **B545**, 43 (2002).
33. CDMS Collaboration, *Phys. Rev. Lett.* **93**, 211301 (2004).
34. S. Coutu et al., *Astropart. Phys.* **11**, 429 (1999).
35. E.A. Baltz et al., *Phys. Rev.* **D65**, 063511 (2002).
36. H.A. Mayer-Hasselwander et al., *Astron. Astrophys.* **335**, 161 (1998).
37. K. Kosack et al., VERITAS Collaboration, astro-ph/0403422; K. Tsuchiya et al., CANGAROO-II collaboration, *Astrophys. J.* **606**, L115 (2004); F. Aharonian et al., H.E.S.S. Collaboration, *Astron. Astrophys.* **425**, L13 (2004).
38. A. Cesarini et al., *Astropart. Phys.* (2004), in press.
39. L. Bergström and P. Ullio, *Nucl. Phys.* **B504**, 27 (1997); P. Ullio and L. Bergström, *Phys. Rev.* **D57**, 1962 (1998).

DARK MATTER SEARCHES

R. BERNABEI

Dipartimento di Fisica, Università di Roma "Tor Vergata"
and INFN, Sezione di Roma II, I-00133, Roma, Italy
e-mail: rita.bernabei@roma2.infn.it

General arguments related to the investigation of a Dark Matter particle component in the galactic halo have been addressed during the lectures as well as latest results obtained by exploiting the annual modulation signature. Only few points are summarized here.

1. Introduction

Although the first evidence that much more than the visible matter should fill our Universe dates back to the beginning of the XX century [1,2], only in the 80's the fact that Dark Matter should be present in our Universe reached a wide consensus. In particular, in the 70's two groups showed that the velocity curves of astrophysical objects in spiral galaxies stay flat even outside the luminous disk [3], crediting the presence of dark matter in the galactic halo. Afterwards, many other experimental evidences for the Dark Universe have been further obtained with improvements of technology definitively pointing out that to explain the observed gravitational effects the mass of the Universe should be much larger than the luminous one [4]. After the '70 many other observations have further confirmed the presence of Dark Matter in the Universe and, at present, the measurements are mainly devoted to the investigation of the quantity, of the distribution (from the cosmological scale down to the galactic one) and of the nature of the Dark Matter in the Universe. In particular, recent measurements of the CMB temperature anisotropy by WMAP [5], analysed in the framework of the *Big Bang* cosmological scenario, support for the density of the Universe a value: $\Omega = 1$, further crediting that most of the Universe is dark. Recently, it has been suggested from observations on the supernovae Ia at high red-shift that about 73% of Ω might be in form of a *dark energy* [6]; further experimental investigations are in progress. However, even in this scenario large space for Dark Matter particles in the Universe exists. In fact, in this scenario the matter density in the Universe would be $\Omega_m \sim 0.3$ [5,7], while the luminous matter can only account for a density $\simeq 0.005$ and the baryonic Dark Matter for $\simeq 0.04$. On the other hand, since the contribution of Dark Matter particles relativistic at the decoupling time cannot exceed 0.01 (by considerations on large scale structure formations [8]), most of the Dark Matter particles in the Universe should be non relativistic at decoupling time; they are named Cold Dark Matter particles (CDM). The CDM candidates have to be neutral, stable or quasi-stable (e.g. with a time decay of order of the age of the Universe) and have to weakly interact with ordinary matter. These features are respected by the axions (also investigated by DAMA/NaI [9]) and by a class of candidates named WIMPs (Weakly Interacting Massive Particles). Since in the Standard Model of particle Physics no suitable particle as CDM candidate exists, a window beyond the Standard Model is investigated. At present, the most widely considered candidate for CDM is the lighest supersymmetric particle named neutralino; in fact, in the supersymmetric theories where the R-parity is conserved, the lighest supersymmetric particle must be stable and can interact neither by electromagnetic nor by strong interactions, otherwise it would be detected in the galactic halo as the ordinary matter is. However, other candidates can also be considered as e.g. a heavy neutrino of the 4th family or a sneutrino (the spin-0 supersymmetric partner of the neutrino) in a multi-component Dark Matter scenario. Moreover, a dominant contribution of a sneutrino candidate remains still possible in supersymmetric models with violation of lepton number, where two mass states and a small energy splitting is present, as reported in ref. [10]; a similar sneutrino can only inelastically scatter off nuclei, after its excitation to the low lying energy level. Other proposed candidates are the mirror Dark Matter particles [11] and the particles from multi-dimensional Kaluza-Klein-like theories. Moreover, in principle even whatever massive and weakly interacting particle, not yet foreseen by theories, can be a good candidate as CDM.

Several observations have pointed out that our Galaxy should also be embedded in a dark halo with mass at least 10 times larger than that of the luminous matter. The CDM particles of this dark halo can be detected either by direct or by indirect methods. In the first case, their elastic interaction on target nuclei can be detected by means of the signal induced by the recoiling nucleus or, in case of inelastic scattering, also by the successive de-excitation gamma's; these searches should be performed deep underground. In the other case, the flux of secondary particles - mainly neutrinos, positrons, gamma's - produced by possible annihilation of CDM particles either in the Sun or in the Earth or in the halo may be detected. Indirect searches can be performed by space experiments or by underwater or underground experiments. Discussions on the relative techniques can be found in refs. [12,13] and references therein.

In the following, we mainly concentrate our attention on the direct detection technique exploited by the DAMA/NaI experiment through the study of the model independent annual modulation signature.

2. A direct detection experiment: DAMA/NaI

The DAMA/NaI experiment was proposed in 1990 [14], designed and realized having the main aim to investigate in a model independent way the presence of a Dark Matter particle component in the galactic halo [15,16,17,18,19,20,21,22,13]. For this purpose, we planned to exploit the effect of the Earth revolution around the Sun on the Dark Matter particles interactions on the target-nuclei of suitable underground detectors. In fact, as a consequence of its annual revolution, the Earth should be crossed by a larger flux of Dark Matter particles in June (when its rotational velocity is summed to the one of the solar system with respect to the Galaxy) and by a smaller one in December (when the two velocities are subtracted). This offers an efficient model independent signature, able to test a large interval of cross sections and of halo densities; it is named *annual modulation signature* and was originally suggested in the middle of '80 in ref. [23].

The annual modulation signature is very distinctive since a WIMP-induced seasonal effect must simultaneously satisfy all the following requirements: the rate must contain a component modulated according to a cosine function (1) with one year period (2) and a phase that peaks roughly around $\simeq 2^{nd}$ June (3); this modulation must only be found in a well-defined low energy range, where WIMP induced recoils can be present (4); it must apply to those events in which just one detector of many actually "fires" ($single-hit$ events), since the WIMP multi-scattering probability is negligible (5); the modulation amplitude in the region of maximal sensitivity must be $\lesssim$7% for usually adopted halo distributions (6), but it can be larger in case of some possible scenarios such as e.g. those in refs. [10,24]. Only systematic effects able to fulfil these 6 requirements and to account for the whole observed modulation amplitude could mimic this signature; thus, no other effect investigated so far in the field of rare processes offers a so stringent and unambiguous signature. With the present technology, the annual modulation signature remains the main signature of a WIMP signal.

The DAMA/NaI experiment [a] was located deep underground in the Gran Sasso National Laboratory of the I.N.F.N., whose main features have been reported in [28,29,30,31]. Considering its main goal, DAMA/NaI ($\simeq$ 100 kg of highly radiopure NaI(Tl)) was designed by employing and further developing all the necessary low-background techniques and procedures. A detailed description of the set-up, of its radiopurity, of its performance, of the used hardware procedures, of the determination of the experimental quantities and of the data reduction was given in refs. [32,18,19,13]. Here only few arguments are addressed. The nine 9.7 kg highly radiopure NaI(Tl) are encapsulated in radiopure Cu housings; moreover, 10 cm long Tetrasil-B light guides act as optical windows on the two end faces of each crystals and are coupled to specially developed low background photomultipliers (PMT). The measured light response is 5.5 – 7.5 photoelectrons/keV depending on the detector. The two PMTs of a detector work in coincidence with hardware thresholds at the single photoelectron level in order to assure high efficiency for the coincidence at few keV level. The energy threshold of the experiment, 2 keV, is determined by means of X-rays sources and of keV range Compton electrons on the basis also of the features of the noise rejection procedures and of the efficiencies when lowering the number of available photoelectrons [32]. The detectors are enclosed in a sealed copper box, continuously maintained in high purity (HP) Nitrogen atmosphere in slightly overpressure with respect to the external environment. A suitable low background hard shield against electromagnetic and neutron background was realized using very high radiopure Cu and Pb bricks [32], Cd foils and 10/40 cm polyethylene/paraffin; the hard shield is also sealed in a plexiglas box and maintained in HP Nitrogen atmosphere. Moreover, about 1 m concrete (made from the Gran Sasso rock material) almost fully surrounds the hard shield outside the barrack and at its bottom, acting as a further neutron moderator.

A three-level sealing system from environmental Radon is effective. In fact, the inner part of the barrack, where the set-up is allocated, has the floor (above the concrete) and all the walls sealed by Supronyl (permeability: $2 \cdot 10^{11}$ cm^2/s [33]) plastic and the entrance door is air-tight. A low

---

[a]We take this occasion to remind that DAMA/NaI has been part of the DAMA project, which is also composed by several other low background set-ups, such as: i) DAMA/LXe ($\simeq$ 6.5 kg pure liquid Xenon scintillator) [25,26]; ii) DAMA/R&D, set-up devoted to tests on prototypes and small scale experiments [27]; iii) the new second generation large mass NaI(Tl) radiopure set-up DAMA/LIBRA (see later); iv) DAMA/Ge detector for sample measurements.

level oxygen alarm informs the operator before entering the inner part of the barrack since the HP Nitrogen which fills both the inner Cu box and the external plexiglas box is released in this closed environment. The Radon level inside the barrack is continuously monitored and recorded with the production data [15,16,32,18,19,13].

On the top of the shield a glove-box (also maintained in the HP Nitrogen atmosphere) is directly connected to the inner Cu box, housing the detectors, through Cu pipes. The pipes are filled with low radioactivity Cu bars (covered by 10 cm of low radioactive Cu and 15 cm of low radioactive Pb) which can be removed to allow the insertion of radioactive sources for calibrating the detectors in the same running condition, without any contact with external environment [32].

The whole installation is air-conditioned and the operating temperature as well as many other parameters are continuously monitored and acquired with the production data. Moreover, self-controlled computer processes automatically monitor several parameters and manage alarms [32,19,13].

The electronic chain and the data acquisition system operative up to summer 2000 have been described in ref. [32], while the new electronics and DAQ installed in summer 2000 have been described in ref. [13].

The DAMA/NaI set-up has exploited the WIMP annual modulation signature over seven annual cycles [15,16,17,18,19,20,21,22,13] and the following part of this paper will summarize the final model independent result, some of the corollary quests for the candidate particle, some implications and some pespectives. In particular, it is worth to remind that - thanks to its radiopurity and features - DAMA/NaI has also investigated other approaches for WIMPs in ref. [34,35] and several other rare processes [36,37,38,39,40,9,41,42].

3. The model-independent result of DAMA/NaI

A model-independent approach on the data collected by DAMA/NaI over seven annual cycles offers an immediate evidence of the presence of an annual modulation of the measured rate of the *single-hit* events in the lowest energy region. In particular, in Fig. 1 – *left* the time behaviour of the residual rate of the *single-hit* events in the cumulative (2-6) keV energy interval is reported. The data favour the presence of a modulated cosine-like behaviour at 6.3 σ C.L. and their fit for this cumulative energy interval offers modulation amplitude equal to (0.0200 ± 0.0032) cpd/kg/keV, a phase $t_0 = (140 \pm 22)$ days and a period $T = (1.00 \pm 0.01)$ year, all parameters kept free in the fit. The period and phase agree with those expected in the

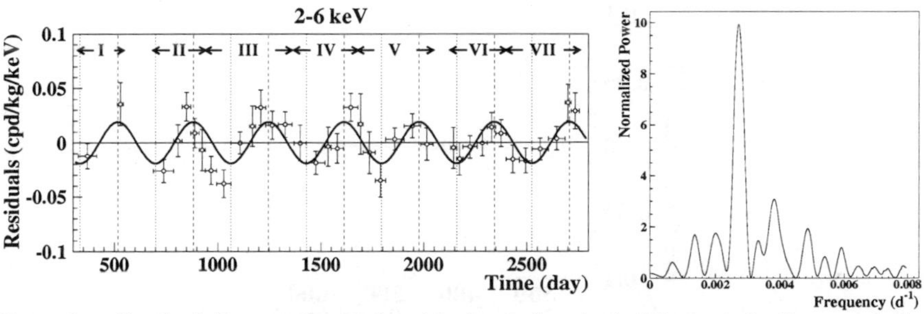

Figure 1. *On the left*: experimental residual rate for *single-hit* events in the cumulative (2–6) keV energy interval as a function of the time over 7 annual cycles (total exposure 107731 kg × day); end of data taking July 2002. The experimental points present the errors as vertical bars and the associated time bin width as horizontal bars. The superimposed curve represents the cosinusoidal function behaviour expected for a WIMP signal with a period equal to 1 year and phase exactly at 2nd June; the modulation amplitude has been obtained by best fit. See ref. [13]. *On the right*: power spectrum of the measured *single-hit* residuals for the cumulative (2–6) keV energy interval calculated including also the treatment of the experimental errors and of the time binning. As it can be seen, the principal mode corresponds to a frequency of $2.737 \cdot 10^{-3}$ d^{-1}, that is to a period of $\simeq$ 1 year.

case of an effect induced by Dark Matter particles of the galactic halo (T = 1 year and t_0 roughly at $\simeq 152.5^{th}$ day of the year). The χ^2 test on the (2–6) keV residual rate disfavours the hypothesis of unmodulated behaviour giving a probability of $7 \cdot 10^{-4}$ ($\chi^2/d.o.f.$ = 71/37). The same data have also been investigated by a Fourier analysis as shown in Fig. 1 – *right*, where a clear peak corresponding to a period of $\simeq$ 1 year is present. Modulation is not observed above 6 keV [13] [b]. Finally, a suitable statistical analysis has shown that the modulation amplitudes are statistically well distributed in all the crystals, in all the data taking periods and considered energy bins. More arguments can be found in ref.[13]. A careful investigation of all the known possible sources of systematic and side reactions has been regularly carried out and published at time of each data release and quantitative discussions can be found in refs. [13,19]. No systematic effect or side reaction able to account for the observed modulation amplitude and to satisfy all the requirements of the signature has been found.

As a further relevant investigation, the *multiple-hits* events collected during the DAMA/NaI-6 and 7 running periods (when each detector was equipped with its own Transient Digitizer with a dedicated renewed electronics) have been studied and analysed by using the same identical hard-

[b] We remind that DAMA/NaI took data up to MeV energy region despite the optimization was done for the keV energy range.

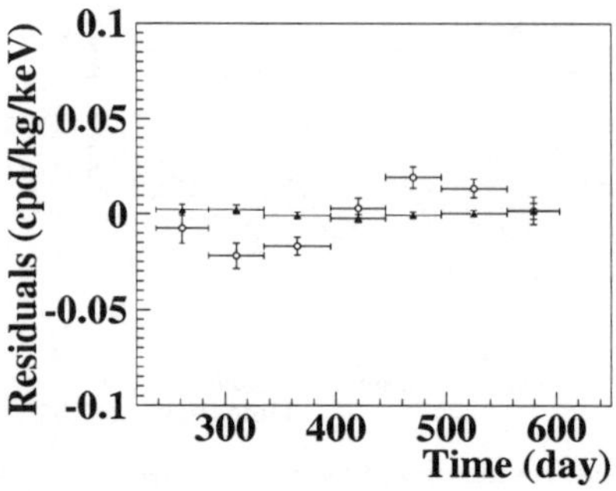

Figure 2. Experimental residual rates over seven annual cycles for *single-hit* events (open circles) – class of events to which WIMP events belong – and over the last two annual cycles for *multiple-hits* events (filled triangles) – class of events to which WIMP events do not belong – in the (2–6) keV cumulative energy interval. They have been obtained by considering for each class of events the data as collected in a single annual cycle and using in both cases the same identical hardware and the same identical software procedures. The initial time is taken on August 7th. See text.

ware and the same identical software procedures as for the case of the *single-hit* events (see Fig. 2). The *multiple-hits* events class – on the contrary of the *single-hit* one – does not include events induced by WIMPs since the probability that a WIMP scatters off more than one detector is negligible. The fitted modulation amplitudes are: $A = (0.0195 \pm 0.0031)$ cpd/kg/keV and $A = -(3.9 \pm 7.9) \cdot 10^{-4}$ cpd/kg/keV for *single-hit* and *multiple-hits* residual rates, respectively. Thus, evidence of annual modulation is present in the *single-hit* residuals (events class to which the WIMP-induced recoils belong), while it is absent in the *multiple-hits* residual rate (event class to which only background events belong). Since the same identical hardware and the same identical software procedures have been used to analyse the two classes of events, the obtained result offers an additional strong support for the presence of Dark Matter particles in the galactic halo further excluding any side effect either from hardware or from software procedures or from background.

In conclusion, the presence of an annual modulation in the residual rate of the *single-hit* events in the lowest energy interval $(2-6)$ keV, satisfying all the features expected for a Dark Matter particle component in the galactic halo is supported by the data of the seven annual cycles at 6.3 σ C.L.. No systematic effect or side reaction able to account for the observed effect has been found. This is the experimental result of DAMA/NaI; it is model-independent. No other experiment, whose result can be directly compared with this one in a model independent way, is available so far in the field of Dark Matter investigation.

4. Some corollary model-dependent quests for a candidate

On the basis of the obtained 6.3 σ model-independent result, corollary investigations can also be pursued on the nature of the Dark Matter particle candidate. This latter investigation is instead model-dependent and – considering the large uncertainties which exist on the astrophysical, nuclear and particle physics assumptions and on the parameters needed in the calculations – has no general meaning (as it is also the case of exclusion plots and of the WIMP parameters evaluated in indirect detection experiments). Thus, it should be handled in the most general way as we have pointed out with time passing [15,16,17,18,19,20,21,22,13].

Candidates, kinds of WIMP couplings with ordinary matter and implications, cross sections, nuclear form factors, spin factors, scaling laws, halo models, priors, etc. are discussed in ref.[13]. The reader can find in this latter paper and in references therein devoted discussions to correctly understand the results obtained in corollary quests and the real validity of any claimed model-dependent comparison in the field. Here, we just remind that the results briefly summarized here are not exhaustive of the many scenarios possible at present level of knowledge, including those depicted in some more recent works such as e.g. refs. [24,43].

DAMA/NaI is intrinsically sensitive both to low and high WIMP mass having both a light (the ^{23}Na) and a heavy (the ^{127}I) target-nucleus; in previous corollary quests for the candidate, dark matter particle masses above 30 GeV (25 GeV in ref.[15]) have been presented [16,18,20,21,22] for few (of the many possible) model frameworks. However, that bound holds only for neutralino when supersymmetric schemes based on GUT assumptions are adopted to analyse the LEP data [44]. Thus, since other candidates are possible and also other scenarios can be considered for the neutralino itself as recently pointed out[c], the present model-dependent lower bound quoted by LEP for the neutralino in the supersymmetric schemes based on GUT assumptions (37 GeV [47]) is simply marked in the following figures. It is worth to note that this model dependent LEP limit – when considered – selects the WIMP-Iodine elastic scatterings as dominant.

For simplicity, here the results of these corollary quests for a candidate particle are presented in terms of allowed regions obtained as superposition of the configurations corresponding to likelihood function values *distant* more than 4σ from the null hypothesis (absence of modulation) in each of

[c]In fact, when the assumption on the gaugino-mass unification at GUT scale is released, neutralino masses down to $\simeq$ 6 GeV are allowed [45,46].

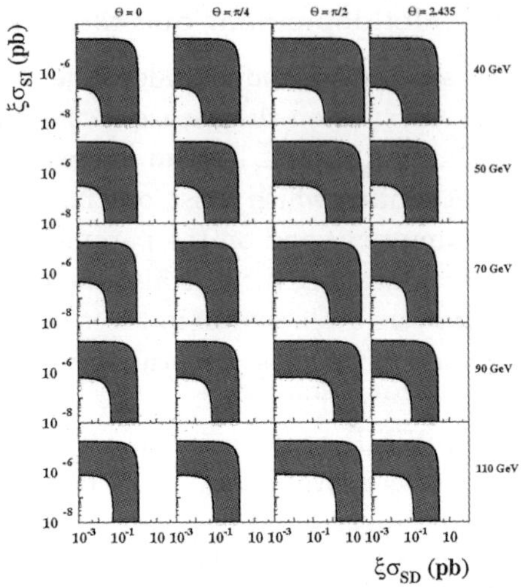

Figure 3. *Case of a WIMP with mixed SI&SD interaction for the model frameworks given in ref.*[13]. Coloured areas: example of slices (of the 4-dimensional allowed volume) in the plane $\xi\sigma_{SI}$ vs $\xi\sigma_{SD}$ for some of the possible m_W and θ values. Inclusion of other existing uncertainties on parameters and models would further extend the regions; for example, the use of more favourable form factors and/or of more favourable spin factors than the ones considered here would move them towards lower cross sections. For details see ref.[13].

the several (but still a limited number) of the possible model frameworks considered in ref.[13]. These allowed regions take into account the time and energy behaviours of the single-hit experimental data and have been obtained by a maximum likelihood procedure (for a formal description see e.g. refs. [15,16,18]) which requires the agreement: i) of the expectations for the modulated part of the signal with the measured modulated behaviour for each detector and for each energy bin; ii) of the expectations for the unmodulated component of the signal with the respect to the measured differential energy distribution and - since ref.[18] - also with the bound on recoils obtained by pulse shape discrimination from the devoted DAMA/NaI-0 data [34]. The latter one acts in the likelihood procedure as an experimental upper bound on the unmodulated component of the signal and – as a matter of fact – as an experimental lower bound on the estimate of the background levels. Thus, the C.L.'s, we quote for the allowed regions, already account for compatibility with the measured differential energy spectrum and with the measured upper bound on recoils. Finally, it is worth to note that the best fit values of cross sections and Dark Matter particle mass span over a large range when varying the considered model framework.

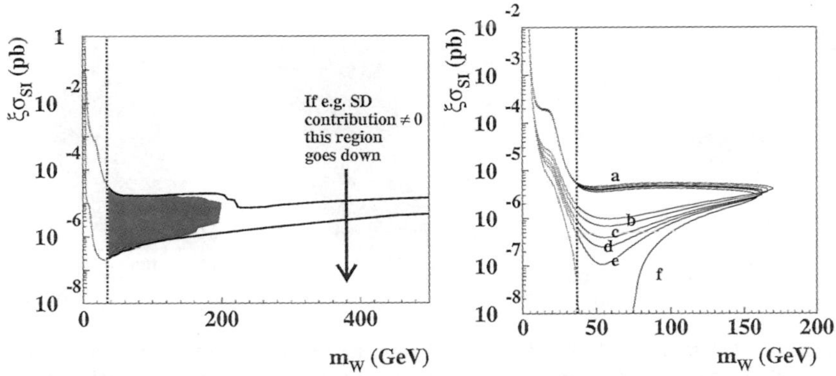

Figure 4. *On the left : Case of a WIMP with dominant SI interaction for the model frameworks given in ref.*[13]. Region allowed in the plane (m_W, $\xi\sigma_{SI}$). The vertical dotted line represents a bound in case of a neutralino candidate when supersymmetric schemes based on GUT assumptions are adopted to analyse the LEP data; the low mass region is allowed for neutralino when other schemes are considered (see text) and for every other dark matter particle candidate. While the area at WIMP masses above 200 GeV is allowed only for few configurations, the lower one is allowed by most configurations (the colored region gathers only those above the vertical line). The inclusion of other existing uncertainties on parameters and models would further extend the region; for example, the use of more favourable SI form factor for Iodine alone would move it towards lower cross sections. *On the right: Example of the effect induced by the inclusion of a SD component different from zero on allowed regions given in the plane $\xi\sigma_{SI}$ vs m_W.* In this example the Evans' logarithmic axisymmetric $C2$ halo model with $v_0 = 170$ km/s, ρ_0 equal to the maximum value for this model and a given set of the parameters' values (see ref.[13]) have been considered. The different regions refer to different SD contributions for the particular case of $\theta = 0$: $\sigma_{SD} = 0$ pb (a), 0.02 pb (b), 0.04 pb (c), 0.05 pb (d), 0.06 pb (e), 0.08 pb (f). Analogous situation is found for the other model frameworks. For details see ref.[13].

Fig. 3, 4, 5 show some of the obtained allowed regions; details and descriptions of the symbols are given in ref.[13]. Here we only remind that $tg\theta$ is the ratio between the Dark Matter particle-neutron and the Dark Matter particle-proton effective spin-dependent coupling strengths and that θ is defined in the $[0,\pi)$ interval. Obviously, larger sensitivities than those reported in the following figures would be reached when including the effect of other existing uncertainties on the astrophysical, nuclear and particle Physics assumptions and related parameters; similarly, the set of the best fit values would also be enlarged as well. For details see ref. [13].

In Fig. 6 the theoretical expectations in the purely SI coupling for the particular case of a neutralino candidate in MSSM with gaugino mass unification at GUT scale released [46] are shown. The marked curve surrounds the DAMA/NaI purely SI allowed region as in Fig. 4 – *left*.

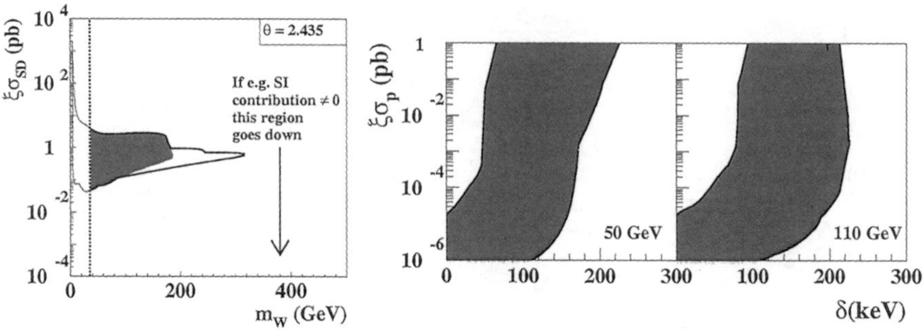

Figure 5. *On the left: Case of a WIMP with dominant SD interaction in the model frameworks given in ref.*[13]. Example of a slice (of the 3-dimensional allowed volume) in the plane (m_W, $\xi\sigma_{SD}$) at a given θ value (θ is defined in the $[0, \pi)$ range); here $\theta = 2.435$ (Z_0 coupling). For the definition of the vertical line and of the coloured area see the caption of Fig. 4. Inclusion of other existing uncertainties on parameters and models (as discussed in ref.[13]) would further extend the SD allowed regions. For example, the use of more favourable SD form factors and/or more favourable spin factors would move them towards lower cross sections. Values of $\xi\sigma_{SD}$ lower than those corresponding to this allowed region are possible also e.g. in case of an even small SI contribution (see ref.[13]). *On the right: Case of a WIMP with* preferred inelastic *interaction in the model frameworks given in ref.*[13]. Examples of slices (coloured areas) of the 3-dimensional allowed volume ($\xi\sigma_p$, δ, m_W) for some m_W values. Inclusion of other existing uncertainties on parameters and models would further extend the regions; for example, the use of more favourable form factors and of different escape velocity would move them towards lower cross sections. For details see ref.[13].

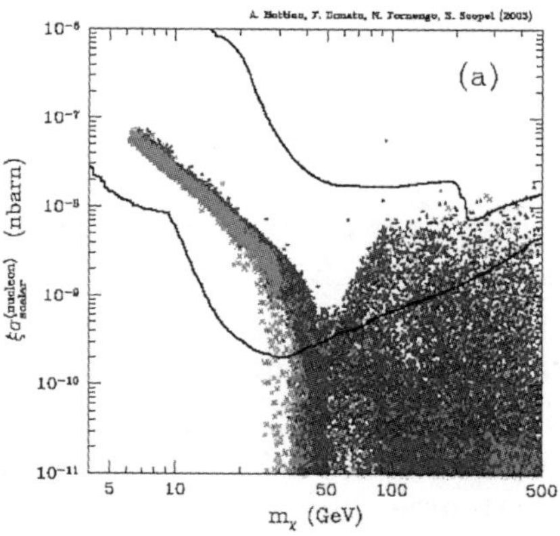

Figure 6. Figure taken from ref.[46]: theoretical expectations of $\xi\sigma_{SI}$ versus m_W in the purely SI coupling for the particular case of a neutralino candidate in MSSM with gaugino mass unification at GUT scale released; the curve is the same as in Fig. 4-*left*.

5. Comparison with other direct and indirect detection experiments

As already mentioned, no other experiment, whose result can be directly compared in a model independent way with that of DAMA/NaI, is available so far in the field of Dark Matter detection.

In fact, most of the activities, started in the 90's, are still at R&D stage and/or have released marginal exposures with the respect to the many years of existence and to the several used detectors. This is the case of CDMS and EDELWEISS experiments, while the Zeplin experiment is more recent [48,49,50]. Since these experiments have claimed to have "excluded" DAMA/NaI, we will briefly point out only few arguments. In particular, Table 1 summarizes some items for comparison.

Firstly, let us preliminarily assume as fully correct the "selected" number of events, the energy threshold, the energy scale, etc. quoted by those experiments (see Table 1) and let us consider if – at least under this hypothesis – their claims might be justified. The answer is obviously not; in fact: i) they give a single model dependent result using ^{nat}Ge or ^{nat}Xe target, while DAMA/NaI gives a model independent result using ^{23}Na and ^{127}I targets; ii) in the single (of the many possible) model scenario, they consider, they "fix" all the astrophysical, nuclear and particle physics assumptions at a single choice; the same is even for the the experimental and theoretical parameters values needed in the calculations. In addition, DAMA/NaI is generally quoted there in an uncorrect, partial and unupdated way and the existing scenarios to which DAMA/NaI is fully sensitive – on the contrary of the others – are ignored.

Let us now briefly comment also some of the experimental aspects. In particular, the counting rate of the Ge bolometers experiments is very high and few/zero events are claimed after applying several strong and hardly safe rejection procedures (involving several orders of magnitude). They usually claim to have an "event by event" discrimination between *noise + electromagnetic background* and *recoil + recoil-like (neutrons, end-range alphas, fission fragments,...)* events by comparing the bolometer and the ionizing signals for each event, but their results are, actually, largely based on huge data selections and on the application of other preliminar rejection procedures (such as e.g. the one on the so-called surface electrons), which are generally poorly described and often not completely quantified. Moreover, most efficiencies and physical quantities entering in the interpretation of the claimed selected events have never been properly accounted; as an example, we mention the case of the bolometer quenching factor of the recoil target nuclei. In fact, for the bolometer signals the quenching factor (on which the energy threshold and the energy scale rely and, hence, also the claimed sensitivity for the given model dependent exclusion plots) is arbitrarily assumed to be exactly equal to one. Up to now, only one

Table 1. Features of the DAMA/NaI results on the WIMP annual modulation signature over the seven annual cycles [13] with those of refs. [48,49,50].

	DAMA/NaI	CDMS-II	Edelweiss-I	Zeplin-I
Signature	Annual modulation	None	None	None
Target-nuclei	^{23}Na, ^{127}I	^{nat}Ge	^{nat}Ge	^{nat}Xe
Technique	well known	poorly experienced	poorly experienced	critical optical liquid/gas interface in this realization
Target mass	$\simeq$ 100 kg	0.75 kg	0.32 kg	$\simeq$ 3 kg
Exposure	$\simeq (1.1 \cdot 10^5)$ kg · day	19.4 kg · day	30.5 kg · day	280 kg · day
Depth of the experimental site	1400 m	780 m	1700 m	1100 m
Software energy threshold	2 keV e.e. (5.5 – 7.5 p.e./keV)	10 keV e.e.	20 keV e.e.	2 keV e.e. (but: $\sigma/E = 100\%$ mostly 1 p.e./keV; [50]) (2.5 p.e./keV for 16 days; [51])
Quenching factor	Measured	Assumed = 1	Assumed = 1 (see also [52])	Measured
Measured event rate in low energy range	$\simeq$ 1 cpd/kg/keV	??, claimed γ's larger than CDMS-I ($\simeq$ 60 cpd/kg/keV, 10^5 events)	$\simeq 10^4$ events total	$\simeq$ 100 cpd/kg/keV
Claimed events after rejection procedures		either 0 or 1	2 (claimed taken in a noisy period)	$\simeq$ 20-50 cpd/kg/keV after rejection and ?? after standard PSD [50,51]
Events satisfying the signature in DAMA/NaI	modulation amplitude integrated over the given exposure $\simeq 10^3$ events	insensitive	insensitive	insensitive
Expected number of events from DAMA/NaI effect		from few down to zero depending on the models (and on quenching factor)	from few down to zero depending on the models (and on quenching factor)	depends on the models (even zero)

measurement has been made available for a given detector [52]; it offers the value: $0.87 \pm 10\%(stat.) \pm 10\%(syst.)$, which is – within the error – compatible with one, but – at the same time – also compatible with much smaller values. Thus, any bolometer result, obtained without considering e.g. the uncertainties about the unknown value of the quenching factor and, hence, about the energy threshold and energy scale, has to be considered partial and arbitrary. For completeness we also mention that the reproducibility of the results over different running periods has not been proved as well as the values of the effective sensitive volumes for the read-outs of the two signals for each event and related quantities; obviously, further uncertainties are present when, as done in some cases, a neutron background modeling and subtraction is pursued in addition.

As regards Zeplin-I [50,51], a very low energy threshold is claimed (2 keV), although the light response is very poor: between $\simeq$ 1 ph.e./keV [50] (for most of the time) and $\simeq$ 2.5 ph.e./keV (claimed for 16 days) [51] [d]. Moreover, a strong data filtering is applied to the high level of measured counting rate (see Table 1) by hardware vetoes, by fiducial volume cuts and, largely, by applying down to few keV a standard pulse shape discrimination procedure, although the LXe scintillation pulse profiles (pulse decay time $<$ 30 ns) are quite similar to the PMT noise events in the lower energy bins and in spite of the poor light response. Quantitative information on experimental quantities related to the used procedures has not yet been given [50,51].

In conclusion, those claims for contradiction have intrinsecally no scientific bases.

On the other hand, some positive hints are present in indirect detection experiments; in fact, an excess of positrons and of gamma's in the space has been reported with the respect to a modellised background; they are not in contraddiction with the DAMA/NaI result. Moreover, recently, it has been suggested [53] that these positive hints and the effect observed by DAMA/NaI can also be described in a scenario with multi-component Dark Matter in the galactic halo, made of a subdominant component of heavy neutrinos of the 4th family and of a sterile dominant component. In particular (see Fig. 7), it has been shown that an heavy neutrino with mass around 50 GeV can account for all the observations, while the inclusion of

[d] For comparison we remind that the data of the DAMA/LXe set-up, which has a similar light response, are analysed by using the much more realistic and safer software energy threshold of 13 keV [26].

possible clumpiness of neutrino density as well as new interactions in the heavy neutrino annihilation, etc. can lead to wider mass ranges: from about 46 up to about 75 GeV (see ref. [53] for details).

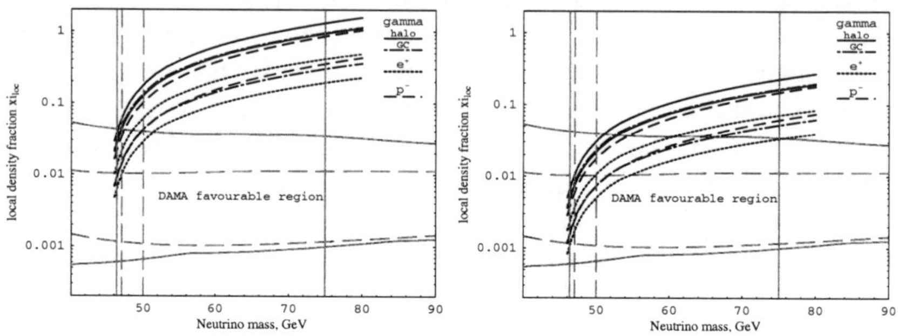

Figure 7. Figure taken from ref.[53]: Case of a subdominant heavy 4th neutrino candidate in the plane local density fraction versus the heavy neutrino mass. The favorable region for this candidate obtained from the DAMA/NaI data (grey dashed line when using the Evan's halo model; solid line when using the other halo models) and the best-fit density parameters deduced from cosmic gamma-radiation (from halo and galactic center), positron and antiproton analysis are shown (left panel). The effect of the inclusion of possible neutrino clumpiness is also reported (right panel). See ref. [53] for details.

6. Conclusions and perspectives

DAMA/NaI has been a pioneer experiment investigating as first the WIMP annual modulation signature with suitable sensitivity and control of the running parameters. During seven independent experiments of one year each one, it has pointed out at 6.3 σ C.L. in a model independent way the presence of a modulation satisfying the many peculiarities of an effect induced by Dark Matter particles in the galactic halo; no systematic effect or side reaction able to account for the observed effect has been found. As a corollary result, it has also pointed out the complexity of the quest for a candidate particle mainly because of the present poor knowledge on the many astrophysical, nuclear and particle physics aspects. At present after a devoted R&D effort, the second generation DAMA/LIBRA (a $\simeq$250 kg more radiopure NaI(Tl) set-up) has been realised and put in operation since March 2003. It will further investigate with increased sensitivity the model independent result of DAMA/NaI and will improve corollary quests on the nature of the candidate particle, trying to disentangle at least among some of the many different possible astrophysical, nuclear and particle physics models as well as to investigate other new possible scenarios. In particular, several arguments will be addressed, such as e.g.: i) possible effects induced on the Dark Matter particles distribution in the galactic halo by contributions from satellite galaxies tidal streams (Sagittarius, Canis Major, ...); ii) possible effects induced on the Dark Matter particles distribution in the galactic halo by the existence of caustics; iii) detection of possible "solar wakes"; iv) possible clumpiness with small scale size; v) the coupling(s) of the Dark Matter particle with the ^{23}Na and ^{127}I and its nature; vi) scaling laws and cross sections. A large work will be faced by DAMA/LIBRA, which is in addition the intrinsecally most sensitive experiment in the field of Dark Matter because of its radiopurity, exposed mass and high duty cycle. These qualities will also allow DAMA/LIBRA to further investigate with higher sensitivity several other rare processes.

Finally, at present a third generation R&D effort toward the possible NaI(Tl) ton set-up, we proposed in 1996 [54], has been funded and related works have already been started.

Complementary informations are also expected fron other serious direct and indirect approaches.

References

1. F. Zwicky, *Helv. Phys. Acta* 6 (1933) 110
2. S. Smith, *Astrophys. J.* 83 (1936) 23
3. V.C. Rubin and W.K. Ford, *Astrophys. J.* 159 (1970) 379; M. Roberts and A.H. Rots, *Astron. Astrophys.* 26 (1973) 483
4. E.W. Kolb and M.S. Turner, "The Early Universe", Addison - Wesley (1989).
5. D. N, Spergel et al., *astro-ph/0302209*
6. A. Riess et al., *Astronom. J.* 116 (1998) 1009; E.D. Perlmutter et al., *Astrophys. J.* 517 (1999) 565
7. C. L. Bennett et al., *astro-ph/0302207*
8. R.A.C. Croft et al., *Phys. Rev. Lett.* 83 (1999) 1092; O. Elgaroy et al. *astro-ph/0204152*
9. R. Bernabei et al., *Phys. Lett.* B515 (2001) 6.
10. D. Smith and N. Weiner, *Phys. Rev.* D **64** (2001) 043502.
11. R. Foot, *hep-ph/0308254*.
12. R. Bernabei, *Prog. in Part. and Nucl. Phys.* **48** (2002) 263.
13. R. Bernabei et al., *La Rivista del Nuovo Cimento* **26** (2003) 1-73 (*astro-ph/0307403*).
14. P. Belli, R. Bernabei, C. Bacci, A. Incicchitti, R. Marcovaldi, D. Prosperi, DAMA proposal to INFN Scientific Committee II, April 24th 1990.
15. R. Bernabei et al., *Phys. Lett.* B **424** (1998) 195.
16. R. Bernabei et al., *Phys. Lett.* B **450** (1999) 448.
17. P. Belli et al., *Phys. Rev.* D **61** (2000) 023512.

18. R. Bernabei et al., *Phys. Lett.* B **480** (2000) 23.
19. R. Bernabei et al., *Eur. Phys. J.* C **18** (2000) 283.
20. R. Bernabei et al., *Phys. Lett.* B **509** (2001) 197.
21. R. Bernabei el al., *Eur. Phys. J.* C **23** (2002) 61.
22. P. Belli et al., *Phys. Rev.* D **66** (2002) 043503.
23. K.A. Drukier et al., *Phys. Rev.* D **33** (1986) 3495. K. Freese et al., *Phys. Rev.* D **37** (1988) 3388.
24. K.Freese et. al. astro-ph/0309279; *Phys. Rev. Lett.* 92 (2004) 11301.
25. P. Belli et al., *Astrop. Phys.* **5** (1996) 217; P. Belli et al., *Nuovo Cimento* C **19** (1996) 537; P. Belli et al., *Phys. Lett.* B **387** (1996) 222; *Phys. Lett.* B **389** (1996) 783 (err.); P. Belli et al., *Phys. Lett.* B **465** (1999) 315; P. Belli et al., *Phys. Rev.* D **61** (2000) 117301; R. Bernabei et al., *New Journal of Physics* **2** (2000) 15.1; R. Bernabei et al., *Phys. Lett.* B **493** (2000) 12; R. Bernabei et al., *Nucl. Instrum & Meth.* A **482** (2002) 728; R. Bernabei et al., *Eur. Phys. J. direct* C11 (2001) 1; R. Bernabei et al., *Phys. Lett.* B **527** (2002) 182; R. Bernabei et al., *Phys. Lett.* B **546** (2002) 23.
26. R. Bernabei et al., *Phys. Lett.* B **436** (1998) 379
27. R. Bernabei et al., *Astrop. Phys.* **7** (1997) 73; R. Bernabei et al., *Nuovo Cimento* A **110** (1997) 189; P. Belli et al., *Astrop. Phys.* **10** (1999) 115; P. Belli et al., *Nucl. Phys.* B **563** (1999) 97; R. Bernabei et al., *Nucl. Phys.* A **705** (2002) 29; P. Belli et al., *Nucl. Instrum & Meth.* A **498** (2003) 352; R. Cerulli et al., *Nucl. Instrum & Meth.* A **525** (2004) 535.
28. M. Ambrosio et al., *Astrop. Phys.* **7** (1997) 109
29. P. Belli et al., *Nuovo Cimento* **101** (1989) 959
30. M. Cribier et al., *Astop. Phys.* **4** (1995) 23
31. C. Arpesella et al., *Health Phys.* 72 (1997) 629
32. R. Bernabei et al., *Nuovo Cimento* A **112** (1999) 545.
33. M. Wojcik: *Nucl. Instrum. & Methods* B61 (1991) 8
34. R. Bernabei et al., *Phys. Lett.* B **389** (1996) 757.
35. R. Bernabei et al., *Nuovo Cimento* A **112** (1999) 1541
36. R. Bernabei et al., *Phys. Lett.* B **408** (1997) 439
37. P. Belli et al., *Phys. Rev.* C60 (1999) 065501.
38. P. Belli et al., *Phys. Lett.* B460 (1999) 236
39. R. Bernabei et al., *Phys. Rev. Lett.* **83** (1999) 4918.
40. F. Cappella et al., *Eur. Phys. J.*-direct C14 (2002) 1.
41. R. Bernabei et al., ROM2F/2004/18, to appear on *Eur. Phys.* J. A
42. R. Bernabei et al., ROM2F/2004/30, submitted for publication.
43. G. Prezeau et al., *Phys. Rev. Lett.* **91** (2003) 231301.
44. D.E. Groom et al., *Eur. Phys. J.* C **15** (2000) 1
45. A. Bottino et al., *Phys. Rev.* D **67** (2003) 063519; A. Bottino et al, hep-ph/0304080; D. Hooper and T. Plehn, MADPH-02-1308, CERN-TH/2002-29, [hep-ph/0212226]; G. Bélanger, F. Boudjema, A. Pukhov and S. Rosier-Lees, hep-ph/0212227.
46. A. Bottino et al., *Phys. Rev.* D **69** (2004) 037302.
47. K. Hagiwara et al., *Phys. Rev.* D **66** (2002) 010001.
48. CDMS collaboration, astro-ph/0405033; *Phys. Rev. Lett.* **84** (2000) 5699

49. EDELWEISS collaboration, in the Proc. of NDM03, Japan (2003); *Phys. Lett.* B **513** (2001) 15
50. N. Smith, talk given at IDM02, York, september 2002
51. R. Luscher , talk given at Moriond, march 2003
52. E. Simon et al., *Nucl. Instrum. & Meth.* A507 (2003) 643.
53. K. Belotsky et al., *hep-ph/0411093.*
54. R. Bernabei et al.,*Astrop. Phys.* **4** (1995) 45; R. Bernabei, "Competitiveness of a very low radioactive ton scintillator for particle Dark Matter search", in the volume *The identification of Dark Matter*, World Sc. pub. 574(1997).

AXION SEARCHES

K. ZIOUTAS
University of Thessaloniki, Greece and CERN, Geneva, Switzerland
E-mail : zioutas@cern.ch

A short introduction to the axion physics is given. Various approaches of detection of axions or other exotic particles with similar couplings are presented. This work is in summary form of the given lectures at this school, and therefore the given references can be consulted for further reading.

1. Introduction

In order to solve the strong CP problem, i.e., why the electric dipole moment of the neutron is some ten orders of magnitude smaller than expected for a strong interaction that should violate CP-symmetry, a new neutral particle with spin-parity 0-, the axion, was invented (see recent ref.'s [1,2]). Following various astrophysical/cosmological observations and earth bound experimental searches as well as theoretical reasoning, the axion rest mass is expected to be in the region ~10 μeV to a few eV. Axions, along with Weakly Interacting Massive Particles (WIMPs), are the two leading particle candidates for the dark matter in the Universe, which is the biggest mystery in all of physics (after the dark energy?).

Energetic axions should also be abundantly produced inside stars, e.g. in the plasma of the solar core [1]. The expected spontaneous decay to two photons ($a \rightarrow 2\gamma$) results in a lifetime much longer than the age of the Universe, which makes axions practically unobservable. However, their interaction with a magnetic field can give rise to the (coherent) oscillation between axions and photons, a process which is being widely used as the working principle of axion detectors [= telescopes].

In theories of large extra-dimensions, the "conventional", almost massless axions become as massive as the reaction energies involved. In the case of the solar axions, the expected mass spectrum of the excited Kaluza-Klein (KK) tower states reaches ~10 keV/c^2 [3],with a relatively short lifetime ($\tau \sim 10^{20}$ s), because of the $\tau \sim m^{-3}$ dependence. The underlying axion-photon-photon coupling constant, $g_{a\gamma\gamma}$, remains the same for the 'conventional' (≈ massless) axion and for the massive KK-axion.

Pioneering contributions to axion physics are associated with the names: PECCEI, QUINN, WILCZEK, WEINBERG, SIKIVIE.

2. Relic axions

If axions exist, they must have been abundantly produced during the Big Bang era as a condensate, and, they can be (part of) the ubiquitous dark matter in the Universe. They can be detected via their conversion to a microwave photon inside a "Sikivie resonant cavity" permeated by a strong magnetic field [2], provided the resonance frequency of the cavity satisfies the condition: $\hbar\omega = m_{axion} c^2$. One should observe an axion signal as an excess of microwave photons at resonance, compared to the background level just a little below/above the axion rest mass. Such cavity experiments aim at axion detection provided axions are the galactic dark matter constituents. At present, two large-scale microwave cavity experiments search for galactic axions. For more detailed information see ref. [2,4].

3. Solar axions

Relativistic axions can be created inside the hot solar core via the conversion of the ~ keV-thermal photons to axions near the electric field of the atomic nuclei. This is the Primakoff effect; the same effect, but time reversed, is used to convert axions back to X-rays in an external magnetic field (B). For more details see ref. [5], about the ongoing CAST experiment (Cern Axion Solar Telescope), with vacuum and/or buffer-gas in the magnetic volume, which is at present the only data taking axion helioscope in the world.

The observed dependence of the solar soft X-ray luminosity on the magnetic field squared, changing smoothly by ~ ± 20% over a solar cycle [6], is intriguing, since the same relation applies also to the expected axion-to-photon oscillation probability inside a transverse magnetic field [5,7]. Therefore, it is an exciting perspective, if the observed $L_x \sim B^2$ relation is the manifestation of an electro-magnetically induced "decay" [9] of axions streaming out of the Sun depending on the local physical conditions, e.g., magnetic field strength, plasma density, etc. [10] indicating also a possible connection to the 11-years cycle and other mysterious observations from the Sun.

4. Massive axions or axion-like particles

The theoretically motivated Kaluza-Klein-axions are taken as a generic example of massive particles, which may be created inside the hot solar core, while a small fraction of them being highly non-relativistic ($\sim 10^{-7}$) can be gravitationally

trapped by the Sun itself in orbits where they accumulate over cosmic times [8]. Their derived density increases rapidly near the solar surface. The spontaneous, or "induced" [9], radiative decay of trapped axions or other particles (from the solar core) with similar properties can give rise to a *self-irradiation* (= self-heating) of the solar atmosphere (see Figure 1) explaining (at least partly) the otherwise unexpected X-ray emission from the solar disk and limb. Thus, the spontaneous radiative decay of bound massive axions can provide the required continuous and steady power input into the unexpectedly hot solar atmosphere.
For more detailed information of other possibly axion-related astrophysical implications, e.g., for the Sun, the Galactic Center and Clusters of Galaxies, see ref.'s [8,10,11,12,13,14].

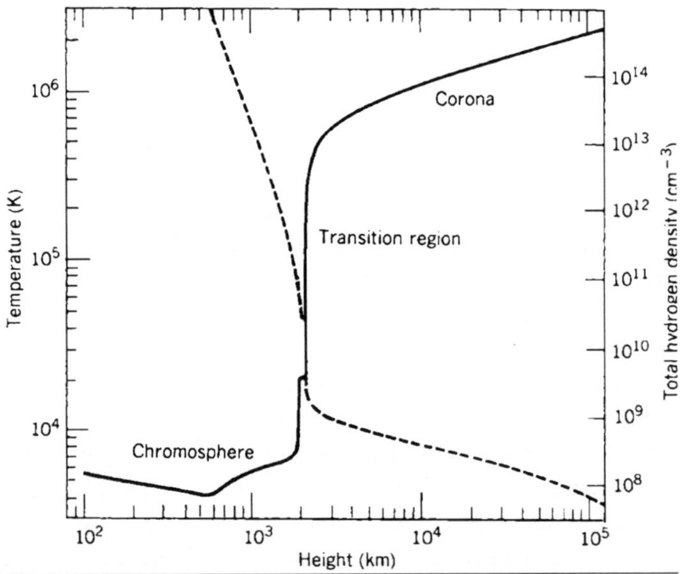

Figure 1. Average temperature and electron number density of the solar atmosphere as a function of height above the photosphere. (http://www.sp.ph.ic.ac.uk/~mkd/AndreHandout.pdf)

5. Underground experiments

For the widely celebrated Primakoff-effect in axion experiments one needs the virtual photon, in order for the axion-to-photon conversion to take place. This virtual photon can be provided also by the Coulomb field of the atomic nuclei of the detector material itself, e.g. Ge, NaI, Si, etc. Some of these detectors are in the form of large single crystals, which allow a Bragg scattering to take place, but with incident solar axions in the energy range below ~10 keV [15,16]. In fact, this energy range coincides with the expected solar axion energy spectrum. Thus, the detector itself becomes the Bragg target and the 4π-detector of the converted axion(-like) particle, when the Bragg condition is satisfied. That is to say, an ideal active target, which provides a very characteristic time-energy dependence of the axion signal, with periods of only a few hours; this can be utilized for particle identification, which is usually missing in dark matter experiments.

Some of the mentioned type of detectors operate since decades underground, in lowest environmental background conditions, which are actually tuned to search for WIMPs, via the mediated nuclear recoil with a deposited kinetic energy up to ~10-100 keV. For example, to search for solar axions, both light and massive ones, or, other particles from the Sun with similar couplings, the energy threshold of the detector must be below a few keV; this is still a very demanding requirement for a large volume detector system. In fact, such searches have been performed already. Though their result is not yet a positive one, such investigations in existing data provided a new limit for the axion-to-photon coupling constant for a rest mass range up to a few keV, which is exceptionally broad. The obtained limits for the axion-to-photon coupling constant -for the same lower axion rest mass range- by the germanium dark matter experiments and also DAMA are much less restrictive than the recently derived limit by CAST [5].
For more details on this type of searches see ref.'s [5,15,16,17].

6. Axions in Laboratory Experiments

6.1. *ν-experiments*

The most recent search for high energy axions in a neutrino experiment was performed by reanalyzing the data taken by the NOMAD experiment at CERN [18]. This is a regeneration-type experiment: the energetic photons produced in 450 GeV proton collisions with the neutrino target, pointing towards the NOMAD detector fiducial volume, can be transformed to axions while crossing the horn magnetic field. Only very feebly interacting particles like neutrinos and axions can penetrate the intervening shielding and reach the detector. Inside its transverse magnetic field, the regeneration of high energy photons from axions interacting with the magnetic field via the Primakoff effect can take place. Note, this is an experiment independent on astrophysical models, since both axion source and detection is supposed to occur in the laboratory. This measurement provided an extended upper limit of the axion-to-photon coupling constant beyond that derived from laser experiments (see below). More details about this otherwise as neutrino designed experiment see ref. [18].

6.2. *Vacuum Birefringence*

Linearly polarized laser light propagating in vacuum inside a strong transverse magnetic field can interact coherently with the field. This is expected to give rise to the optical production and detection of dark matter particles, e.g. the pseudoscalar axion [19]. Again, the Primakoff effect can be at work, resulting to: a) a rotation of the polarization plane induced by the production of a massive particle which couples to two photons, and b) an ellipticity induced by the retardation between the two components of the electric field vector of the laser beam by the virtual production of a massive particle coupling to two photons. Apparently, the PVLAS experiment [20] observes an effect, which is far in excess of the QED expectation via the vacuum-polarization loop. Interestingly, a first signature was observed also some 15 years earlier in a similar type of measurement [21]. In any case, an interpretation in terms of axions requires a coupling strength axion-to-photon far larger than any existing limit derived from earth bound experiments and from astrophysical considerations. For more details on such experiments see ref.'s [19,20,21].

6.3. *Shining Through Wall*

The configuration of such an experiment is in principle similar to that of the axion search with the NOMAD experiment (see above). In fact, here one uses again as photon source a pulsed laser, which propagates inside the magnetic field pipe following some 10^5 reflections on special mirrors outside the magnet. The other side of the intervening shielding ("wall") can only be reached by penetrating particles, e.g. axion or axion-like particles, which can be detected by being back-converted to photons inside the second magnet downstream the initial laser beam on the other side of the wall (see Figure 2). The observation of regenerated photons with the same energy as that from the laser and inside the allowed time window, will be the direct signature of some new particle being created and detected in the lab, i.e. in a model independent way. For further reading the recent ref. [22] can be consulted.

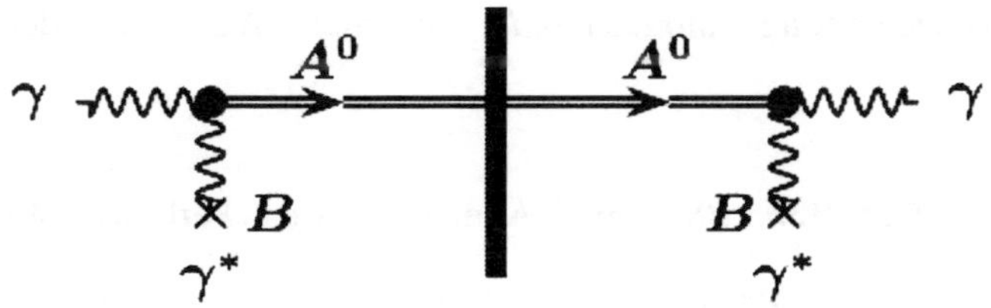

Figure 2. Schematic view of axion production through photon conversion to axion (A^0) inside a magnetic field (*left*), subsequent travel through a wall, and regeneration to a photon (*right*) [22]. γ is the real photon, and γ^* represents the virtual photon from the magnetic field (B).

6.4. *Atomic/Nuclear Transitions*

Since an axion can replace the photon in a magnetic dipole transition, atomic [23] as well as nuclear M1 transitions can be utilized to search for axions. For example, the photoelectric effect can take place also with exotic particles like axions with sufficient energy ("axioelectric effect") [24]. The Sun is the strongest nearby source of various nuclear transitions/reactions, which are associated with an M1 transition. It has been proposed that almost monochromatic axions should be produced with 14.4 keV total energy due to decay of the (thermally excited) 14.4 keV state of ^{57}Fe nuclei in the solar core [25]. Such axions, streaming out of the Sun, can resonantly excite the same isotope of an earth-bound detector. Such a search for photons of the same energy has been performed already with detectors utilizing the resonant absorption by the ^{57}Fe nuclei of the detector material [26]. This type of experiments are complementary to the suggested Mössbauer experiment, where the radioactive source has to be highly shielded [27].

7. Conclusions

The presented ongoing searches for axions or other axion-like particles are complementary to each other with the potential for exciting new results. The celebrated Primakoff-effect dominates most axion experiments as the underlying detection mechanism. The origin of various solar as well as astrophysical-cosmological observations might be due to axions, massive or light ones, which seem to have been overlooked in the past. Then, some of the mentioned fine tuned earth bound axion experiments might be at work already on astrophysical scales. This holds in particular, for the ongoing CAST experiment at CERN and the PVLAS laboratory experiment, whose signatures are based on oscillations between axions (or other particles with similar couplings) and photons. Thus, important evidence has been sitting unnoticed *under our noses* 1 AU for decades or even much longer.

References

1. G.G. Raffelt, Phys. Rep. **333** (2000) 593, Ann. Rev. Nucl. Part. Sci. **49** (1999) 163.
2. R. Bradley et al., Rev. Mod. Phys. **75** (2003) 777.
3. K.R. Dienes, E. Dudas, T. Gherghetta, Phys. Rev. **D62** (2000) 105023; L. DiLella, A. Pilaftsis, G. Raffelt, K. Zioutas, Phys. Rev. **D62** (2000) 125011.
4. K. Yamamoto et al., hep-ph/0101200.
5. S. Andriamonje et al., The CAST collaboration, Phys. Rev. Lett. (2005) *in press* (hep-ex/0411033); see also http://www.cern.ch/CAST.
6. E.E. Benevolenskaya, A.G. Kosovichev, J.R. Lemen, P.H. Scherer, G.L. Slater, ApJ. **571** (2002) L181.
7. K. Zioutas et al., Nucl. Instrum. Meth. **A425** (1999) 480.
8. L. DiLella, K. Zioutas, Phys. Lett. **B531** (2002) 175, and, Astroparticle Physics **19** (2003) 145.
9. D.H.H. Hoffmann, B. Lakic, Y. Semertzidis and K. Zioutas, *in preparation.*
10. K. Zioutas, K. Dennerl, L. DiLella, D.H.H. Hoffmann, J. Jacoby, Th. Papaevangelou, ApJ. **607** (2004) 575.
11. R. Horvat, M. Krcmar, B. Lakic, Phys. Rev. **D69** (2004) 125011.
12. K. Zioutas, D.H.H. Hoffmann, K. Dennerl, Th. Papaevangelou, Science **306** (2004) 1485.
13. J. Hogan, New Scientist **182** (#2443) (2004) 8.
14. D.H.H. Hoffmann, J. Jacoby, K. Zioutas, Astropart. Phys. **20** (2003) 73.
15. E.A. Paschos, K. Zioutas, Phys. Lett. **B323** (1994) 367.
16. R.J. Creswick, F.T. Avignone III, H.A. Farach, J.I Collar, A.O. Gattone, S. Nussinov, K. Zioutas, Phys. Lett. **B427** (1998) 235.
17. F.T. Avignone III et al., The SOLAX Collaboration, Phys. Rev. Lett. **81** (1998) 5068; A. Morales et al., The COSME Collaboration, Astropart. Phys. **16** (2002) 325; R. Bernabei et al., The DAMA Collaboration, Phys. Lett. **B515** (2001) 6.
18. P. Astier et al., The NOMAD Collaboration, Phys. Lett. **B479** (2000) 371.
19. L. Maiani, R. Petronzio, E. Zavattini, Phys. Lett. **B175** (1986) 359.
20. G. Cantatore et al., presented at IDM 2004, Edinburgh, England, 4-10 September 2004.
21. Y. Semertzidis et al., Phys. Rev. Lett. **64** (1990) 2988.
22. A. Ringwald, Phys. Lett. **B569** (2003) 51.
23. K. Zioutas, Y. Semertzidis, Phys. Lett. **A130** (1988) 94.
24. S. Dimopoulos, G.D. Starkman et al., Phys. Lett. **B168** (1986) 145.
25. S. Moriyama, Phys. Rev. Lett. **75** (1995) 3222.
26. M. Krcmar et al., Phys. Lett. **B442** (1998) 38.
27. A. De Rujula, K. Zioutas, Phys. Lett. **B217** (1989) 354.

MAGNETIC MONOPOLE SEARCHES

G. GIACOMELLI AND L. PATRIZII

INFN and Phys. Dept. of the University of Bologna, v.le B.Pichat 6/2, I-40127 Bologna, Italy
E-mail: giacomelli@bo.infn.it, patrizii@bo.infn.it

In these lecture notes we discuss the status of the searches for classical Dirac Magnetic Monopoles (MMs) at accelerators, for GUT superheavy MMs in the penetrating cosmic radiation and for Intermediate Mass MMs. Also the searches for nuclearites and Q-balls are considered.

1. Introduction

The concept of magnetic monopoles (MMs) goes back to the origin of magnetism. At the beginning of the 19th century there were discussions concerning the magnetic content of matter and the possible existence of isolated magnetic charges. In 1931 Dirac introduced the MM in order to explain the quantization of the electric charge [1]. He established the relation between the elementary electric charge e and a basic magnetic charge g: $eg = n\hbar c/2 = ng_D$, where n is an integer, $n = 1, 2, ..$; $g_D = \hbar c/2e = 68.5e$ is the unit Dirac charge. The existence of magnetic charges and of magnetic currents would symmetrize in form Maxwell's equations, but the symmetry would not be perfect since $e \neq g$ (but the couplings could be energy dependent and could merge in a common value at high energies) [2]. There was no prediction for the MM mass; a rough estimate, obtained assuming that the classical monopole radius is equal to the classical electron radius yields $m_M \simeq \frac{g^2 m_e}{e^2} \simeq n\ 4700\ m_e \simeq n\ 2.4\ GeV/c^2$. From 1931 searches for *"classical Dirac monopoles"* were carried out at every new accelerator using simple setups, and recently also large collider detectors [3–7].

Electric charge is naturally quantized in Grand Unified Theories (GUT) of the basic interactions; they imply the existence of *GUT monopoles* with calculable properties. The MMs appear in the Early Universe at the phase transition corresponding to the breaking of the unified group into subgroups, one of which is U(1) [8]. The MM mass is related to the mass of the X, Y carriers of the unified interaction, $m_M \geq m_X/G$, where G is the dimensionless unified coupling constant at energies E $\simeq m_X$. If $m_X \simeq 10^{14} - 10^{15}$ GeV and $G \simeq 0.025$, $m_M > 10^{16} - 10^{17}$ GeV. This is an enormous mass: MMs cannot be produced at any man–made accelerator, existing or conceivable. They may have been produced only in the first instants of our Universe.

Larger MM masses are expected if gravity is brought into the unification picture, and in some SuperSymmetric models.

Multiply charged Intermediate Mass Monopoles (IMMs) may have been produced in later phase transitions in the Early Universe, when a semisimple gauge group yields a U(1) group [9]. IMMs with $m_M \sim 10^7 \div 10^{13}$ GeV may be accelerated to relativistic velocities in one galactic magnetic field domain. Very energetic IMMs could yield the highest energy cosmic rays [10].

The lowest mass MM is stable, since magnetic charge is conserved like electric charge. Thus the poles produced in the Early Universe should still exist as cosmic relics; their kinetic energy was affected by the Universe expansion and by travel through galactic and intergalactic magnetic fields.

GUT poles are best searched for underground in the penetrating cosmic radiation (CR). IMMs may be searched for at high altitude laboratories.

In this lecture we review the experimental situation on MM searches and briefly discuss the searches for nuclearites [11] and Q-balls [12].

1.1. *Properties of magnetic monopoles*

The main properties of MMs are obtained from the Dirac relation.
- If n =1 and the basic electric charge is that of the electron, then the *basic magnetic charge* is $g_D = \hbar c/2e = 137e/2$. The magnetic charge is larger if $n > 1$ and if the basic electric charge is $e/3$.
- In analogy with the fine structure constant, $\alpha = e^2/\hbar c \simeq 1/137$, the *dimensionless magnetic coupling constant* is $\alpha_g = g_D^2/\hbar c \simeq 34.25$; since it is > 1 perturbative calculations cannot be used.
- *Energy W acquired in a magnetic field B*: $W = ng_D B\ell = n$ 20.5 keV/G cm. In a coherent galactic–length ($\ell \simeq 1$ kpc, $B \simeq 3$ μG), the energy gained by a MM with $g = g_D$ is $W \simeq 1.8 \times 10^{11}$ GeV. Classical poles and IMMs in the CR may be accelerated to relativistic velocities. GUT poles should have low velocities, $10^{-4} < \beta < 10^{-1}$.
- *MMs may be trapped in ferromagnetic materials* by an image force, which could reach values of ~ 10 eV/Å.
- Electrically charged monopoles (dyons) may arise as quantum–mechanical

excitations or as M–p, M-nucleus composites.

- The interaction of a MM magnetic charge with a nuclear magnetic dipole could lead to the formation of a M–nucleus bound system. A monopole–proton bound state may be produced via radiative capture. Monopole–nucleus bound states may exist for nuclei with large gyromagnetic ratios.
- *Energy losses of fast poles.* A fast MM with magnetic charge g_D and velocity $v = \beta c$ behaves like an electric charge $(ze)_{eq} = g_D \beta$, Fig. 1.
- *Energy losses of slow poles* ($10^{-4} < \beta < 10^{-2}$) may be due to ionization or excitation of atoms and molecules of the medium ("electronic" energy loss) or to recoiling atoms or nuclei ("atomic" or "nuclear" energy loss). Electronic energy loss predominates for $\beta > 10^{-3}$.
- *Energy losses at very low velocities.* MMs with $v < 10^{-4}c$ may lose energy in elastic collisions with atoms or with nuclei. The energy is released to the medium in the form of elastic vibrations and/or infra–red radiation [13].

Fig. 1 shows the energy loss in liquid hydrogen of a $g = g_D$ MM vs β [4].

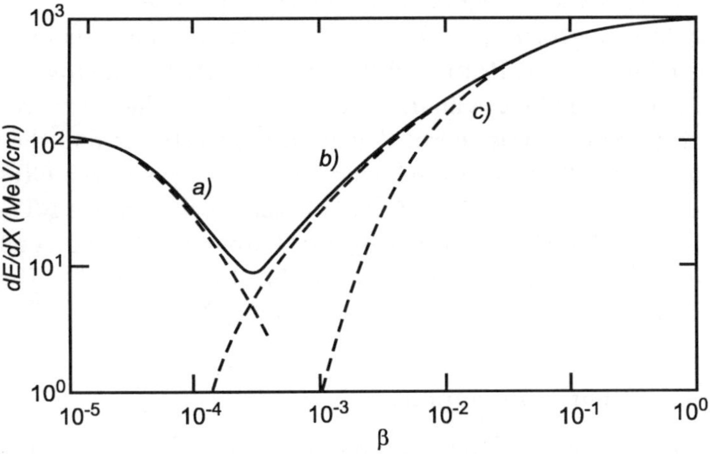

Figure 1. The energy losses, in MeV/cm, of $g = g_D$ MMs in liquid hydrogen vs β. Curve a) corresponds to elastic monopole–hydrogen atom scattering; curve b) to interactions with level crossings; curve c) describes the ionization energy loss.

- *Energy loss of MMs in celestial bodies.* For $\beta < 10^{-4}$ the dE/dx in the Earth is due to pole–atom elastic scattering, eddy currents, and nuclear stopping power. MMs may be stopped by celestial bodies if they have:
Moon: $\beta \leq 5 \times 10^{-5}$, Earth: $\beta \leq 10^{-4}$, Sun: $\beta \leq 10^{-3}$.

1.2. *Monopole detectors*

Monopole detectors are based on MM properties given by Dirac's relation.
- *Superconducting induction devices are sensitive to MMs of any velocity* [3]. A moving MM induces in a ring an electromotive force and a current change (Δi). For a coil with N turns and inductance L, $\Delta i = 4\pi N n g_D / L = 2\Delta i_o$, where Δi_o is the current change corresponding to a change of one unit of the flux quantum of superconductivity. This method of detection is based only on the long–range electromagnetic interaction between the magnetic charge and the macroscopic quantum state of a superconducting ring.
- *Scintillation counters* for MMs have a threshold $\beta \sim 10^{-4}$, above which the light signal is larger than that of a minimum ionizing particle [13,14].
- *Gaseous detectors* of various types have been used. MACRO used a gas mixture of 73% helium and 27% n–pentane [14]. This allows exploitation of the Drell [15] and Penning effects [3]: a MM leaves a helium atom in a metastable state (He*) with an excitation energy of $\simeq$ 20 eV. The ionization potential of n–pentane is $\simeq$ 10 eV; the excited energy of the He* is converted into ionization of the n–pentane molecule (Penning effect).
- *Nuclear track detectors (NTDs).* The formation of an etchable track in a NTD is related to the Restricted Energy Loss (REL), the fraction of the energy loss localized in a cylindrical region of 10 nm diameter around the particle trajectory. It was shown that both the electronic and the nuclear energy losses are effective in producing etchable tracks in the CR39 NTD which has a threshold at $z/\beta \simeq 5$ [16]; it is the most sensitive NTD and it allows to search for MMs with $g = g_D$ for β around 10^{-4} and $> 10^{-3}$, the whole β-range of $4 \times 10^{-5} < \beta < 1$ for MMs with $g \geq 2g_D$ [13]. The Lexan and Makrofol polycarbonates are sensitive for $z/\beta \geq 50$ [17].

2. "Classical Dirac monopoles"

- *Accelerator searches.* If MMs are produced at high–energy accelerators, they would be relativistic and would ionize heavily. Examples of *direct searches* are the experiments performed with scintillators or NTDs. Experiments at the Fermilab $\overline{p}p$ collider established cross section limits of $\sim 2 \times 10^{-34}$ cm^2 for MMs with $m_M < 850$ GeV [18]. Searches at e^+e^- colliders excluded masses up to 45 GeV and later in the 45-102 GeV range ($\sigma < 5 \times 10^{-37}$ cm^2). Recently few high energy general purpose detectors used some subdetectors to search for Dirac MMs [7].

Fig. 2 summarizes the cross section limits vs MM mass obtained by direct and indirect experiments (solid lines and dashed lines) at the Fermilab

$\bar{p}p$ collider, e^+e^- colliders, the ISR pp collider [4]. Most searches are sensitive to poles with magnetic charges $g = ng_D/q$ with $0.5 < n < 5$.

Examples of indirect searches are those performed at the CERN SPS and at Fermilab: the protons interacted in ferromagnetic targets, later the targets were placed in front of a superconducting solenoid with a field $B > 100$ kG, large enough to extract and accelerate the MMs, to be detected in scintillators and in NTD sheets [3]. An indirect experiment performed at the $\bar{p}p$ Tevatron collider, assumed that produced MMs could stop, be trapped and bound in the matter surrounding a collision region [5]. Small Be and Al samples were passed through the 10 cm diameter bore of two superconducting coils, and the induced charge measured by SQUIDs. Limits $m_M > 285$ GeV were published for $g = g_D$ poles. It is difficult to establish the validity of the hypotheses made to interpret these results.

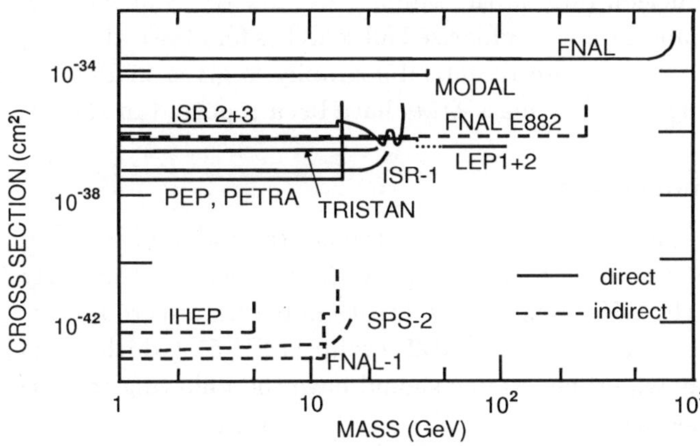

Figure 2. Classical Dirac MMS cross section upper limits vs MM mass obtained from direct accelerator searches (solid lines) and indirect searches (dashed lines).

- *Multi–γ events.* Five peculiar photon showers found in emulsion plates exposed to high–altitude CRs, are characterized by an energetic narrow cone of tens of photons, without any incident charged particle [19]. The total energy of the photons is $\sim 10^{11}$ GeV. The small radial spread of photons suggested a c.m. $\gamma = (1-\beta^2)^{-1/2} > 10^3$. The energies of the photons are too small to have π^o decays as their source. One possible explanation: a high–energy γ–ray, with energy $> 10^{12}$ eV, produced a pole–antipole pair, which suffered bremsstrahlung and annihilation producing the final multi–γ

events. Searches for multi-γ events were performed in pp collisions at the ISR at $\sqrt{s} = 53$ GeV, at the $\bar{p}p$ 1.8 TeV collider and in e^+e^- collisions at LEP (Fig. 2). The D0 experiment searched for γ pairs with high transverse energies; virtual pointlike MMs may rescatter pairs of nearly real photons into the final state via a box monopole diagram; they set a 95% CL limit of 870 GeV [5]. At LEP the L3 coll. searched for $Z \to \gamma\gamma\gamma$ events; no deviation from QED predictions was observed, setting a 95% CL limit of 510 GeV [5]. Many authors studied the effects from virtual monopole loops [2,20]. The authors of Ref. [6] criticized the underlying theory and believe that no significant limit can be obtained from present experiments.
- *Searches in bulk matter.* Classical MMs could be produced by CRs and could stop at the Earth surface, where they may be trapped in ferromagnetic materials. Bulk matter searches used hundreds of kg of material, including meteorites, schists, ferromanganese nodules, iron ore and others. A superconducting coil through which the material was passed, yielded a monopole/nucleon ratio in the samples $< 1.2 \times 10^{-29}$ at 90% CL [3].

Ruzicka and Zrelov summarized all searches for classical poles performed before 1980 [21]. A more recent bibliography is given in Ref. [22]. Possible effects arising from low mass MMs have been reported in Ref. [23].

3. GUT monopoles

As already stated, GUT theories of the electroweak and strong interations predict the existence of superheavy MMs produced in the Early Universe (EU) when the GUT gauge group breaks into separate groups, one of which is U(1). Assuming that the GUT group is SU(5) (which is excluded by proton decay experiments) one should have the following transitions:

$$SU(5) \underset{10^{-35}s}{\overset{10^{15}\ GeV}{\longrightarrow}} SU(3)_C \times [SU(2)_L \times U(1)_Y] \underset{10^{-9}s}{\overset{10^{2}\ GeV}{\longrightarrow}} SU(3)_C \times U(1)_{EM} \quad (1)$$

MMs would be generated as topological point defects in the GUT phase transition, about one pole for each causal domain. In the standard cosmology this leads to too many poles (the monopole problem). Inflation would defer the GUT phase transition after large supercooling; in its simplest version the number of generated MMs would be very small. However the flux depends critically on several parameters, like the pole mass, the reheating temperature, etc. If the reheating temperature is large enough one would have MMs produced in high energy collisions, like $e^+e^- \to M\bar{M}$.

Fig. 3 shows the structure of a GUT MM: a very small core, an electroweak region, a confinement region, a fermion–antifermion condensate (which may contain 4–fermion baryon–number–violating terms); for $r \geq 3$ fm it behaves as a point particle generating a field $B = g/r^2$ [24].

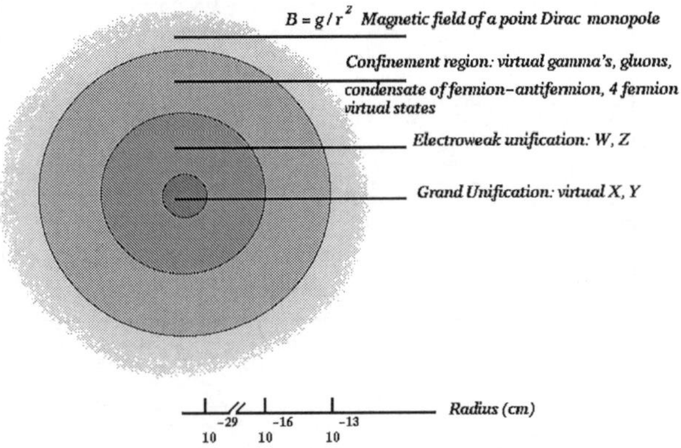

Figure 3. Structure of a GUT pole. The 4 regions correspond to: (i) Grand Unification ($r \sim 10^{-29}$ cm; inside this core one finds virtual X, Y particles); (ii) electroweak unification ($r \sim 10^{-16}$ cm; inside one finds virtual $W^{\pm}$ and Z^0); (iii) confinement region ($r \sim 10^{-13}$ cm; inside one finds virtual γ, gluons, fermion-antifermion pairs and possibly 4-fermion virtual states); (iv) for $r >$ few fm one has the field of a point magnetic charge.

A flux of cosmic GUT MMs may reach the Earth with a velocity spectrum in the range $4 \times 10^{-5} < \beta < 0.1$, with possible peaks corresponding to the escape velocities from the Earth, the Sun and the Galaxy. Searches for such MMs in the CR performed with superconducting induction devices yielded a combined 90% CL limit of 2×10^{-14} cm^{-2} s^{-1} sr^{-1}, independent of β [4]. Direct searches were performed above ground and underground [4,25–27]. MACRO performed a search with different types of detectors (liquid scintillators, limited streamer tubes and NTDs) with an acceptance of $\sim$ 10,000 m^{2}sr for an isotropic flux. No MM was detected; the 90% CL flux limits, shown in Fig. 4 vs β for $g = g_D$, are at the level of 1.4×10^{-16} cm^{-2} s^{-1} sr^{-1} for $\beta > 4 \times 10^{-5}$ [25]. The figure shows also the limits from the Ohya [26], Baksan, Baikal, and AMANDA experiments [27].

The interaction of the GUT monopole core with a nucleon can lead to a reaction in which the nucleon decays (monopole catalysis of nucleon decay), f. e. $M + p \rightarrow M + e^+ + \pi^0$. The cross section for this process is very small, of the order of magnitude of the core size; but the catalysis process could proceed via the Rubakov-Callan mechanism with a σ of the order of the strong interaction cross section [28]. MACRO performed a dedicated search

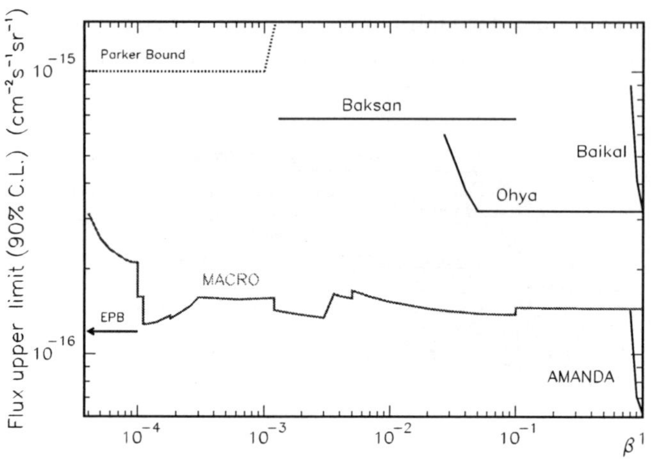

Figure 4. The 90% CL MACRO direct upper limits vs β for GUT $g = g_D$ poles in the penetrating CR, and direct limits from other experiments (see text).

for nucleon decays induced by the passage of a GUT pole in the streamer tube system. The flux limits obtained, $3-8\times10^{-16}$ cm^{-2} s^{-1} sr^{-1}, depend on the MM velocity and on the catalysis cross section [29]. Previous limits were at levels 10^{-15} cm^{-2} s^{-1} sr^{-1} [29], except the Baikal limit which is 6×10^{-17} cm^{-2} s^{-1} sr^{-1} for $\beta \simeq 10^{-5}$ [27].

Indirect GUT MM searches used ancient mica, which has a high threshold. It is assumed that a pole passing through the Earth captures an Al nucleus and drags it through subterranean mica causing a trail of lattice defects, which survive as long as the mica is not reheated. Only small sheets were analyzed (13.5 and 18 cm^2), but should have been recording tracks for $4 \div 9 \times 10^8$ years. The flux limits are 10^{-17} cm^{-2} s^{-1}sr^{-1} for $10^{-4} < \beta < 10^{-3}$ [30]. There are reasons why these indirect experiments might not be sensitive: if MMs have a positive electric charge or protons attached, then Coulomb repulsion could prevent capture of heavy nuclei.

4. Cosmological and astrophysical bounds

Rough upper limits for a GUT monopole flux in the CR were obtained on the basis of cosmological and astrophysical considerations.

- *Limit from the mass density of the universe:* it is obtained requiring that the present MM mass density be smaller than the critical density ρ_c of the universe. For $m_M \simeq 10^{17}$ GeV one has the limit: $F = \frac{n_M c}{4\pi}\beta <$

$3 \times 10^{-12} h_0^2 \beta$ ($\mathrm{cm}^{-2}\mathrm{s}^{-1}\mathrm{sr}^{-1}$). It is valid for poles uniformely distributed in the universe. If poles are clustered in galaxies the limit is larger [3].
- *Limit from the galactic magnetic field (Parker limit).* The ~ 3 μG magnetic field in our Galaxy is probably due to the non–uniform rotation of the Galaxy, which generates a field with a time–scale of the order of the rotation period of the Galaxy ($\tau \sim 10^8$ yr). An upper bound for the MM flux is obtained by requiring that the kinetic energy gained per unit time by MMs be less than the magnetic energy generated by the dynamo effect: $F < 10^{-15}$ cm^{-2} s^{-1} sr^{-1} [31]; taking into account the almost chaotic nature of the field, with domains of $\ell \sim 1$ kpc, the limit becomes mass dependent [31]. An extended "Parker bound", obtained by considering the survival of an early seed field [32], yields $F \leq 1.2 \times 10^{-16} (m_M / 10^{17} GeV)$ cm^{-2} s^{-1} sr^{-1}.
- *Limit from the intergalactic (IG) magnetic field.* If $B_{IG} \sim 3 \times 10^{-8}$ G with a regeneration time $\tau_{IG} \sim 10^9$ y, a more stringent bound is obtained; the limit is less reliable because the IG field is less known.
- *Limits from peculiar A4 stars and from pulsars* may be stringent, but the assumptions made are not clear (see the pulsar PSR 1937+214) [3,4].

5. Intermediate mass magnetic monopoles

IMMs may appear as topological point defects at a later time in the Early Universe; f.e. the SO(10) GUT group would not yield directly a U(1) group

$$SO(10) \underset{10^{-35} s}{\overset{10^{15}\ GeV}{\longrightarrow}} SU(4) \times SU(2) \times SU(2) \underset{10^{-23} s}{\overset{10^{9}\ GeV}{\longrightarrow}} SU(3) \times SU(2) \times U(1) \qquad (2)$$

This would lead to MMs with masses of $\sim 10^{10}$ GeV; they would survive inflation, be stable, "doubly charged" ($g = 2g_D$) and do not catalyze nucleon decay [9]. The structure of an IMM would be similar to that of a GUT MM, but the core would be larger (since R $\sim 1/m_M$) and the outer cloud would not contain 4–fermion baryon–number–violating terms.

Relativistic IMMs, $10^7 < m_M < 10^{13}$ GeV, could be present in the cosmic radiation, could be accelerated to large γ in one coherent domain of the galactic field. Thus one would have to look for $\beta \geq 0.1$ MMs.

Detectors at the Earth surface could detect MMs coming from above if they have $m_M > 10^5 - 10^6$ GeV [13]; lower mass MMs may be searched for with detectors located at high mountain altitudes, balloons and satellites.

Few experimental results are available. Fig. 5 shows the situation on the flux upper limits for IMMs [4]. The Cherenkov neutrino telescopes under ice and underwater are sensitive to fast ($\gamma >> 1$) MMs coming from above.

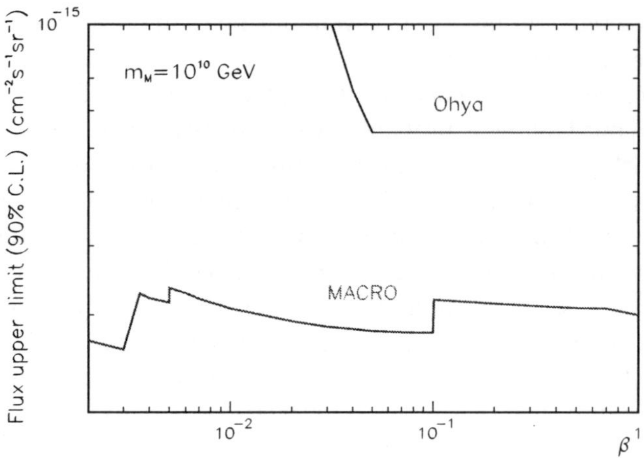

Figure 5. Experimental 90% CL upper limits for a flux of IMMs with mass $m_M = 10^{10}$ GeV plotted versus β.

The SLIM experiment, which searches for IMMs with NTDs at the Chacaltaya high altitude lab (5290 m a.s.l.) [33], is sensitive to $g = 2g_D$ MMs in the whole range $4 \times 10^{-5} < \beta < 1$.

6. Nuclearites and Q-balls

Strange Quark Matter (SQM) should consist of aggregates of *u, d* and *s* quarks in almost equal proportions; the number of *s* quarks should be lower than the number of *u* or *d* quarks and the SQM should have a positive integer charge. The overall neutrality of SMQ is ensured by an electron cloud which surrounds it, forming a sort of atom (see Fig. 6). SQM should have a constant density $\rho_N = M_N/V_N \simeq 3.5 \times 10^{14}$ g cm^{-3}, larger than that of atomic nuclei, and it should be stable for all baryon numbers in the range between ordinary heavy nuclei and neutron stars (A $\sim 10^{57}$). Lumps of SQM with baryon number $A < 10^6 - 10^7$ are usually called "strangelets"; the word "nuclearite" was introduced to indicate large lumps of SQM which could be present in the CR [11]. SQM lumps could have been produced shortly after the Big Bang and may have survived as remnants; they could also appear in violent astrophysical processes, such as in neutron star collisions. SQM could contribute to the cold dark matter. The main energy loss mechanism for low velocity nuclearites is elastic or quasi-elastic collisions with the ambient atoms. The energy loss is large; therefore nuclearites should be easily detected in scintillators and CR39 NTDs [34] . Nuclearites should have typical galatic velocities, $\beta \sim 10^{-3}$, and for masses larger than 0.1 g could traverse the earth. Most nuclearite searches were obtained as byproducts of CR MM searches; the flux limits are similar to those for MMs.

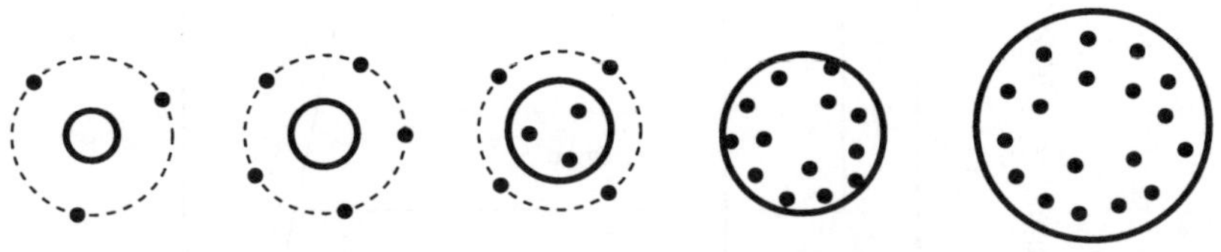

Figure 6. Nuclearite structure. Dimensions of the quark bag (radius R_N) and of the core+electron system; the black points are the electrons (the border of the core + electron cloud for small masses is indicated by the dashed lines). For masses smaller than 10^9 GeV, the electrons are outside the quark bag, the core+electron system has size of $\sim 10^5$ fm; for $10^9 < M_N < 10^{15}$ GeV the e^- are partially inside the core, for $M_N > 10^{15}$ GeV all electrons are inside the core.

The most relevant direct flux limits for nuclearites come from three large area experiments: the first two use CR39 NTDs; one experiment was performed at mountain altitude (Mt. Norikura at 2770 m a.s.l.) [35], the 2nd at the depth of 10^4 g cm^{-2} in the Ohya mine [26]; the third experiment, MACRO, at an average depth of 3700 hg cm^{-2}, used liquid scintillators besides NTDs [36]. A 4th experiment (SLIM) is deployed at high altitudes. Indirect searches with old mica samples could yield the lowest limits, but they are affected by several uncertainties. Some exotic cosmic ray events were interpreted as due to incident nuclearites, f. e. the "Centauro" events and the anomalous massive particles, but the interpretation is not unique [37]. Supermassive nuclearites (M $\sim$ 1 ton) passing through Earth could induce epilinear earthquakes [11,38]. Fig. 7 shows a compilation of limits for a flux of downgoing nuclearites compared with the dark matter (DM) limit, assuming a velocity at ground level $\beta = 10^{-3}$, corresponding to nuclearites of galactic or extragalactic origin. The MACRO limit is extended above the DM bound to show the transition to an isotropic flux for $M_n > 0.1$ g ($\sim 10^{23}$ GeV). Some possible positive indications are discussed in Ref. [37].

Q-balls should be aggregates of squarks $\tilde{q}$, sleptons $\tilde{l}$ and Higgs fields [12]. The scalar condensate inside a Q-ball core has a global baryon number Q (and may be also a lepton number). Protons, neutrons and may be

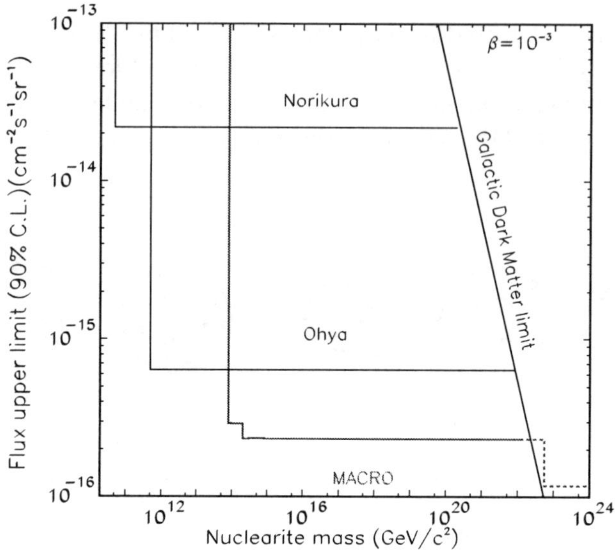

Figure 7. 90% CL flux upper limits versus mass for nuclearites with $\beta = 10^{-3}$ at ground level. These nuclearites could have galatic or extragalatic origin. The limits are from Refs. [26,35,36].

electrons could be absorbed in the condensate. There could exist neutral and charged Q-balls. Supersymmetric Electrically Neutral Solitons (SENS) are generally massive and may catalyse proton decay. SENS may obtain a positive electric charge absorbing a proton in their interactions with matter yielding SECS (Supersymmetric Electrically Charged Solitons), which have a core electric charge, have generally lower masses and the Coulomb barrier could prevent the capture of nuclei. SECS have only integer charges because they are color singlets. A SENS which enters the earth atmosphere could absorb a nitrogen nucleus which would give it the positive charge of +7 (SECS with $z = 7$). Other nuclear absorptions are prevented by Coulomb repulsion. If the Q-ball can absorb electrons at the same rate as protons, the positive charge of the absorbed nucleus may be neutralized by the charge of absorbed electrons. If, instead, the absorption of electrons is slow or impossible, the Q-ball carries a positive electric charge after the capture of the first nucleus in the atmosphere. Q-balls may be cold DM candidates. SECS with $\beta \simeq 10^{-3}$ and $M_Q < 10^{13}$ GeV could reach an underground detector from above, SENS also from below. SENS may be detected by their continuons emission of charged pions (energy loss $\sim$ 100

GeV $g^{-1}cm^2$), SECS may be detected by scintillators, NTDs and ionization detectors.

Note that we did not consider here the possibility of strongly interacting, colored, MMs, nuclearites [41] and Q-balls.

7. Conclusions. Outlook

Direct and indirect accelerator searches for classical Dirac MMs placed limits at the level $m_M > 850$ GeV with cross section upper values as shown in Fig. 2. Future improvements may come from experiments at the LHC [42].

Many searches were performed for GUT poles in the penetrating cosmic radiation. The 90% CL flux limits are at $\sim 1.4 \times 10^{-16}$ cm^{-2} s^{-1} sr^{-1} for $\beta \geq 4 \times 10^{-5}$. It may be difficult to do much better since one would require refined detectors of considerably larger areas.

Present limits on Intermediate Mass Monopoles with high β are relatively poor. Experiments at high altitudes and at neutrino telescopes should improve the situation. In particular stringent limits may be obtained by large neutrino telescopes for IMMs with $\beta > 0.5$ coming from above.

As a byproduct of GUT MM searches some experiments obtained stringent limits on nuclearites and on Q-balls. Future experiments at neutrino telescopes and at high altitudes should perform searches for nuclearites and Q-balls of smaller masses.

We acknowledge the cooperation of many colleagues, in particular S. Cecchini, M. Cozzi, M. Giorgini, G. Mandrioli, V. Popa, M. Spurio, and others. We thank ms. Giulia Grandi for typing the manuscript.

References

1. P.A.M. Dirac, Proc. R. Soc. London 133(1931)60; Phys. Rev. 74(1948)817.
2. A. De Rujula, Nucl. Phys. B435(1995)257.
3. G. Giacomelli, Riv. Nuovo Cimento 7(1984)N.12, 1.
4. G. Giacomelli et al. hep-ex/011209; hep-ex/0302011; hep-ex/0211035.
5. G.R. Kalbfleisch, Phys. Rev. Lett. 85(2000)5292. K.A.Milton et al. hep-ex/0009003. B. Abbott et al., Phys. Rev. Lett. 81(1998)524. M. Acciarri et al., Phys. Lett. B345(1995)609.
6. L. Gamberg et al., hep-ph/9906526.
7. Private communication by M. Cozzi.
   K. Kinoshita et al., Phys. Rev. D46(1992)R881.
8. G.'t Hooft, Nucl. Phys. B29(1974)276. A.M. Polyakov, JETP Lett. 20(1974)194. N.S. Craigie et al., Theory and Detection of MMs in Gauge Theories, World Scientific, Singapore (1986).

9. G. Lazarides et al., Phys. Rev. Lett. 58(1987)1707.
   T. W. Kephart and Q. Shafi, Phys. Lett. B520(2001)313.
10. P. Bhattacharjee and G. Sigl, Phys. Rept. 327(2000)109 and refs. therein.
11. E. Witten, Phys. Rev. D30(1984)272.
    A. De Rujula and S. Glashow, Nature 31(1984)272.
12. S. Coleman, Nucl. Phys. B262(1985)293.
    A. Kusenko and A. Shaposhnikov, Phys. Lett. B418(1998)46.
13. J. Derkaoui et al., Astrop. Phys. 9(1998)173; Astrop. Phys. 9(1999)339.
14. S. Ahlen et al., Phys. Rev. Lett. 72(1994)608. M. Ambrosio et al., Astrop. Phys. 6(1997)113; Nucl. Instr. Meth. A486(2002)663; Astrop. Phys. 4(1995)33; Astrop. Phys. 18(2002)27.
15. G.F. Drell et al., Nucl. Phys. B209(1982)45.
16. S. Cecchini et al., Nuovo Cim. A109(1996)1119.
17. S. Cecchini et al., 22th ICNTS, Barcelona, Spain, 2004.
18. M. Bertani et al., Europhys. Lett. 12(1990)613.
19. M. Schein et al., Phys. Rev. 99(1955)643.
20. I. F. Ginzburg and A. Schiller, Phys. Rev. D60(1999)075016.
21. J. Ruzicka and V.P. Zrelov JINR-1-2-80-850(1980).
22. G. Giacomelli et al., hep-ex/0005041.
23. V.A. Skvortsov et al., 29th EPS Plasma Conf., ECA 26B, D–5.013 (2002).
24. D. Bakari et al., hep-ex/0004019.
25. M. Ambrosio et al., MACRO Coll., hep-ex/0207020, Eur. Phys. J. C25(2002)511; Phys. Lett. B406(1997)249; Phys. Rev. Lett. 72(1994)608.
26. S. Orito et al. ("Ohya"), Phys. Rev. Lett. 66(1991)1951.
27. E.N. Alexeyev et al. ("Baksan"), 21st ICRC 10(1990)83.
    V.A. Balkanov et al. ("Baikal") Nucl. Phys. B(Proc. Suppl.) 91(2001)438.
    P.Niessen et al., 27st ICRC 3(2001)1496.
28. V.A. Rubakov, JETP Lett. B219(1981)644.
    G.G. Callan, Phys. Rev. D26(1982)2058.
29. M. Ambrosio et al., Eur. Phys. J. C26(2002)163.
30. P. B. Price, Phys. Rev. D38(1988)3813.
    D. Ghosh and S. Chatterjea, Europhys. Lett. 12(1990)25.
31. E.N. Parker, Ap. J. 160(1970)383.
    M.S. Turner et al., Phys. Rev. D26(1982)1296.
32. F.C. Adams et al., Phys. Rev. Lett. 70(1993)2511.
33. D. Bakari et al., hep-ex/0003028. S. Cecchini et al. 28st ICRC 3(2003)1657; Nucl. Phys. B(2004) in press.
34. M. Ambrosio et al., Eur. Phys. J. C13(2000)453.
35. S. Nakamura et al., Phys. Lett. B263(1991)529.
36. G. Giacomelli, hep-ex/0210021.
37. M. Rybczynski et al., hep-ph/0410064
38. D. P. Anderson et al., astro-ph/0205089
39. D. Bakari et al., Astrop. Phys. 15(2001)137.
40. J. Arafune et al., hep-ph/0005103.
41. S.D. Wick et al., astro-ph/0001233.
42. Proposal MOEDAL at the LHC, CERN/LHCC 98-5

COSMIC RAYS AT EXTREME ENERGIES

R. CESTER

Universitá di Torino and INFN
Via P. Giuria 1, Torino 10125, Italia
E-mail: Cester@to.infn.it

This paper briefly reviews the status of research on High Energy Cosmic Rays, in particular those populating the highest part of the energy spectrum ($E_{CR} > 510^{18} eV$) and believed to be of extra-galactic origin. An outlook on a new generation of experiments hopefully capable of answering open questions on production, acceleration and propagation mechanisms of these particles, concludes the paper.

1. Introduction

The discovery of Cosmic Rays (C.R.)dates back to the beginning of the 20th century. In 1936 the Nobel prize was awarded jointly to V. Hess [1] and to C.D.Anderson [2], to Hess for his 1912 experiments that proved the existence, in the highest layers of the atmosphere, of radiation coming from outer space, and to Anderson who in 1932 discovered, within the flux of secondary C.R. reaching earth, a new component of light, positive particles (positrons) soon after identified as the antiparticles of electrons.

After Anderson experiment, C.R. became a fertile ground of search for new particles with the rate of discoveries increasing rapidly as detection and measuring techniques improved, until, in the middle 1950s, it became possible to continue the search at particle accelerators. In the range of energy that can be reached at accelerators, C.R. are not competitive in the study of elementary particles and their interactions, however an energy window for the discovery of new phenomena still exists above 2×10^{15} eV.

A complementary and fundamental aspect in the study of C.R. is the measurement of the characteristics of the primary component reaching earth from outer space, to infer from these measurements informations on their sources. A major problem in the attempt to unravel this puzzle lies in the fact that we have for each incoming particle only three observables: mass, energy and direction, the measurement of which is far from trivial.

In this paper I will mainly concentrate on primary C.R. of energy above

10^{15} eV. At these energies it is not possible to measure directly the characteristics of the incoming particles that must be reconstructed from the study of secondaries reaching ground. A powerful handle to reconstruct energy and direction and to identify the type of primary particles comes from the study of electromagnetic air showers (EAS) generated from interactions of the incoming particles in the higher layers of the atmosphere and from the observation of muons from charged secondary particles decays, that penetrate down to ground. EAS were first detected in 1934 by B.Rossi [3] and systematically studied by P.Auger [4] who, with a comparatively large array of counters and good timing coincidence-circuits, succeeded in detecting showers of energy up to 10^{15} eV.

As experiments continued to yield evidence of the existence of C.R. of very high energy (up to 10^{20} eV and more), speculations on their sources, the acceleration processes, and on the effects of propagation through galactic and intra-galactic media were brought forward in an attempt to con struct a coherent picture of the phenomena.

This paper is organized in two parts. In the first I will attempt to follow the flow of ideas that in the last 50 years have contributed to what understanding we now have of the origin of high energy C.R.[5]. In the second, I will review experimental results and give an overview of the new more powerful generation of experiments exploring the highest energies, now in data taking or under construction.

2. C.R. from the Galaxy

The measured energy spectrum of C.R. is shown in Fig.1. For energies greater than 10^{10} eV, where the influence of the solar system can be ignored, and smaller than 10^{15} eV, the spectrum is well described by a power law:

$$\frac{dJ(E)}{dE} \propto E^{-2.7}, \tag{1}$$

The average concentration of elements in the Galaxy and in C.R. are quite similar; however there are some differences, the most pronounced one for light nuclei, Li, Be and Bo that are abundant in C.R. and almost absent in the galaxy. All the differences can be explained by the fact that C.R., while propagating in the galaxy, collide with particles (for the most part protons) of the interstellar medium (ISM). A quantitative analysis of the chemical and isotopic composition of C.R. reaching the earth indicate that the thickness of interstellar medium traversed by C.R. before reaching the earth is of the order of 5 to 10 g/cm^2 [6].

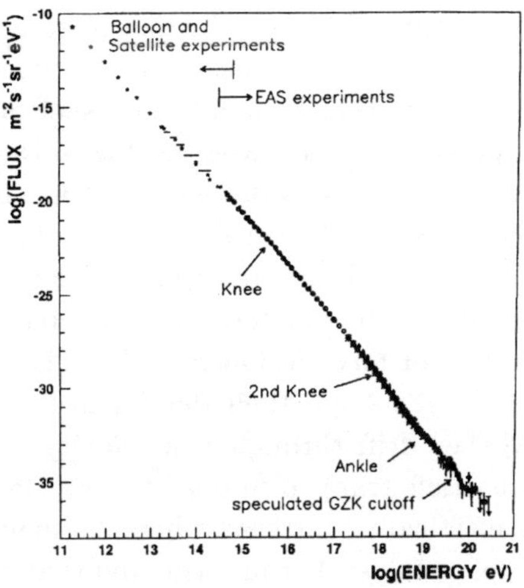

Figure 1. Measured C.R. spectrum.

C.R. appear to be isotropic and to carry an energy density of about 1 eV per cm^3, about twice that of star light. If they are uniformly distributed over the volume of the galaxy halo ($V_h \sim (10\ parsecs)^3 \sim 10^{68} cm^3$), the total energy carried by these particles is of the order of $W_{C.R.} \sim 10^{56}$ erg. Such huge energy implies that a powerful acceleration mechanism must be active in our galaxy.

2.1. *Fermi Acceleration mechanisms*

Fermi started studying C.R. in 1946. At the time it was already known that the majority of C.R were nuclear particles with a power-law energy spectrum with exponent -2.9, a value remarkably close to the present one.

From an estimate of the average density of ISM and the known cross-section for nuclear interactions, Fermi [7] calculated the lifetime of C.R. to be: $\tau_{nucl} \sim 7 \times 10^7$ years, much shorter than the universe lifetime. This implies continuous creation. If one assumes C.R. to be created at a constant rate, then the flux of such particles with age between t and $t+dt$ is:

$$dJ(t) \propto dt \times e^{-\frac{t}{\tau_{nucl}}}, \tag{2}$$

But how do these particles get their energy and what is the acceleration process that leads to a power-law spectrum? This law requires a very specific acceleration mechanism capable of indefinitely increasing the particle energy. Fermi observed that, for the principle of equipartition of energy, this can be achieved in successive collisions with extremely large moving objects. The next step was that of identifying such objects. In Fermi's words: " **...the main process of acceleration is due to the interaction of cosmic particles with wandering magnetic fields which, according to Alfven, occupy the interstellar space..**" Alfven *clouds* drift through galactic space in stable configurations of highly irregular magnetic fields and " **.. to each line of force one should attach a material density due to the mass of the matter to which the line of force is linked...** ". The dimension of these clouds is of several light-years and their density is 10 to 100 times higher than that of the ISM they drift through at a velocity $V \sim 3 \times 10^6\ cm/sec$, ($\beta = \frac{V}{c} \sim 10^{-4}$). A particle trapped in one of these clouds, spirals around the field lines being elastically scattered when the magnetic field changes abruptly. With a simple analysis Fermi estimated that a particle fractional energy increases on average by: $\frac{\Delta E}{E} = \frac{4}{3}\beta^2$ in each collision with an Alfven cloud. After n collisions:

$$n \times \frac{\Delta E}{E} \sim \int_{E_0}^{E} \frac{dE}{E} = \int_{E_0}^{E} dlnE = ln\frac{E}{E_0} = n \times \frac{4}{3}\beta^2 \approx \frac{t}{\tau_{coll}}\beta^2 , \qquad (3)$$

where E_0 is the injection energy and τ_{coll} is the mean time interval between collisions. Combining the equation $E(t) = E_0 \times e^{\frac{t}{\tau_{coll}}\beta^2}$ with eq.2, the differential flux of C.R. as a function of energy is derived:

$$dJ(E) \propto dE \times E^{-\alpha} \qquad with \qquad \alpha = \frac{4}{3}\frac{\tau_{coll}}{\tau_{nucl} \times \beta^2} + 1 , \qquad (4)$$

As seen, the Fermi stochastic acceleration process leads naturally to a power-law spectrum; unfortunately the mechanism is highly inefficient, since the fractional energy increase in a collision is proportional to β^2 (hence the name *Fermi acceleration mechanism of second order*) and $\beta \approx 10^{-4}$.

It was only in the late 1970's that Fermi's basic ideas were extended and a more efficient *first order Fermi acceleration mechanism* was proposed. By that time supernova explosions had been studied in detail and it was suggested that C.R. in the galaxy could be generated and accelerated in this process[8]. In fig.2 the first order acceleration processes is compared to the second order one. The acceleration mechanism is basically the same but the particles are now accelerated in shock waves propagating out in the explosion with velocity $\beta \sim 0.1$, three orders of magnitude larger than that of Alfven clouds, and $\frac{<\Delta E>}{E}$ is proportional to β (*Fermi acceleration mechanism of first order*). It is estimated that C.R. emitted in a supernova explosion can reach energies up to $\sim 10^{13}$ eV and that multiple interactions

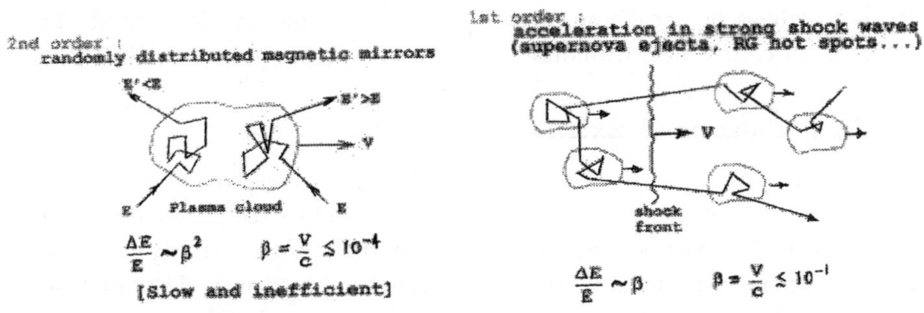

Figure 2. Fermi acceleration mechanism of second (left) and first (right) type.

of C.R.wandering through the galaxy, with supernova remnants can extend their energy up a few orders of magnitude[9].

2.2. *C.R. motion in the ISM*

Fig.3 shows an example of C.R. trajectories in the galactic magnetic fields which has a uniform component $B_0 \sim 2\mu G$ along the spiral arms and a component with random direction, B_r, of the same order of magnitude [10]. The C.R. particles spiral around the field lines with a Larmor radius:

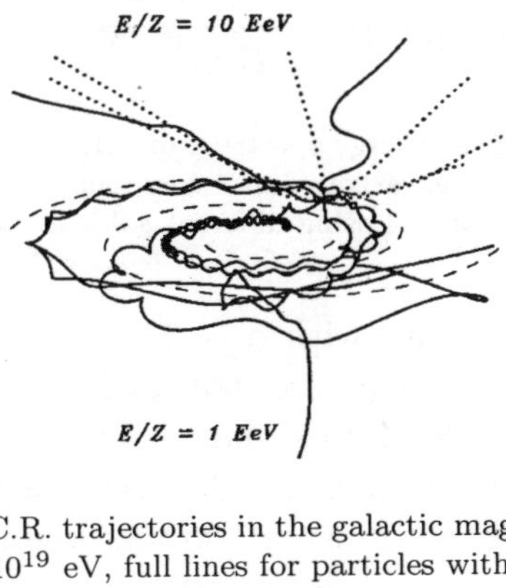

Figure 3. Simulation of C.R. trajectories in the galactic magnetic field; dotted lines are for particles with E/Z = 10^{19} eV, full lines for particles with E/Z = 10^{15} eV

$$r_L(pc) = \frac{E_{15}}{B_0(\mu G)Z}\,, \tag{5}$$

(with E measured in (10^{15} eV) and r_L measured in parsecs) and scatter on the random irregularities. The long random walk of C.R. in the galaxy explains why, when reaching earth, C.R. are isotropically distributed. This effect is, of course, less pronounced for C.R. of higher energies (fig.3).

3. From Galactic to Extra-Galactic C.R.

In fig.4 the flux of C.R. with $E > 10^{13}$ eV, as measured by recent experiments, is plotted as a function of energy. The flux is rescaled by E^3 to enhance the features showing-up above 10^{15} eV[11]. As we have seen, for C.R. energies up to 10^{15} eV the differential spectrum is described by a power law $\frac{dJ(E)}{dE} \propto E^{-\alpha}$ with $\alpha = 2.7$. The first anomaly (*the knee*) appears at $E \sim 3 \times 10^{15}$ eV where α changes from 2.7 to ~ 3, followed by a second change in steepness (*the second knee*) at $E \sim 4 \times 10^{17}$ eV where α reaches a value ~ 3.3. At $E \sim 5 \times 10^{18}$ eV the spectrum shape changes again,

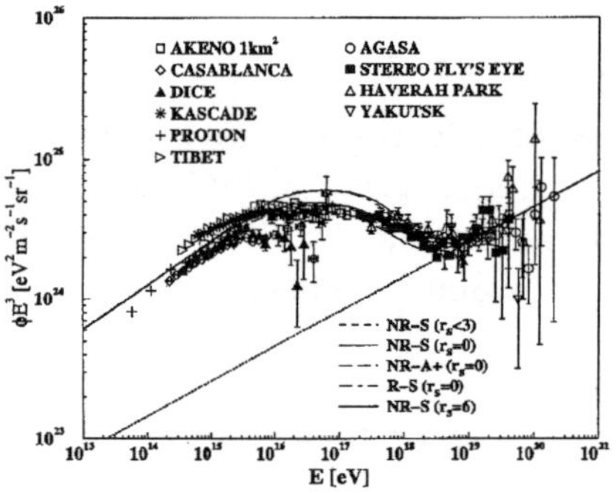

Figure 4. Measurements of the energy spectrum of C.R. (rescaled by a factor E^3). The dotted line shows the onset of an extragalactic component. Also shown are predictions of transport calculations performed using different galactic models.

and can be fit to a power law with $\alpha \sim 2.7$ (*the ankle*). The *knee* structures are explained by the conjecture that for E/Z above a threshold value, C.R. start escaping from the Galaxy halo, causing the observed reduction in their local density. The phenomenon has been successfully modeled with detailed simulations of C.R. transport through the magnetic fields permeating the galaxy, assuming a realistic distribution of supernova sources[12]. Since the magnetic field effect depends only on E/Z, lighter nuclei will start escaping first so that we expect to observe a change in composition as the energy increases. Fig.5 shows how the measured composition evolves with energy. Recent results appear to agree well with the prediction of the model. Above the *ankle* C.R. originating in the Galaxy are no longer

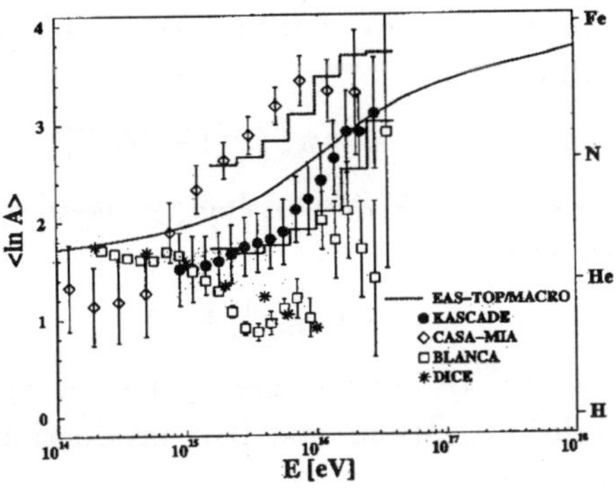

Figure 5. Measured composition ($< ln\ A >$) as a function of energy compared to the result of a transport calculation (full line).

confined and the hardening of the spectrum suggests the onset of a new component probably of extra-galactic origin. The C.R. spectrum extends up to energies $\geq 10^{20} eV$ and the crucial issue becomes that of identifying sources where energies of such magnitude could be reached.

4. Ultra High Energy Cosmic Rays (UHECR)

As we see in fig.4 the flux of C.R. cuts off at about 10^{20} but it is not clear if this is due to a real effect or a consequence of the limited aperture of experiments at an energy where the flux is at most of one event per $sterad \times km^2 \times century$. Even with the new large-aperture experiments under construction, if a cut-off is observed, we will need auxiliary informations to establish if this is due to lack of high-power sources or to the degrading effects of interactions in the inter-galactic medium, discussed in the next section.

4.1. *The Greisen-Zatseping-Kuz'min (GZK) effect*

In 1965 Penzias and Wilson[13] discovered quite accidently that the Universe is uniformly filled with soft electromagnetic radiation, the so called Cosmic Microwave Background (CMB) with a blackbody spectrum peaked at 6×10^{-4} eV and a density of about 400 photons/cm^3. The existence of CMB had actually been predicted as a relic of an early-universe time when H atoms were formed and neutral matter decoupled from radiation [14].

Soon after the discovery, Greisen and independently Zatseping and Kuz'min [15] noticed that C.R. will interact with these CMB photons and loose energy in the process. These authors were referring specifically to a highly inelastic reaction:

$$p + \gamma \rightarrow \Delta^{+} \rightarrow \pi^{+}(\pi^{0}) + N(P)\,, \tag{6}$$

with a threshold at $\sim 4 \times 10^{19}$ eV and an energy loss of $\sim 20\%$ of the proton energy. Other reactions are:

$$p + \gamma \rightarrow P + e^{+} + e^{-}\,, \tag{7}$$

with a threshold at $\sim 10^{18}$ eV and an energy loss of only $\sim 0.1\%$ and, for heavy nuclei, photo-disintegrations and pair production, with the largest energy loss for the process:

$$A + \gamma \rightarrow (A - 1) + N\,, \tag{8}$$

These predictions seemed to contradict the findings of the Vulcano Range experiment [16] that in 1962 had detected an event with energy E $> 10^{20}$ eV. However this apparent contradiction could be explained if the source of such event was at a relatively small distance from our observation point. Recent calculations, which include the effect of all components (microwave, infrared and radio) of intergalactic background radiation, have determined the energy dependence of the attenuation length of different C.R. particles[5] (see fig.6) and indicate that, if the Vulcano Range event was initiated by a primary proton,it must have come, with high probability, from a distance < 50 Mpc. As we will discuss in detail in the second part of this paper, after 45 years of the detection of the Vulcano Range event and in spite a wide experimental effort, we still do not have a reliable estimate of the flux of events with energy $\geq 10^{20}$ eV. It is hoped that the new generation of experiments will clarify the matter and identify the sources of such events.

4.2. *The sources*

There are two classes of models to explain the existence of C.R. with E$>$ $10^{19} eV$ (UHECR). The first one (Bottom-up models) assumes that they are produced and accelerated in astronomical objects where catastrophic processes with large energy transfers are taking place. The second one (Top-down models) assumes that UHECR get their energy from the decay of some yet undiscovered high-mass particle.

4.2.1. *Bottom-up models*

I discuss first Bottom-up models as they are less controversial. In one scenario the acceleration process occurs in shock waves projected out in

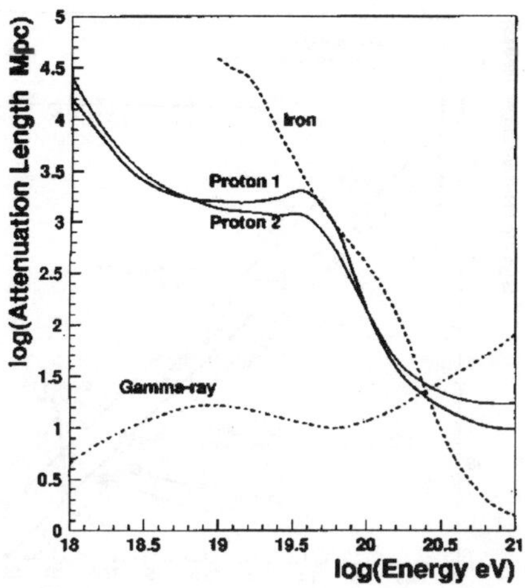

Figure 6. Attenuation length of proton, iron and gamma-ray primary C.R. in the intergalactic background radiation.

extra-galactic astronomical explosions far more powerful than the galactic supernova ones, and that UHECR acquire energy through a Fermi mechanism where the relevant parameters reach extreme values. A simple analysis helps identifying such parameters and their range of values[17].

To be effectively accelerated particles must be trapped in the magnetic field (B) of the acceleration region, hence the characteristic length (L) of this region must be larger than twice the particle Larmor radius r_l:

$$L(pc) >> 2 \times r_l(pc) \approx 2 \times \frac{E_{15}}{Z \times B_T(\mu G)}, \tag{9}$$

where B_T is the component of the magnetic field normal to the particle direction and E_{15} is the particle energy in units of 10^{15} eV. Moreover, as we have seen, in a first order Fermi-type process, the energy increment depends linearly on the velocity β of the shock wave, so that we may expect the maximum energy attainable to be:

$$E_{15} = k\frac{B_T(\mu G) \times L \times Z \times \beta}{2}, \tag{10}$$

with k always < 1. The relevant quantities in this problem are therefore the magnetic rigidity $B_T \times L$ and the velocity of the shock, β. Particles may also be accelerated to high energy directly by extended electric fields. As is well known, such extended electric fields can be generated in the rotation of magnetized conductors. In rapidly rotating neutron stars, the conditions

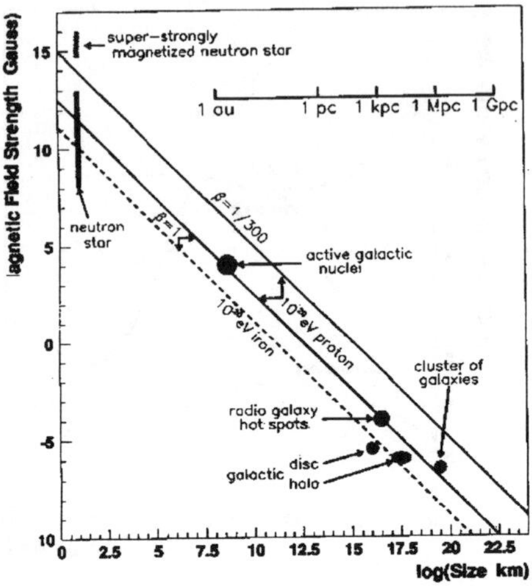

Figure 7. Size and magnetic-field strength of acceleration sites where CR could reach energies above $10^{20}eV$. The two lower lines refer to ultra-relativistic shock waves ($\beta \sim 1$) (full line for protons and dotted line for iron nuclei), the upper line refers to protons in shock waves with $\beta = \frac{1}{300}$. Drawn on the plot are candidate astronomical sites.

are met for such an acceleration mechanism to take place. In Fig.7 known astronomical objects where UHECR could be produced and accelerated, are drawn on a log/log plot of B vs L [18]. Only in regions above the diagonal lines can particles be accelerated to $E \geq 10^{20}eV$.

The lower-right part of the plot is populated by extra-galactic objects. I list some of the most interesting below, referring[5] the reader to the specialized literature for a deeper understanding of their characteristics.

Radio-Galaxies Hot Spots are gigantic jets ejected at relativistic speed from an Active Galactic Nucleus (AGN) where shock waves of few kiloparsec, with magnetic fields up to hundreds of $\mu Gauss$ could be generated.

Cosmological Gamma Ray Bursts (GRB) are thought to be produced when massive stars or binary systems collapse into black holes. Energy is released almost instantaneously in an expanding fireball where the γ emitting region moves relativistically with a Lorentz factor of several hundreds. It is conjectured that in such environment, protons are accelerated to the highest energies, through a second order Fermi mechanism, within regions expanding with a velocity V $\sim c$ in which strong, random magnetic-fields are thought to be present. A common origin for the two phenomena is suggested by a remarkable coincidence between the energy flow of UHECR and GRB[19]. These sites and AGNs are favored candidates as sources of protons of energies up to 10^{21} eV.

Colliding Galaxies: when two galaxies collide, the converging flows could contain shock fronts capable of accelerating iron nuclei up to 10^{20} eV.

Clusters of galaxies: particles could be accelerated to high energy by accretion shocks formed by the in-falling flow toward a cluster of galaxies.

Nearby Galaxies: It can not be excluded that UHECR come from 'normal' galaxies where the level of activity, star formation and magnetic fields are higher than in our galaxy. Then the C.R. flux at emission should be proportional to the distribution of luminous matter and their mass composition similar to the one of our galaxy.

In the upper-left part of fig.7 are galactic **Neutron stars** where rapidly rotating, strong magnetic-fields can generate electromotive forces capable of accelerating Iron nuclei to ultra-high energy[20].

I conclude by noting that what we will observe with our experiments depends crucially not only on the source accelerating power but also on its distance (d) from the solar system and on the distribution of sources in the universe. Particles coming from objects close-by, on the universe scale of distances ($d \leq 50 Mpc$), will not be substantially affected by the interaction with the CMB even if their energy is above the GZK threshold. The hypothesis that these objects are sources of UHECR can then be tested by measuring the pointing direction of the incoming particles which should correlate with the position of the source since over distances d $\leq 50 Mpc$ particles will not be significantly deviated by the rather weak intergalactic magnetic-fields. If, on the other side, the sources are distant objects ($d >> 50 Mpc$), we do not expect to detect particles with energy above $\sim 6 \times 10^{19}$ eV irrespective of their energy at the source. Moreover since particles of this energy have attenuation paths in the inter-galactic media of $\sim$ 1000 Mpc (see fig.4) they will wonder around, scattered by magnetic fields irregularities. We therefore expect that when reaching our galaxy they will have lost memory of the initial direction. It is also very important to determine the mass of the incoming particles, protons or heavier nuclei, since this might help differentiating between possible sources.

4.2.2. *Top-down models*

Far more speculative are the models that assume that UHECR come from the decay of super-heavy particles predicted by theory. First because no experimental hint of such particles exists so far, and second because to explain the observed flux of UHECR, their density and decay life-time must be chosen 'ad-hoc'. A common feature is that, given their large mass, they would accumulate in the galaxies halo so that their decay products would reach the earth unaffected by CMB. Candidate objects are: early-universe relic super-heavy particles, topological defects created in the phase-transition of the early universe, magnetic monopoles and supersymmetric hadrons[5]. A strong signature for some of these models is that γs and neutrinos would account for a large fraction of the decay products reaching the solar system.

Finally, I touch on yet an other model[21] that would predict a large flux of high energy γ rays and neutrinos from the production-decay chain:

$$p + ... \rightarrow \nu + ... \qquad with \qquad E_\nu \sim 4.10^{21}\,, \tag{11}$$

$$\nu + \bar{\nu}_{rel} \rightarrow Z^0 \rightarrow kP + n\pi^0 + m\pi^{+-} .... \rightarrow k \times P + 2n \times \gamma + 3m \times \nu..\,, \tag{12}$$

The Z^0 decay branching ratios are well known and a ratio of k/n/m/= 2/10/17 is expected. The high-energy incoming neutrino interacts resonantly with an anti-neutrino of the *cosmic neutrino background* (a relic of the early stages of the universe) to form a Z^0 particle. Given the values of the neutrino and of the Z^0 masses, the energy of the incoming neutrino is derived by simple kinematics. Once again the major problem here is that of finding a source sufficiently powerful to accelerate the particles that initiate the production-decay chain.

5. Experimental Overview

High energy C.R. entering the atmosphere undergo nuclear interactions producing a large number of hadrons. While propagating through the atmosphere, particles interact repeatedly producing a high multiplicity cascade. The main components reaching ground are the barionic one, in a small angular cone approximately along the direction of the incoming C.R. and, distributed over a wider area, muons from charged π decays and electrons and γs in electromagnetic showers.

Electromagnetic showers are initiated by the decay of π^0's to 2γs which undergo conversion to e^+e^- pairs; electrons, in turn, emit γs in the *Bremsstrahlung* process. The interplay of the two processes leads to particles multiplication and to the degrading of particles energy until a point is reached where the electron energy falls below the *Bremsstrahlung* threshold and the multiplication process stops. Electrons loose their residual energy interacting with atoms and molecules of the atmosphere and the shower tapers off. The shower can be parametrized by the atmospheric depth of the first interaction X_0, the depth at maximum, X_{max}. and the number, N_{max} of electrons at X_{max}.

5.1. *Detection and Analysis methods*

The properties of the UHECR have been studied in two ways: a) with large arrays of ground-based detectors that measure the flux of muons and/or the energy carried by the electromagnetic component and b) with optical detectors measuring the yield of fluorescence photons emitted along the electromagnetic shower path[22].

5.1.1. *Arrays of Surface Detectors (SD)*

The use of these detectors has two major advantages: a) they are little affected by external conditions, and therefore can run continuously and, b) they can be built of identical modules that, to instrument large areas, must be simple, low cost and stable. Two types of modules have been frequently used: a) scintillator/absorber sandwiches to count muons (*μ detectors*), and b) water Cherenkov tanks to measure the energy deposited by electrons and muons (*H_2O Cher.*). The main disadvantage of this technique lies in the fact that all informations on the characteristics of primary C.R. come from a snapshot of the shower at ground level. As a consequence the validity of the results are strongly dependent on the ability to model accurately the development of showers through the atmosphere.

Events are accepted if they meet a trigger condition that requires a minimum number of modules to have fired (typically 3 or 4). Hence the energy range of acceptance of the detector is bound on the low side by the choice of the array lattice spacing. On the upper side it is the total exposure, the product of the detector aperture times the effective running time, that sets a bound to the energy range.

Measurement of the primary CR direction: The direction of the axis of the shower measures the primary CR arrival direction. If at least three non-collinear ground stations record the shower, its direction can be computed using detectors position and arrival times.

Energy measurement: the determination of the primary C.R. energy. is based on the measurement of a single parameter, the signal density ρ at an optimal distance l from the shower core (typically 500 to 1000 m), This method [23] relies on the fact (born out by simulations) that the value of $\rho(l)$

correlates strongly with the primary energy, weakly to the particle identity. and is quite insensitive to fluctuations in the development of the shower. The weak point of the method lies in the fact that the conversion factor between $\rho(l)$ and primary energy must come from simulations.

Identification of the primary: for each event the most effective indicators of the primary identity are the rise-time of the signal detected far from the shower core and the muon content of the showers. As an example, a Fe nucleus will produce $\sim 70\%$ more muons than a proton of the same energy. Showers initiated by γs will have a low muon content.

5.1.2. *Fluorescence Fly's Eye Detectors (FD)*

When a Cosmic Ray interacts in the upper levels of the atmosphere, originating a shower that propagates down to earth, the charged particles in the shower excite the Nitrogen molecules and ions which decay emitting light in the near-ultraviolet wavelength range. The process has a very low efficiency with only 5×10^{-5} of the shower energy carried by fluorescence photons. The light is emitted isotropically with an yield proportional to the particle ionization loss, hence, for relativistic particles, to track length, and almost independent of the atmospheric depth. Detecting this signal one can perform a measurement of the shower energy profile using the atmosphere as a calorimeter. The signal is however so weak that this technique can only be applied to study very high energy showers when background light levels are low, typically in clear, moonless nights.

Each Fluorescence Detector *eye* is built out of telescopes. The conceptual scheme of a telescope is very simple(see fig.8): a light collecting element (i.e. a spherical mirror) defines the aperture and focuses the light collected onto a camera, an array of phototubes positioned approximately on the mirror focal surface. To reduce background, the incoming light is filtered to cut out the unwanted components of the night-sky light spectrum. As the shower comes into the field of view of one telescope, an image is formed on the camera that tracks the trajectory of the shower as it develops through the atmosphere. From each PM, amplitude and timing of the signal are read out. From these data the following characteristics of the shower can be reconstructed:

Direction of the shower axis. The reconstruction of the shower geometry starts with the determination of the shower-detector plane (SDP) obtained by a fit to trial configurations of the triggered pixels directions, weighted by signal amplitude. A precision of $\sim 0.25^0$ on the direction of the

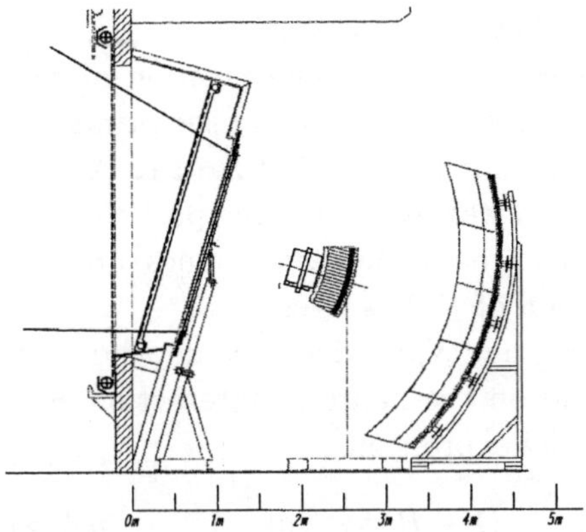

Figure 8. Drawing of a Fluorescence Detector telescope for the Auger experiment.

normal to the SDP can be achieved. If the shower is detected by two eyes (*stereo event*), the direction of the shower is precisely reconstructed from the intersection of the SDP's determined from the two eyes. For *monocular events*, that is for events seen by one eye only, it is not always possible to reconstruct unambiguously the direction of the shower axis within the SDP. **Energy measurement**: The light yield reaching the detector, scaled to ionization energy and corrected for geometrical factors and for the losses due to absorption and scattering in the atmosphere, measures the shower-energy: It gives a lower bound to the primary energy since $\sim 10\%$ of the shower energy is carried by neutral particles and part is lost in the ground. This effect can be corrected for without introducing a large systematic error. Dangerous sources of systematics are: a) the yield of fluorescence light emitted by Nitrogen and its dependence on the atmosphere parameters is still not precisely known[24], b) the presence of a Cherenkov-light component in the detected signal will affect the energy integral and distort the shower profile[5]. The Cherenkov light beamed directly into the detector acceptance can be estimated from the known angular distribution of electrons in the shower, and subtracted out. It is however more difficult to evaluate the fraction of Cherenkov light scattered into the detector acceptance by the aerosol molecules in the atmosphere,

c) an insufficient understanding of the characteristics of the atmosphere along the light path between the shower and the detector is potentially the worst cause of systematic errors in the determination of the shower parameters. An accurate monitoring of the atmosphere parameters and of the aerosol component is therefore essential.

Identification of the primary: The position of the shower maximum (X_{max}) is a good indicator of the nature of the primary particle. It changes over a range of 100 gm/cm^2 for nuclei of A=1 to A = 56 and, of course, is quite different for γs and weakly interacting particles. To have a reasonable sensitivity to the nuclear mass one needs to measure the position of shower maximum to a precision of $\leq$ 20 gm/cm^2.

It should be noted that X_{max} does not depend only on A but also on the shower energy and on the hadronic interaction characteristics. It can be approximated by the expression[5]:

$$X_{max} = (1 - B).\chi_0.(ln(\frac{E}{\epsilon}) - < lnA >), \tag{13}$$

where B (always < 1) carries the information about the hadronic interactions cross sections and particles multiplicities. χ_0 and ϵ are radiation length and critical energy of air. The rate of change of X_{max} with the logarithm of energy (*Elongation rate*):

$$D_e = \frac{\partial X_{max}}{\partial lnE} = (1 - B).\chi_0.(1 - \frac{\partial < lnA >}{\partial lnE}), \tag{14}$$

(or the more commonly used $D_{10} = 2.3D_e$), is constant if the mass compositions is constant, in which case the only parameter to be determined is B. A change in D at a given energy, signals a change in the average mass composition at that energy

5.2. *Experimental Results*

In Table 1 I have summarized some of the characteristics and results of UHECR experiments[25,26,27,28,29,30]. All entries refer to experiments located in the northern hemisphere, except Auger South. I have omitted the only other southern experiment, SUGAR[31] - Sidnay University Giant Array - which yielded few results.

There is not a good agreement in the measured energy spectrum between experiments using ground arrays (1st part of the Table) and experiments using fluorescence detectors (2nd part of the Table) as shown in fig.9. Over the full range of UHECR energies, the flux of AGASA is higher than the flux measured by the FD experiments. It has been estimated that an energy miscalibration of $\sim$ 30% [32]could explain the difference. Also the rate of events above $E > 10^{20}$ eV disagrees. Summing separately over the two sets of experiments, we find for the ratio of the (*number of events with energy above* 10^{20} *eV/exposure*), respectively (2.26 ± 0.54) $(m^2 \times sterad \times sec)^{-1}$

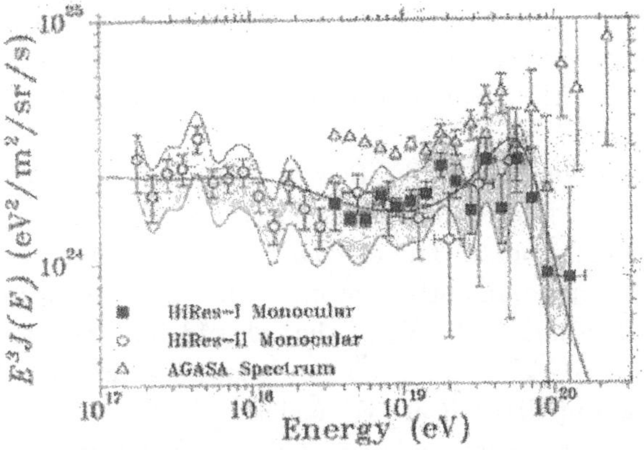

Figure 9. Comparison between the energy spectrum of UHECR as measured by Agasa, and the combined HiRes-I and HiRes-II results. The shaded band reflects the estimate of systematic errors by HiRes.

for SD and $(0.26^{+0.25}_{-0.14})$ $(m^2 \times sterad \times sec)^{-1}$ for FD. If we assume that this is due to a systematic error on the energy measurement, rather than to a $\sim 3.5\sigma$ statistical fluctuation, we would be tempted to privilege the FD result since it depends less on simulations. In this case one could conclude that there is evidence of a GZK cutoff. However, as we saw, the FD energy measuring technique has its own problems. Moreover, at the highest energies, there is considerable uncertainty on the HiRes aperture[33]. All experiments have searched for anisotropies in UHECR incoming directions. AGASA has reported small-angle correlations between doublets (5)

Table 1. Experimental overview: Volcano Range (V.R.),Haverah Park (H.P.),AGASA and Yakutsk (Ykt) use Ground Arrays while Fly's Eye (Fleye) and Hires use Fluorescence Detectors).

Exp.	Status	Detect type	Exposure (10^{16} $m^2.ster.s$)	SD Area (Km^2)	FD eye Aperture $\Delta\phi.(\theta_1 - \theta_2)$	N_{events} ($> 10^{19.6}$ /$> 10^{19.8}$ eV)	N_{events} ($> 10^{20}$ eV)
V.R.	ended	μ	0.2	8		6/-	1
H.P.	"	H_2OCh	0.73	12		27/-	4
AGASA	"	μ	5.1	100		72/24	11
Ykt	Data	μ	~1.5	18 (10)		14/-	1
FlEye	ended	FD mn	2.6			24/-	1
"	"	FD st	0.46			2/-	0
HiresI	Data	FD mn	7.6		$2\pi(3^0 - 17^0)$	-/10	2
"II	"	FD mn	1.0		$2\pi(3^0 - 31^0)$	-/3	0
Aug.S	Constr.	Hybrid		3000	$\pi(1.7^0 - 31.3^0)$	-	~30/yr
T.A.	Constr.	Hybrid		760	$2\pi.(2^0 - 32^0.)$	-	~ 10/yr

and triplets (1) of events[34]. However a recent analysis [35] has cast doubts on the significance of the signal. Other reported correlations of UHECR incoming directions with known astronomical objects have not, so far, found confirmation.

The mass assignment of UHECR is also an open question, with different methods giving different answers and the conclusions being based on shower development models. The *elongation rate* by FD detectors suggests that UHECR are predominantly protons. The same conclusion is reached measuring the fluctuations in X_{max} that should be larger for lighter nuclei. In contrast, AGASA, from muons densities at ground, finds that a consistent fraction of UHECR are Fe nuclei[32].

Attempts have also been made to set limits to the fraction of γs in the primary flux at and above the GZK limit. At present the limits are still rather loose due to the small number of events. Simulations [36] show that *elongation rate* is a sensitive tool to detect very high energy γs and will certainly be exploited by future experiments.

5.3. *The new generation of experiments*

In the last two raws of Table 1 are listed some of the characteristics of two experiments now under construction that will, hopefully, soon produce data capable of substantially improving our understanding of UHECR. These experiments were designed when it was already clear that two essential conditions should be met: a) the aperture of the detector systems had to be large enough to collect a statistically significant number of events in the GZK cutoff energy region (possibly within the life-span of a physicist), and b) guarantee control of the systematics on energy measurements.

The design of the **Auger Observatory** includes two equal sites, one in the South and one in the North hemisphere to provide full sky coverage. Construction of the South Observatory, situated in the district of Malargue, Argentina, (35.1^0 - 35.6^0 South, 69.0^0 - 69.6^0 West) is well underway, while decisions on the Northern site are still pending. The experiment is built as a *Hybrid Detector* with two components: a) a 3000 Km^2 SD array of 1600 H_2O Cherenkov counters, distributed over a hexagonal lattice with 1.5 Km spacing between modules and, b) 4 fluorescence *eyes* overlooking the SD array. The number of FD eyes and their location on the site are chosen so that all showers of energy $\geq 10^{19} eV$ that hit the SD, will be seen by at least one eye, when the FD is operational. Details on the detectors characteristics can be found in a recent paper [37].

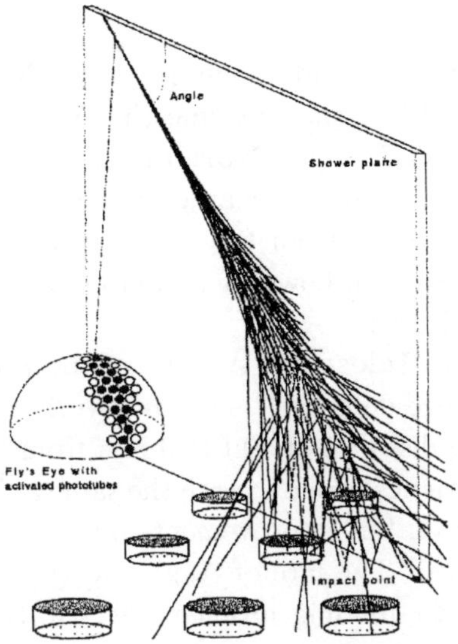

Figure 10. Artist view of an Hybrid detector.

As we have seen, a ground array has 100% duty cycle while the duty cycle of FD's is not more than 10%. Therefore the merit of an hybrid detector is that of providing a subset of *golden events* that can be used to understand the systematics of the energy and direction measurements of both SD and FD. The cross-check is particularly important if looked upon as a training procedure that will validate the measurements done on the majority of the events where only SD information is available. Furthermore, for this subset of events, the identity of primary particles can be more reliably determined exploiting the combined set of tools; X_{max} from FD , μ density, signal rise-time etc. from SD data.

At the time of writing $\sim$ 600 H_2O Cherenkov detectors have been deployed and are routinely taking data, as are 2 of the 4 FD eyes. It is anticipated that by the middle of 2005 a set of data equivalent to that collected in 12 years running of AGASA, will be available.

The Telescope Array (TA) Hybrid Detector, the first stage of a program to explore the north-hemisphere sky, is under construction in the West desert of Utah (USA). The SD Array will have 576 scintillator detectors for muon counting, spaced by 1.2 Km, and will be surrounded by three FD eyes of 40 telescopes each, with full azimuth coverage. It is expected that data taking will start in 2007.

5.4. *Future Outlook*

Programs for the next two decades include: Auger North, the Fluorescence Telescope Array and EUSO, the Extreme Universe Space Observatory.

The base-line design of **Auger North** foresees an observatory with the same area and the same detector characteristics as Auger South to allow a direct comparison of data from the two sky hemispheres. The results obtained in Auger South will however undoubtedly have an impact on the final design of Auger North.

The final goal of the **Telescope Array** collaboration is that of building a very large fluorescence detector[38] system covering an area of about 200 km^2 with 10 eyes at a distance of 30 to 40 Km from each other. The basic construction characteristics of one *eye* are the same as in the hybrid detector now being built in Utah, but improvements on the optics and on read-out electronics are proposed. The fate of Auger North and of the fluorescence TA, both now stalled by funding problems, will in the end depend on the physics scenario that Auger South and Hybrid TA will uncover in the next few years.

By far more innovative is the **EUSO** experiment which has been designed to measure EAS with a satellite-borne fluorescence detector [39]. A space mission requires a compact, radiation resistant detector with low power consumption, and capable to operate reliably over a long period of time. The present design of the detector, based on an aggressive R.&D. program, features a wide angle ($\pm 30^0$), high resolution ($\sim 0.1^0$) telescope. This resolution is reached with a fine segmentation of the detector area ($> 10^5$ pixels) and corresponds to a spatial resolution on the ground of $\sim 1Km$ and a resolution on the primary C.R. direction of $\sim 1^0$, comparable to that of ground-based detectors. The FD will be installed in a payload facility aboard the International Space Station (ISS) and will then look down at the earth atmosphere from a $\sim$450 Km high ISS orbit. The geometrical aperture of $\sim$500000 km^2 sr will be reduced by a factor $\sim$ 10 due to the limited observational duty cycle. Even so it will be about $\sim$ 7 times that of the Auger South ground-array, with uniform sky coverage. Uniform sky coverage is very important in the study of large scale anisotropies associated with the galactic structure and super-galactic plane. Euso will monitor an atmospheric volume of $\sim$ 1500 Giga-tons of air and therefore has a strong capability to detect down-going UHE neutrinos interacting in the atmosphere or up-going ones interacting in the upper layers of the earth crust.

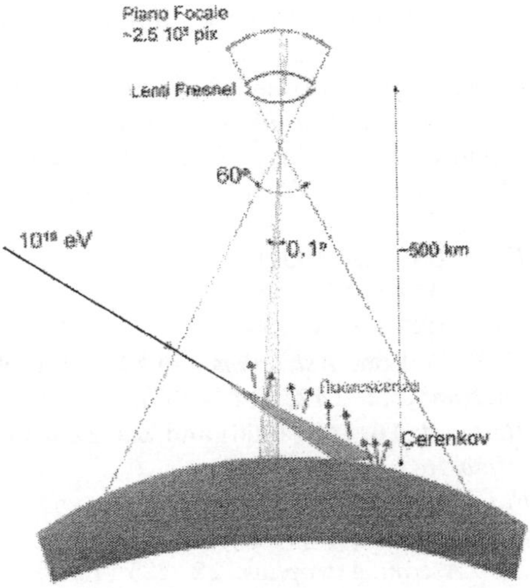

Figure 11. Schematic view of Euso experiment.

There are a number of reasons why it is important to detect UHE neutrinos: a) they are the only messengers from optically thick sources, opaque to other particles and b) they will provide a signature for sources where high energy π meson are produced, since the charged pion decay-chain ends into 3 neutrinos (ν *and* $\overline{\nu}$s) and one $e^{\pm}$. If nothing else, we should see neutrinos from proton interactions on CMB. It should also be noted that, due to vacuum oscillation, if neutrinos are produced in distant objects, they should be almost equally distributed between the three flavors when they reach earth. Auger[40] and the large FD of TA[41] both claim to be, under favorable neutrino flux conditions, sensitive to neutrino-initiated EAS, however, a systematic study of these events with an adequate statistics, will probably have to wait for EUSO. Unfortunately at present the space mission is uncertain and might be delayed until the end of the next decade.

References

1. V.Hess, *Phys.Zs.* **13**, 1084 (1912).
2. C.D.Anderson, *Phys.Rev.* **43**, 491 (1932).
3. B.Rossi, *Phys.Rev.* **45**, 212 (1934).
4. P.Auger, *Rev.Mod.Phys.* **11**, 288 (1939).
5. For a complete review of the subject and list of references see M.Nagano and A.A. Watson*Rev.Mod.Phys* **72**, 689 (2000).

6. V.S.Berezinsky et al.*Astrophysics of Cosmic Rays, North Holland,Amsterdam* (1990).
7. E.Fermi, *Phys.Rev.* **75**, 1169 (1949) and *N.C. Suppl.* **6**, 317 (1949)
8. R.D.Blanford and J.P.Ostriker *Astrophys.J.Lett.***221**, L29 (1978)
9. W.H.Ip and W.I.Axford *Proceedings of the International Simposium on AstrophysicalAspects of the MostEnergetic C.R.* edited By M.Nagano and F.Takahara (W.S. Singapore) 273 (1991)
10. D.Harari et al., *JHEP* **08**, 022 (1999).
11. E.Roulet et al., *JHEP* **0212**, 032 (2002).
12. J.Candia et al., *JHEP* **0212**, 033 (2002).
13. A.A.Penzias and R.W.Wilson, *Astrophys.J.* **142**, 419 (1965).
14. R.H.Dicke et al., *Astrophys.J.* **142**, 414 (1965).
15. K.Greisen, *Phys.Rev.Lett.* **16**, 748 (1966) and Z.T.Zatsepin and V.A.Kuz'min *Exsp.Teor.Phys.Pis'ma Red.* **4**, 144 (1966)
16. J.Linsley et al., *Phys.Rev.Lett.* **10**, 146 (1963).
17. L.Drury, *Contemp.Phys.* **35**, 232 (1994).
18. A.M.Hillas, *Ann.Rev.Astron.Astrophys.* **22**, 425 (1984).
19. E.Waxman, *Phys.Rev.Lett.* **75**, 386 (1995).
20. A.V.Olinto et al. *Proceedings of the 26tth ICRC, Salt Lake City* vol.**4**, 361 (1999)
21. S.Yoshida et al., *Phys.Rev.Lett.* **81**, 5505 (1998).
22. A.N.Bunner,Ph.D.thesis (Cornell University) 1964
23. A.M.Hillas, *Acta Phys.Scient.Hungar.* **29**, Suppl.3, 355 (1970).
24. M.Nagano et al., *Astropart.Phys.* **22**, 235 (2004).
25. J.Linsley (Volcano Range), *Proceedings of the 8tth ICRC, Jaipur* (Tata Inst.for Fundam.Research) vol.**4**, 77 (1963)
26. LaurenceM.A. et al.(Haverah Park), *Phys.Rev.Lett* **63**, 1121 (1989).
27. M.Takeda et al.(AGASA), *Astropart.Phys.* **19**, 447 (2003).
28. V.P. Knurenko et al.(Yakutsk), *ArXiv:astro-ph.*/0411484 v1 (17 Nov 2004).
29. D.Bird et al.(Fly's Eye), *Astropart.J.* **511**, 739 (1999).
30. R.U.Abbasi et al.(HiRes), *Phys.Rev.Lett.* **92**, 151101-1 (2004).
31. L.J.Kevley et al.(SUGAR), *Astropart.Phys.* **5**, 69 (1996).
32. A.A. Watson, *Proceedings of the conference on 'Thinking and Observing the Universe, Sorrento-Italy,* 22 (Sept.2003).
33. E.Bergman et al., *Proceedings of the 28th ICRC, Tsukuba* **1**, 397 (2003).
34. M.Teshima et al., *Proceedings of the 28th ICRC, Tsukuba* **1**, 437 (2003).
35. C.B.Finley and S.Westerhoff, *ArXiv:astro-ph.*/0303484
36. D.Heck (private communication]
37. The Auger Collaboration, *Nucl.Instr.and Meth.A* **523**, 50 (2004).
38. M.Sasaky, *Proceedings of the 285th ICRC, Durban* **5**, 369 (1997).
39. Euso Proposal
40. K.S. Capelle et al. *Astropart.Phys.* **8**, 321 (1998).
41. M.Sasaki et al., *ArXiv:astro-ph.*/0204167, 15 Jul 2002

PHOTONS AND ANTIMATTER IN SPACE

G. BARBIELLINI AND F. LONGO
Department of Physics, University of Trieste
and INFN, Sezione di Trieste,
via Valerio 2, I-34100, Trieste, Italy

The existence of antimatter has been predicted by Dirac in 1928. Only 4 years later Anderson confirmed experimentally the anti-electron (positron) production. Antimatter is present in our Galaxy in a very little percentage. Indirect evidence collected by high energy photon experiments disfavours large amount of antimatter in the Universe. The present interest on space experiments for antimatter is focused on understanding several aspects of cosmic-ray physics and possible signals of dark matter annihilation on ordinary matter and antimatter. High energy gamma-ray astrophysics is essential to understand galactic and extragalactic sources emitting through non-thermal mechanisms.

1. Introduction

Dirac effort to reconcile Quantum Mechanics with Special Relativity was awarded with the discovery of negative energy electron states that Dirac interpreted as a Fermi gas with all states occupied until $-\infty$. Dirac proposal on the existence of the Fermi sea suggested that the energy transfer by photon to this sea could bring a negative energy electron state to positive energy leaving a hole in the sea. The hole behaves like a positively charged brother of the electron.

Later this formulation of the antimatter has been reviewed but the strong belief of Dirac in the theory stimulated experimental researches. The first evidence of the anti-electron existence was published by Anderson in 1932. The event was an electron-positron production by an energetic photon on the Coulomb field of a lead nucleus (see fig. 1). Anderson was awarded with the Nobel Prize in Physics in 1936 for the discovery of the positron[1].

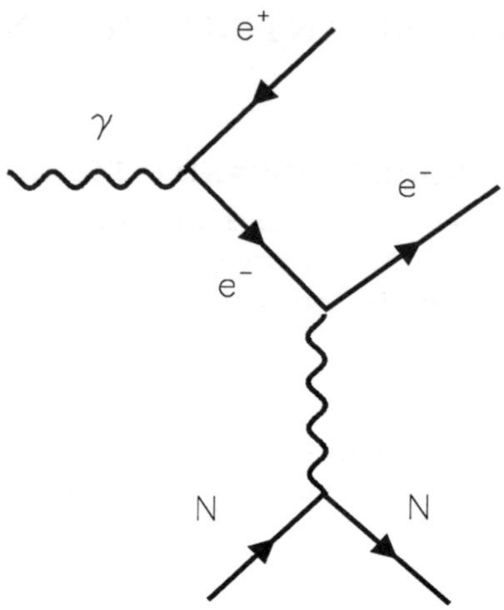

Figure 1. Feynman diagram for the pair production process in the electric field of the target nucleus.

2. Antimatter in Space

The center of our Galaxy is a strong source of positrons. As discovered by the OSSE experiment on the Compton Gamma-Ray Observatory (CGRO)[2] (see figure 2) and recently confirmed by SPI on board INTEGRAL,[3] a strong photon line with central energy of 511 keV is emitted from the Galactic Center. This line is the evidence of positron annihilation with galactic electrons according to a time inverted Feyman diagram with respect to that reported in figure 1. The annihilation is mainly through two body decay. The production rate of positrons, if annihilation and production are in equilibrium is $1.3 \cdot 10^{43}$ e^{+} s^{-1}. The real nature of this high production is still to be understood.

Hypernovae or Type Ic Supernovae are good candidates for positron production in the MeV energy range[4]. A recent paper suggests Gamma-ray bursts explosions as possible origin for galactic positrons[5].

If high energy positrons or antiprotons are experimentally investigated, the field of research moves from nuclear stellar explosion to high energy cosmic ray protons interacting with Galactic interstellar matter (ISM), mainly Hydrogen. High energy protons and heavier ions fill our Galaxy forming

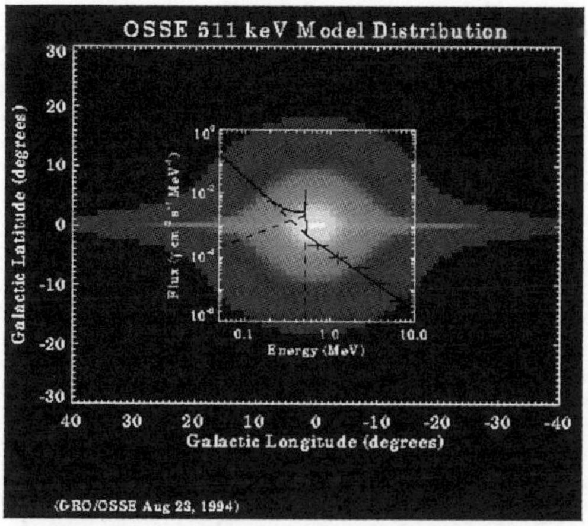

Figure 2. This map represent the intensity of gamma-ray emission from positron-electron annihilation in the plane of our Galaxy near the Galactic center. The emission is at 511 keV, which is the rest-mass energy of the electron and positron. The map is of a model that fits the OSSE 511 keV observations. OSSE has discovered that the radiation is mostly contained in a region of about 10 degrees diameter centered on the center of the Galaxy. The line plot superimposed on the map represents an OSSE observation of the 511 keV emission line. Picture taken from the web site http://cossc.gsfc.nasa.gov/

the flux of cosmic rays. Fig. 3 shows the energy spectrum of the cosmic ray flux at the top of the atmosphere[6].

The main proton component produces stable antiparticles through the reactions between cosmic ray and ISM.

$$pp \to pp + p\overline{p} + n(\pi) \tag{1}$$

$$pp \to pp + n(\pi) \tag{2}$$

$$\pi^+ \to \mu^+ \nu_\mu$$

$$\mu^+ \to e^+ \nu_e \overline{\nu_\mu}$$

Reaction 1 has a production threshold $E_{pp} \sim 7$ GeV, while reaction 2 has a lower threshold. Precision measurements on the expected galactic antimatter production is important to set the limits on other exotic sources like extragalactic antimatter or antimatter from annihilation of non baryonic matter. Among the particle candidates for the non-visible matter in the Universe (Dark Matter) there are supersymmetric partners of known

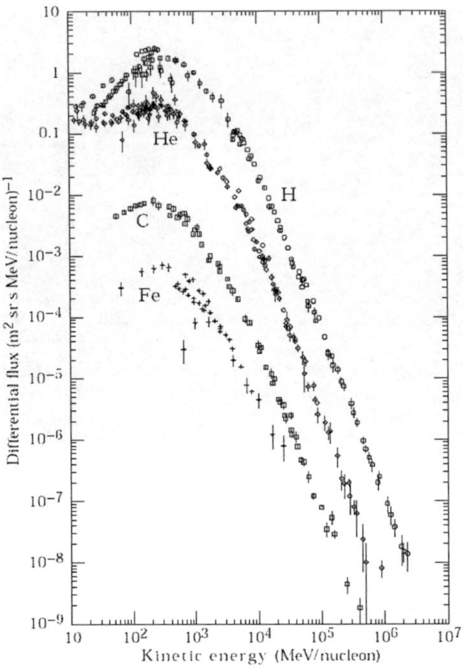

Figure 3. Cosmic ray energy spectrum at the top of the atmosphere.

particles. Their annihilation could produce an excess on the antimatter content in our Galaxy[7].

The measurements of the antimatter content in the proximity of the Earth has been until now mainly realized through sophisticated techniques similar to those in use in the Accelerator Laboratories, put on board of stratospheric balloons (see fig. 4).

In the measurements of the cosmic antimatter flux is crucial to know the contamination from the local source of antimatter, for example due to interactions of cosmic rays with the atmosphere. Recent results are obtained by the BESS collaboration[8].

Results collected by different experiments[9] on antiproton flux are shown in fig. 5. From the data collected until now it is clear that to set more stringent limits higher statistical level is required.

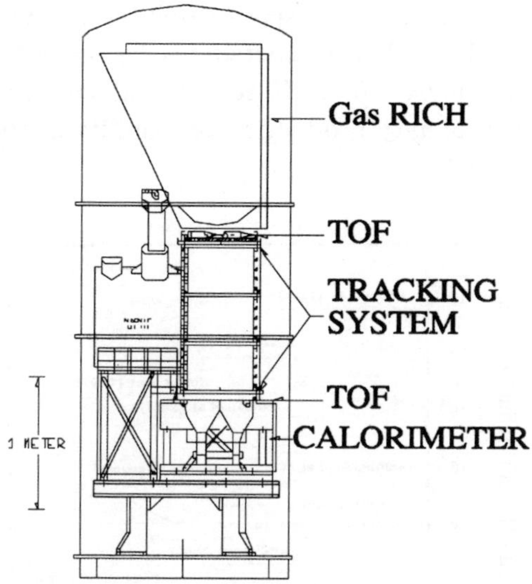

Figure 4. The Caprice apparatus in the 1998 configuration.

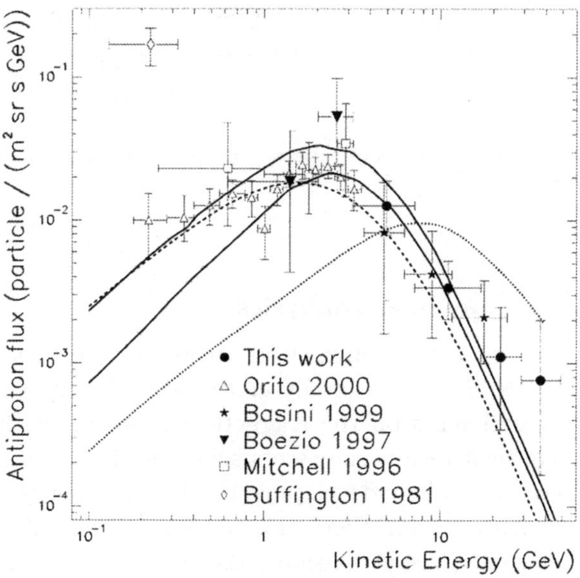

Figure 5. Experimental antiproton spectrum. Taken from Boezio et al. (2001)

Two major missions on board space satellites will fullfill this purpose, PAMELA[10] and AMS.[11] The predicted results for PAMELA and AMS missions are shown in fig. 6[12]. It reports the sensitivity for detecting antihelium, which secondary production probability is predicted[13] to be of the order of less than 10^{-13}.

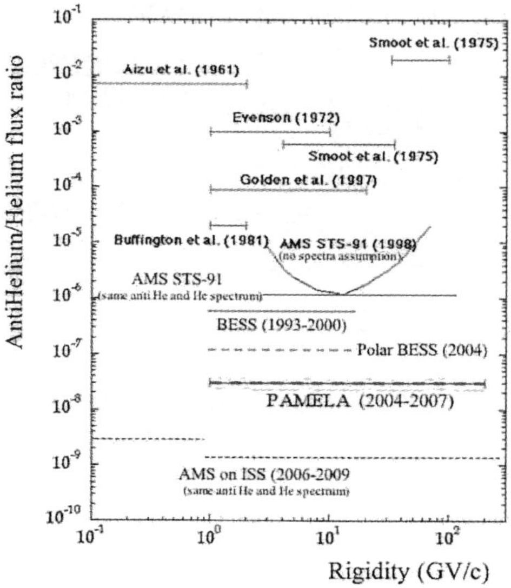

Figure 6. Actual measurements and expected sensitivity for anti-helium/helium ratio detection.

3. High energy gamma astrophysics

The study of the Universe through electromagnetic radiation in the energy range from MeV to TeV is generally called High Energy Gamma Astrophysics. The experiments on the energetic Universe and the theoretical interpretations of their results is a recent scientific effort and it is partly related to the space technology development. The high energy emission from different sources shows a non-thermal behaviour. The study of these phenomena and their energy spectra shows that the origin of the high energy photons is probably due to shock particle acceleration processes. They are associated with the acceleration and annihilation of relativistic charged particles at energies beyond those achieved in manmade accelerators. Gamma-ray astronomy is the astronomy of the "non-thermal universe" where energy is released on a massive scale during the explosive deaths of stars (supernovae or hypernovae), the particle winds and shocks driven by neutron stars spinning on their axes thousands of times per second, or in the superluminal jets of active galaxies powered by supermassive black holes. High energy gamma rays thus offer a unique insight into the most violent regions of our universe[14].

Our own Galaxy is the main gamma-ray source in the 100 MeV band due to interaction of cosmic-rays with the ISM. Figure 7 shows the sky seen in gamma-rays by the EGRET experiment[15].

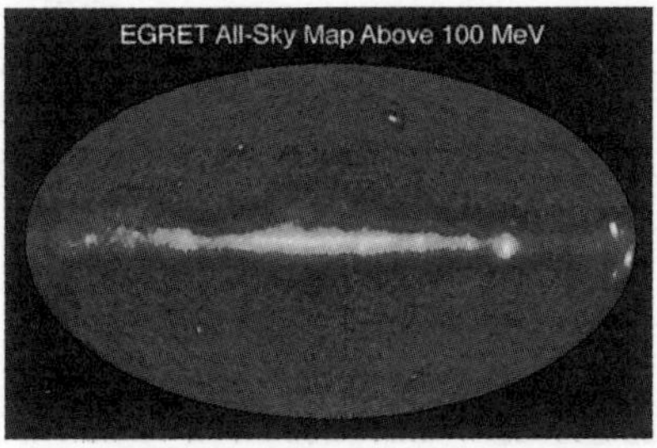

Figure 7. High Energy Gamma-ray Sky seen by EGRET

This area of research opens a large field of investigations. A classical list of astrophysically interesting subjects is given in table 1.

Photons in this energy band radiated by Galactic and Cosmological sources are absorbed by the atmosphere so their flux measurement is either measured by space based experiments or by detectors that taking advantage from the high energy photon interaction with the atmosphere itself, measure the electromagnetic shower development.

The present day space and ground based technology divide the High Energy Gamma Astrophysics in two broad energy ranges: 50 MeV - 50 GeV for space detectors (see fig. 8) and 0.1 to 50 TeV for ground based detectors (see fig.9).

Two experiments are foreseen to fly in the following years, AGILE[16] and GLAST[17]. They could take advantage of the continuous improvement of the high energy gamma-ray astronomy on ground done by experiments such as MAGIC[18] and HESS[19].

Source	Physical Interest
Active Galactic Nuclei (AGN)	AGN Duty Cycle Photon Emission Mechanism Jet Origin Radio Plasmoid emission
Gamma Pulsar (PSR)	Outer Gap or Polar Cap model Accurate Timing Population Studies ms PSR searches
Unidentified Gamma Ray Sources	Identification Cosmic Rays origin Gamma-ray plerions Supernova Remnants gamma detection New source classes
Gamma-Ray Bursts	Acceleration mechanism Gamma-ray afterglows High Energy prompt component
Astroparticle Physics	Dark Matter Lorentz Invariance test

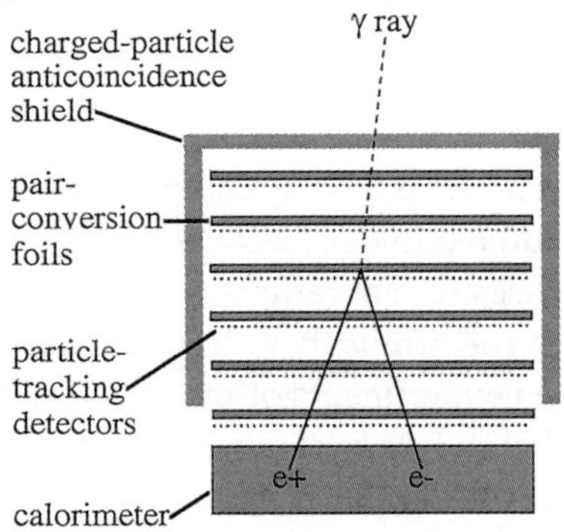

Figure 8. Pair-conversion telescope main elements. Picture taken from the web site http://cossc.gsfc.nasa.gov/

4. Conclusion

In the following years a new era for the study of the high energy universe study is foreseen. New experiments are ready to be launched. Several pending astrophysical misteries will hopefully be solved such as the nature

Figure 9. Whipple Imaging Air Cherenkov Telescope. Picture taken from the web site http://cossc.gsfc.nasa.gov/

of GRB, the origin of Cosmic Rays and a different insight in the Dark Matter nature. To the experiments the last word.

References

1. C. D. Anderson, Physical Review **43**, 491 (1933)
2. R. L. Kinzer *et al.*, The Astrophysical Journal, **559**, 282 (2001)
3. G. Weidenspointner, *et al.*, in Proc. of the 5th INTEGRAL Workshop, The INTEGRAL Universe, arXiv/astro-ph 0406178
4. M. Cassé *et al.*, The Astrophysical Journal, **602**, L17 (2004)
5. E. Parizot *et al.*, arXiv/astro-ph 0411656, Astronomy & Astrophysics accepted
6. J. A. Simpson, Ann. Rev. Nucl. and Part. Sci. **33**, 323 (1983)
7. A. Morselli and P. Picozza, in Proceedings of the Fourth International Workshop on the Identification of Dark Matter, 501 (2003)
8. K. Yamato *et al.*, Proc. 28th Intl. Cosmic Ray Conf., 1785 (2003)
9. M. Boezio *et al.*, Astrophysical Journal **561** 787 (2001)
10. http://wizard.roma2.infn.it/pamela/
11. http://ams.cern.ch/
12. P. Picozza and A. Morselli, J.Phys. G**29**, 903 (2003)
13. P.Chardonnet *et al.*, Technical Report, ENS-LAPP-A-643-97
14. http://www1.ast.leeds.ac.uk/research/vhe-cr.html
15. S. D. Hunter *et al.*, The Astrophysical Journal, **481**, 20 (1997)
16. http://agile.mi.iasf.cnr.it
17. http://glast.gsfc.nasa.gov
18. http://wwwmagic.mppmu.mpg.de/
19. http://www.mpi-hd.mpg.de/hfm/HESS/

ASTROPARTICLE PHYSICS FROM SPACE

STEFANO CECCHINI
IASF/CNR, Sez. di Bologna, I-40129 Bologna, Italy
INFN, Sez. di Bologna, I-40127 Bologna, Italy
E-mail: cecchini@bo.infn.it

The purpose of this lecture is to present an overview of future space experiments at the intersection of astronomy and particle physics. We concentrate our attention on observations that cover three themes: dark energy, dark matter and birth of the Universe.

1. Introduction

By the end of the 20th century it became clear that the standard model of particle physics plays a prominent role in cosmology. Particle physics investigate the smallest scales of the Nature and cosmology probes the largest scales of our Universe. But the two fields are in fact unified because it is possible to understand the origin and evolution of large-scale structures of the Universe only by understanding its "initial conditions" [1]. The idea of the Big Bang has been built on the predicted effects of certain particle fields and potentials *ab initio* of the cosmic expansion [2].

The study of galaxies, galaxy clusters and anisotropies in the cosmic microwave background radiation (CMBR) may reveal insights into events that occurred in the early Universe. By confronting observations with predictions we can get new insight on particle physics at energies not attainable on the Earth. The great success of the observation of the primordial adiabatic perturbations, produced during inflation, seems to provide consistent scenarios/models for the origin of our Universe [3].

As it usually happens, however, these findings have posed more questions than they have solved. The 21st century opens with the necessity of new observations at the intersection of physics and astronomy that require large investments and program plans. To make further progress on understanding the underlying physics require multiple, complementary and independent approaches.

Among the many themes that represent the frontier for the discovery of the physics of the Universe we will present in the followings the ones that have in our opinion a higher priority: Dark matter, Dark Energy and Birth of the Universe.

The choice and presentation of the planned or undergoing missions in connection of one of the above themes is somewhat arbitrary as the observations made by a single experiment will give information that will help in putting constrains in all these research fields.

2. Why observations from space

To perform experiments in space represents a great challenge and has large costs. These latter come mainly from manufacturing cost as high redundancy is required to operate instruments in low gravity and low pressure ambient, and high radiation fields. To that one has to add launch and operational costs. The Hubble Space Telescope (HST) - 2.4m mirror diameter, at 550km Earth orbit with 28.6° inclination - is still the only telescope in space, small compared to the dimensions of actual existing ground telescopes, because of these intrinsic difficulties.

Moreover the preparation and development of the instrument is usually long, so the technology actually used in space risks to be old.

There are, however, advantages.

The absence of atmosphere allows observations not possible from Earth because at wavelengths largely absorbed [4]. Space optical telescopes do not suffer for the terrestrial light background that nowadays represents a serious limit for astronomical observations. For that reason astronomers tend to place observatories where artificial lights are absent. These sites, however, are few and almost saturated. Others limitations for ground based observatories come from the weight of the large mirrors needed to increase the collecting area for looking at fainter objects and from the errors introduced by the mechanics used to move them. The new "active optics" techniques, however, have greatly improved the situation as well as regarding the influence of the air turbulence.

The International Space Station (ISS) may provide unprecedented opportunities for hosting and operating large observatories and for launching special mission in deep space. It has been already envisaged to use the ISS for assembling in low gravity large structures and detectors that will be impossible to put in space by a single rocket launch scheme (e.g. XEUS [5], SEU [6]).

Other advantages for many types of instruments are the possibility to operate the detectors at low temperatures and consequently to work with low noise; the possibility to survey the whole sky with a unique instrument, to have long base interferometer observations, to perform in situ measurements, to use multi-detector satellite that can observe simultaneously at different wavelengths.

3. What is Dark Matter?

Astronomers have shown that the objects in the Universe from galaxies a million time smaller than our to the largest clusters of galaxies are held together by a form of matter that neither emits nor absorbs light – it is "dark" - and that interacts very weakly with ordinary matter. Its nature is a complete mystery.

This matter probably consists of one or more as-yet-undiscovered elementary particles, and whose aggregation would produce the gravitational attraction leading to the formation of galaxies and large-scale structures in the Universe. These particles may also be streaming through our Earth-bound laboratories.

The fact that most of the mass in the Universe is non luminous became evident about 65 years ago when F. Zwicky [7] noticed that the speed of galaxies in large clusters is too great to keep them gravitationally bound together unless they weight over 100 times more than one would estimate on the basis of the number of stars in the cluster, $\Omega_m \cong 0.1\text{-}0.3$ (in units of the critical density [8]).

Also it was known for a long time that if there was matter beyond the luminous one, the time required for structures we observe to form would be very short, thereby requiring fluctuations in the CMBR considerably larger than those observed.

The most robust observational evidence involves galactic dynamics: there is not enough luminous matter observed in spiral galaxies to account for their rotation curve, a fact that imply the existence of a diffuse halo of dark matter [9,10]. Summing the contributions from all galaxies one can infer that dark matter associated with galaxies contributes $\Omega_{halo} \geq 0.1$. On the other hand Big-Bang nucleosynthesis suggests a baryon density Ω_b£ 0.1 [11]. Thus the bulk of the halo must be non-baryonic. The CMBR measurements [12] have given independent and precise confirmation.

So what is it?

Neutrinos of mass ***O***(10eV) could provide the right dark-matter density, but N-body simulations of structure formation in a neutrino-dominated Universe do not succeed in reproducing observations [12].

Numerical simulations indicate that long-lived, cold, collisionless particles are favoured candidates for dark matter (Cold Collisionless Dark Matter-CCDM scenario) [13]. Among them: WIMPs (Weakly Interactive Massive Particles) and Axions [14]. The CDM paradigm, however, is not without problems; some inconsistencies seem to exist at sub-galactic scales [15,16].

Many other possible scenarios have been proposed and the way to distinguish among them is discussed in [17].

There are 4 different strategies for studying CDM:

- Gamma rays or neutrino resulting from annihilation of DM particle-antimatter counterpart (e^+ and anti-p)
- Astronomical observations through dynamical studies of the motions of stars, galaxies and X-ray emitting hot gas in clusters of galaxies (e.g. gravitational lensing)
- Direct detection of DM particles through highly specialized instruments designed to detect directly the extremely weak signal of rare DM interaction with massive detectors
- Create DM particles by colliding ordinary particles in HE accelerators

The first two can be exploited from space (see also [18]).

Instruments and satellites able to address the first strategy are already under construction - for example AMS, PAMELA and GLAST - and are discussed in an extensive way in [19,20].

Regarding the second strategy two instruments/missions that are planned for the near future: New Generation Space Telescope (NGST) and Constellation X will be presented here.

The NGST [21] is a telescope larger and more powerful than the HST, and it is scheduled for launch late in this decade. Its main characteristics are reported in Table 1. It will operate from a special orbit around the L2 point of the Sun-Earth system. Being so far from the Sun it will be cooled so that its instruments will have an extraordinary sensitivity in the infrared region making possible to observe object 400 times fainter than ground based telescope. By this way it will

Table 1. The main requirements for the design of NGST

Characteristic parameters	**Attainable values**
Schedule	Launch readiness: 06/2010
Wavelength range	0.6 to > 10_m
Sensitivity (10^5 s integration)	Roughly from 1.3 nJy at 0.7_m to 800nJy at 20_m
Spatial resolution & stability	0.15" at 1_m < 2% RMS variation about the mean over 24-hour period
Telescope separated FOVs	0.6-5_m $\geq$ 21 sq arcmin ^{2} >5_m $\geq$ 5 arcmin ^{2}
Sky coverage	100% in 1 yr 48.9 % at any give time 100% visible for at least 69 contiguous days
Mission life	Minimum 5 years Consumables sized for 10 years
Orbit	Orbit around L2 point
Overall observing efficiency	> 70%
Total mass	5400 kg payload mass maximum

be possible to extend our observations to the early Universe at an age between 10^6 y and few 10^9 y; study the dynamics of galaxies and clusters of galaxies.

Constellation-X [18] is a high spectra resolution, large aperture X-ray mission currently under design. It should be able to study black holes and others features of compact objects. Furthermore it should allow examining the dynamics within the cluster of galaxies and where they have formed and evolved.

Figure 1. Artist's impression of the Constellation-X Observatory

Table 2. Baseline characteristics for Constellation-X mission

Minimum effective area	1,000 cm^2 from 0.25 keV to 10 keV 15,000 cm^2 at 1.25 keV 6,000 cm^2 at 6.0 keV 1,500 cm^2 from 10 keV to 40 keV
Minimum telescope angular resolution	15 arcsec HPD from 0.25 to 10 keV 1 arcmin HPD from 10 keV to 40 keV
Spectral resolving power (E/ΔE)	>~ 300 from 0.25 keV to 6.0 keV 1,500 from 6.0 keV to 10 keV > 10 from 10 keV to 40 keV
Band Pass	0.25 to 40 keV
Diameter Field of View	SoftXT > 2.5 arc min > 30 Å~ 30 array (5" pixels) HardXT > 8 arc min
Mission Life	> 4 years (full capability)
Redundancy/Reliability	No one on-orbit failure to result in loss of more than 25% of the mission science

As what regards the issue of the dark matter the mission can determine the shape of the mass distribution in all three dimensions through measurements of the X-ray emitting gas of a galaxy. The hot gas is a tracer for dark matter because gas, like usual matter, is attracted by the dark matter gravitational pull. The existence of a stronger force leads to faster moving and thus hotter gas.

From the observations of the X-ray emission, it will be able to measure the temperature and density distribution of the gas as well as its rotational velocity. Due to its great spectral resolution (5 eV) the relative line-strength measurements together with their displacements will reveal the existence and amount of dark matter halos.

In Figure 1 the possible configuration of the 4 telescopes that will operate in close proximity to each other and orbiting all together around the L2 point. The baseline mission characteristics are reported in Table 2.

4. What is the nature of Dark Energy?

Dedicated space-based experiment to precisely determine the nature of Dark energy and its evolution over the history of the universe is a critical centrepiece to future astroparticle programs.

Three different types of recent observations on high redshift supernovae [23,24], Cosmic Microwave Background Radiation (CMBR) [12] and on large-scale galactic distribution [25] indicate that in our Universe it exists *something else*, besides radiation and matter and it causes the acceleration of the expansion of the Universe (Figure 2). The nature of this component, named Dark Energy (DE) is still unknown. One possible explanation is that DE is the modern version of the Einstein positive cosmological constant (L), interpreted as a vacuum energy density [26]. Alternatively DE can be originated by self-interacting scalar field [27] or by exotic physics of extra-dimensions [28].

Strategic measurements are:

- Study of the evolution of galaxy clusters, through X-ray surveys at $z = 2\text{-}3$, observations of Sunyaev-Zeldovich effect and gravitational lensing. For the purpose a new class of optical/IR/X-ray telescope with high spatial and spectral resolution are necessary. Already we have discussed two missions that have these characteristics.
- Measurements of CMBR fluctuation (spectrum). A new class of CMBR anisotropy and polarization observatories both from ground and space are now planned and will be discussed in the next paragraph.
- Use of observations of SNIa at $0.5 < z < 1.8$ and count of galaxy clusters. For the first a new class of wide field telescopes to discover and follow several 1000 SNe (this will also be possible with NGST).

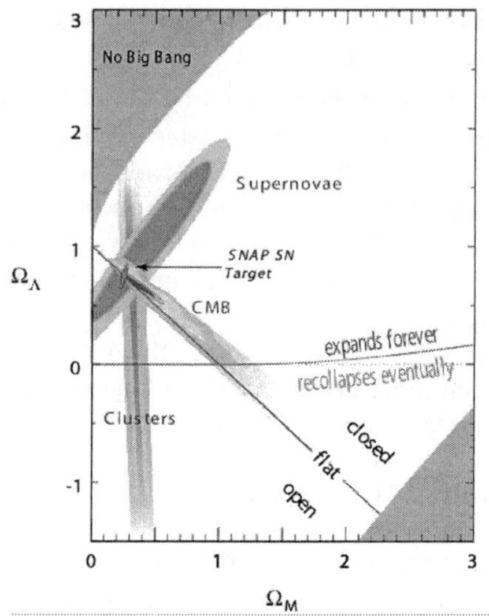

Figure 2. Cosmological parameters Ω_M and Ω_Λ estimated through the observation of galaxy clusters, SNIa and CMBR. These results rule out simple flat cosmology and support DE. Also shown is the confidence region for the parameter estimate from the proposed SNAP satellite. (from [30])

SNAP [29] is a proposed space mission specifically designed to precisely and accurately measuring the distance-redshift relation of Type Ia supernovae and carrying out a deep, wide area weak gravitational lensing survey.

SNIa can be considered as "standard candles" [30,3] so by observing their dimming with distance we can measure the cosmic expansion rate of the Universe over the last 10 billion years to 1% accuracy.

Space is the ideal site for such observations that require high resolution and stability of the imaging, as well as infrared measurements to obtain accurate photometric redshifts and greater redshift depth. Precise colour measurements are impossible from the ground beyond z = 0.7 independent of exposure time and mirror aperture.

SNAP will be also able to map the geometric and dynamic effects of dark energy through the growth history of large-scale structure using weak gravitational lensing.

Another proposal is DUO (Dark Universe Observatory) [31] a 2-year mission that will perform two X-surveys: one will cover the same region of the Sloan Digital Survey (roughly 6000 deg^2) detecting about 8000 clusters of galaxies, complete to a redshift of 0.7; the other will go deeper, detecting around 1800 cluster with redshift up to 1.5 in a region of 150 deg^2 around the South galactic pole. By combining the X-ray observation with the CMBR data it will be possible to put stronger constrain in the DE content of the Universe.

Table 3 - Characteristics of SNAP at the proposal stage

Telescope aperture	2 meter
Optics	Diffraction limited, f/10 0.1" pixel scale
Field of view	0.7 sq. degree instrumented equal CCD, NIR coverage
Wavelength coverage	0.35 – 1.7 μm
Orbit	Elliptical high Earth orbit or L2
Pointing stability	Within 0.02 arcsec Focal plane feedback
Imaging camera	0.5 10^9 pixel imager 9 fixed filter CCD detectors: • high resistivity p-channel • high QE from 0.35-1.0 μm • low noise HgCdTe infrared devices: • high QE from 0.9 – 1.7 μm • low noise
Integrated field spectrograph	0.35 – 1.7 μm low resolution R 100

5. How did the Universe begin?

One of the greatest successes of cosmology over the past two decades has been the development and initial testing of the inflationary paradigm, which provides an explanation for the large size and uniformity of the Universe as well as the origin of the lumpiness that led to galaxies and clusters of galaxies.

This has been mainly due to the assessment of the nature of the CMBR by the *Relict* [32] and COBE [33] satellites, and after them by a number of ground and balloon experiments till the recent WMAP [12] mission that have allowed to explore and put significant constraints to the multidimensional parameter space of available cosmological models.

The spectrum of the CMBR is well characterized as an ideal blackbody, or thermal source with T = (2.728±0.004) K [34]. Moreover this radiation is isotropic on both large and small angular scales to a higher degree (1 part in 10^5) than other source in the Universe.

The principal observable of the CMBR is the power spectrum of the temperature fluctuations across the sky. The shape of this power spectrum, the relative positions and amplitudes of the Doppler peaks, are very sensitive to initial conditions and to curvature, dark energy, baryons density and matter density [31,32]. In order to better constrain the information on the cosmological parameters however it is necessary to still improve the determination of the anisotropy angular power spectrum in the multipole range 4-3000 [31].

One space mission already ongoing is *Planck* that is planned for a 2007 launch by the European Space Agency [32]

Its instruments will image the anisotropies of the CMBR with high sensitivity over a broad wavelength band – 9 channels from 30 GHz to 857 GHz using radio receiver and bolometer arrays in the focus of the telescope with unprecedented spatial resolution – 10 arcmin. Simulated power-spectrum determinations WMAP and *Planck* missions show the great improvement achievable with the latter one (Figure 3a).

The main target for further analysis is "polarization" of the CMBR. Besides conveying independent information on the cosmological model and helping in removing residual degeneracy between various parameters (different set of parameters can fit the same observations) large angle polarization data allow to explore the thermal history of the Universe at low redshift or even to discover gravitational waves from inflation. This last objective implies that only full-sky experiments, based on space can provide such type of observations, ground based or balloon experiments being intrinsically unable to access such scales.

The major experimental issues are:

<u>Sensitivity</u>: The level of the signal is expected to be 2-3 orders of magnitude lower than the anisotropy signal posing a big challenge for the detection. Detectors are already photon noise limited so that one needs to operate a large number of detectors simultaneously or very long integration time.

<u>Foregrounds</u>: polarized radiations from other emission mechanisms (e.g. synchrotron) or point sources have to be removed; this implies the use of multi-frequency measurements.

<u>Systematic effects</u>: imperfect modelling of the detector, use of mirrors for making observations at small angular scales can introduce spurious contributions and leakage of signal from the two components (E, B) of the polarization field.

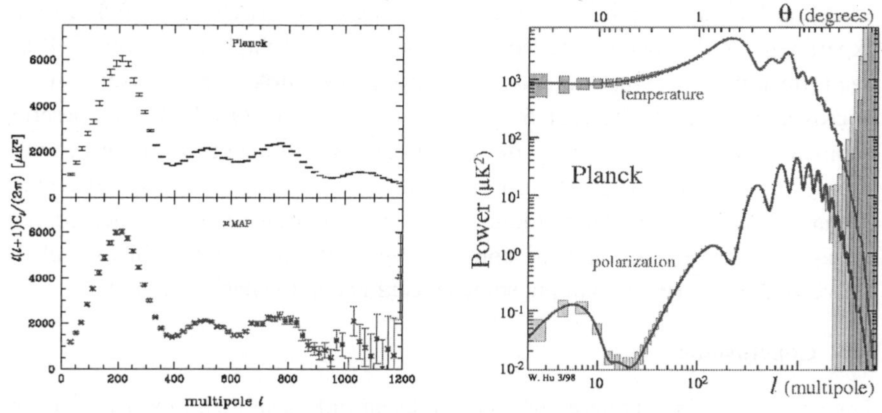

Figure 3. (a-left) Estimated error bars for power spectrum determination by WMAP and *Planck* when the spectra are binned with a width δl=20 [37]; (b-right) E-Polarization spectrum compared to temperature spectrum and sensitivity for Planck mission [35].

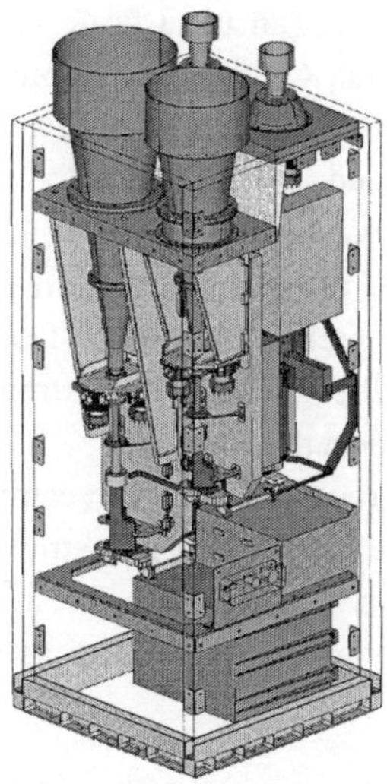

Figure 4. The drawing of the SPOrt payload. On-axis optics (corrugated feed horns) is used to minimize the spurious polarization.

Planck will be able to greatly improve our knowledge on the polarization of the CMBR if able to control all the above aspects (Figure 3b) being the instrument not optimized for such measurement.

Another approach has been developed by SPOrt, an ASI-funded experiment to be flown on the ISS Columbus External Payload Facility [38]. It has been designed to simultaneously measure the Stokes parameters of the microwave sky at 3 different frequencies (22, 33 and 90 GHz). By operating independently from anisotropy the instrument is optimized to minimize every source of systematics, for long-term stability and observing time efficiency (Figure 4). Having angular resolution of 7° its main objective is the detection of large-scale polarization and the constraining of the reionization history of the Universe the experiment will put some constraints on the nature of the DE.

6. Conclusions

The space experiments here reported represent only a limited number of all the astronomy space missions. Several others have been planned to explore fundamental physics from space (e.g. STEP the "equivalence principle" test and LISA, searching for gravitational waves); others are in development (e.g. GAlaxy Evolution Explorer (GALEX) for observing galaxies in UV and studying how galaxies evolve and change). Finally other missions are only "concepts" like *Generation-X* (an ultra large aperture X-ray telescope able to image 1000 times deeper than *Chandra* and obtain spectra of sources 100-1000 fainter than *Constellation-X*), MAXIM (Micro-Arcsecond X-ray Imaging Mission) even more advanced and CMBPol for the detection of the B-mode of the CMBR polarization, a post-Planck Mission.

For many missions the progressive pattern towards launch depends on the rate of advancement of the technology at the frontier. It is however difficult to predict this evolution and the scientific goals may meanwhile change and new ones can emerge and transform the scenario.

This kind of activity requires large investments in technology and, more than that, it requires at the same time investment in user groups, theory, and simulations and data archival activities. These other programs elements are necessary for success of every major facility as much as the development and design of new missions.

References

1. M.S. Turner and J.A. Tyson, *Rev. Mod. Phys.*, **71**, S145 (1999)
2. M. Trodden and S.M. Carroll, *astro-ph/0401547*
3. S. Dodelson, *Modern Cosmology* (Academic Press, 2003)
4. G. Giacconi, H. Gursky and L.P. van Spoeybroeck, *Ann. Rev. Astron. Astrophys.,* **6**, 375 (1968).

5. G.G.C. Palumbo, in *Proc. 6th School Non-Accelerator Particle Physics* (R.A. Carrigan, Jr *et al.* Eds), World Scientific (Singapore, 2002) 233; http://astro.estec.esa.nl/XEUS/
6. T.Y. Takahashi et al., *AIP Conf Proc.,* **458**, 315 (2000); http://www.subaru.naoj.org/
7. F. Zwicky, *Physica Acta*, **6**, 124 (1933).
8. K.A. Olive and J.A. Peacock, *Phys. Rev.,* **D66**, 010001-152 (2002).
9. K. Begeman *et al.*, MNRAS, **249**, 523 (1991).
10. M. Persic and P. Salucci, in *Proc. 4th School Non-Accelerator Particle Physics* (E. Bellotti *et al.* Eds), World Scientific (Singapore, 1996) 79.
11. S. Esposito et al., *Nucl. Phys.,* **B568**, 421 (2000).
12. D. N. Spergel et al., *Astrophys. J. Suppl.*, **148**, 161 (2003).
13. M. Kamionkowski and A. Kosowsky, *Ann. Rev. Nucl. Part. Sci.*, **49**, 77 (1999); *hep-ph/0210370.*
14. J. Ellis, *Phil. Trans. Roy. Astron. Soc. London* **A361**, 2607 (2003)
15. J. F. Navarro et al., *Astrophys. J.*, **490**, 493 (1997).
16. J. F. Navarro, *Phil. Trans. Roy. Astron. Soc. London* **A361**, 2515 (2003)
17. J.P. Ostriker and P. Steinhardt, *Science*, **300**, 1909 (2003).
18. J.L. Feng, *astro-ph/0405479.*
19. S. Cecchini and T. Chiarusi, in *Proc. 6th School Non-Accelerator Particle Physics* (R. A. Carrigan *et al.* Eds.), World Scientific (Singapore, 2002) 175.
20. G. Barbiellini and F. Longo, these proceedings.
21. http://www.esa.int/
22. http://constellation.gsfc.nasa.gov/
23. A. Riess et al., *Astron. J.,* **116**, 1009 (1998); *astro-ph/0402512*
24. S. Perlmutter et al., *Astrophys. J.,* **483,** 565 (1999)
25. G. Efstathiou et al., *MNRAS*, **330**, 29 (2002)
26. Ya. Zeldovich, *Sov. Phys. Usp.***11,** 381 (1968)
27. B. Ratra and P.J.E Peebles, *Phys. Rev.* **D37**, 3406 (1988)
28. G. Dvali and M. Turner, *astro-ph/0301510*
29. http://snap.lbl.gov/
30. S. Perlmutter, *Physics Today*, **56 (4),** 53 (2003)
31. http://duo.gsfc.nasa.gov
32. A. A. Klypin et al., *Soviet. Astron. Lett.* **13**,104 (1987); I.A. Strukov et al., *Soviet. Astron. Lett.* **13**, 65 (1987)
33. G. F. Smoot et al., *Astrophys. J. Lett.,* **396**, L1 (1992)
34. D. J. Fixsen et al., *Astrophys. J.,* **473**, 576 (1996)
35. W. Hu and S. Dodelson, *Ann. Rev. Astron. Astrophys.* **40**, 171 (2002); http://background.uchicago.edu/ whu
36. http://astr.estec.esa.nl/SA-general/Projects/Planck/
37. S. Church et al., *astro-ph/0111203*
38. S. Cortiglioni et al., *New Astron.* **9**, 297 (2003)

THEORY OF GRAVITATIONAL WAVES

J. C. MILLER

SISSA and INFN, Via Beirut 2-4, 34014 Trieste, Italy
and
Department of Physics (Astrophysics), University of Oxford,
Keble Road, Oxford OX1 3RH, England
E-mail: miller@sissa.it

These lecture notes introduce the basic ideas of gravitational waves and outline how they are treated within general relativity. Their interaction with matter and generation by astronomical objects are discussed.

1. Basic ideas

As an entry into our discussion of gravitational radiation, we start by asking the question: How does information about a change in a gravitational field propagate?

In Newtonian theory, the gravitational potential Φ is given by

$$\nabla^2 \Phi = 4\pi G \rho \,, \tag{1}$$

in the presence of matter (where ρ is the mass density), or

$$\nabla^2 \Phi = 0 \tag{2}$$

in vacuum. The acceleration of a test particle moving freely in the field produced is then given by

$$\frac{\mathrm{d}^2 x^i}{\mathrm{d}t^2} = -\frac{\partial \Phi}{\partial x^i} \,. \tag{3}$$

Suppose now that there is a change in the matter distribution giving rise to the field. According to the above equations, the effect of this as felt in the acceleration of the test particle, would occur *instantaneously* even if the test particle were very distant from the source, and this corresponds to an instantaneous transmission of information about the change in the source which is contrary to the ideas of relativity theory.

In general relativity (GR), all physical laws must be written in a *covariant* form, i.e. in terms of scalars, 4-vectors and 4-tensors. In line with this, the equation of motion needs to be written in a four-dimensional form (rather than a three-dimensional form as above), and the Laplacian needs to be replaced by a four-dimensional operator. If the modification were simply to replace ∇^2 by the d'Alembertian:

$$\Box = -\frac{1}{c^2}\frac{\partial^2}{\partial t^2} + \nabla^2 \,, \tag{4}$$

then it is easy to see that the speed of propagation of information would become *finite*, going to the speed of light, c, in vacuum. The concept of gravitational waves then arises as the finite-speed carriers of information about changes in the source of the gravitational field. In analogy with electromagnetism, these waves carry *energy* away from the source as well as *information* about the field changes.

2. What does GR tell us about gravitational waves?

General relativity is a geometrical theory describing gravity in terms of curvature of spacetime and, within this context, gravitational waves appear as *ripples* in spacetime.

The *source* of the gravitational waves might be either a strong-field object (such as a dynamically-changing black hole or neutron star) or a weak-field object (such as a normal stellar binary system). Clearly, strong-field objects will normally give rise to the larger-amplitude gravitational waves although this is not certain because the amplitude also depends on the degree of dynamical motion and asymmetry of the object concerned. Here, we will be focusing on a *weak-field* approach which will certainly be appropriate for regions far enough away from the source and can also be appropriate for the source itself in some cases. Treating these problems in strong field is much more complicated and many important features appear already in a weak-field treatment.

In weak-field situations, the wavelength of gravitational waves is usually much shorter than the length-scale associated with the curvature of the background space-time and so it is an excellent approximation to treat the waves as a perturbation about *flat space*. The general *metric line element* is

$$ds^2 = g_{\mu\nu}\, dx^\mu\, dx^\nu \tag{5}$$

where ds is the space-time interval, $\mathbf{x} = (x^0, x^1, x^2, x^3)$ is the space-time position vector and $g_{\mu\nu}$ is the metric tensor. (We will be using the conventions that Greek indices range from 0 to 3, referring to both time and space dimensions, and Latin indices range from 1 to 3, referring to spatial dimensions only.) In the weak-field limit we write

$$g_{\mu\nu} = \eta_{\mu\nu} + h_{\mu\nu} \tag{6}$$

where $\eta_{\mu\nu}$ is the metric tensor for flat space and $h_{\mu\nu}$ is a small perturbation.

The gravitational field equation in general relativity is the *Einstein field equation*:

$$R_{\mu\nu} - \tfrac{1}{2} g_{\mu\nu} R = \frac{8\pi G}{c^4} T_{\mu\nu} \tag{7}$$

where: $R_{\mu\nu}$ is the *Ricci tensor* (contraction of the Riemann curvature tensor),
R is the *Ricci scalar* (contraction of $R_{\mu\nu}$),
$T_{\mu\nu}$ is the *Energy-momentum tensor* (telling us about the source).

We will now write this out and *linearize* it, retaining only first order terms in h. (We use the convention of taking $c = G = 1$ which allows masses and times to be expressed in terms of lengths). First, we need to calculate Ricci tensor:

$$R_{\mu\nu} \equiv R^{\alpha}_{\mu\alpha\nu} \tag{8}$$

$$= \Gamma^{\alpha}_{\mu\nu,\alpha} - \Gamma^{\alpha}_{\mu\alpha,\nu} + \Gamma^{\alpha}_{\beta\alpha}\Gamma^{\beta}_{\mu\nu} - \Gamma^{\alpha}_{\beta\nu}\Gamma^{\beta}_{\mu\alpha} \tag{9}$$

where a *comma* indicates a standard partial derivative (for example, $\Gamma^{\alpha}_{\mu\nu,\alpha} = \partial\Gamma^{\alpha}_{\mu\nu}/\partial x^{\alpha}$), a *repeated index* implies summation over that index and the Γs (the *Christoffel symbols*) are given by

$$\Gamma^{\mu}_{\alpha\beta} = \tfrac{1}{2}\eta^{\mu\nu}(h_{\alpha\nu,\beta} + h_{\beta\nu,\alpha} - h_{\alpha\beta,\nu}) \tag{10}$$

$$\rightarrow \tfrac{1}{2}(h_{\alpha}{}^{\mu}{}_{,\beta} + h_{\beta}{}^{\mu}{}_{,\alpha} - h_{\alpha\beta}{}^{,\mu}) \tag{11}$$

(note that indices are here raised and lowered with $\eta^{\mu\nu}$). Then

$$R_{\mu\nu} \rightarrow \Gamma^{\alpha}_{\mu\nu,\alpha} - \Gamma^{\alpha}_{\mu\alpha,\nu} \tag{12}$$

$$= \tfrac{1}{2}(h_{\mu}{}^{\alpha}{}_{,\nu\alpha} + h_{\nu}{}^{\alpha}{}_{,\mu\alpha} - h_{\mu\nu,\alpha}{}^{\alpha} - h_{,\mu\nu}) \tag{13}$$

where $h \equiv h_{\mu}{}^{\mu} = \eta^{\mu\nu} h_{\mu\nu}$. Contracting $R_{\mu\nu}$ gives the Ricci scalar:

$$R = \eta^{\mu\nu} R_{\mu\nu}\,. \tag{14}$$

The Einstein equation

$$R_{\mu\nu} - \tfrac{1}{2} g_{\mu\nu} R = 8\pi T_{\mu\nu} \tag{15}$$

then becomes

$$h_{\mu\alpha,\nu}{}^{\alpha} + h_{\nu\alpha,\mu}{}^{\alpha} - h_{\mu\nu,\alpha}{}^{\alpha} - h_{,\mu\nu} - \eta_{\mu\nu}(h_{\alpha\beta}{}^{,\alpha\beta} - h_{,\beta}{}^{\beta}) = 16\pi\, T_{\mu\nu}\,. \quad (16)$$

On the left-hand-side we see four-dimensional second derivatives of $h_{\mu\nu}$, which describes the field, including

$$h_{\mu\nu,\alpha}{}^{\alpha} \equiv \Box h_{\mu\nu}\,. \quad (17)$$

It is convenient to define

$$\bar{h}_{\mu\nu} \equiv h_{\mu\nu} - \tfrac{1}{2}\eta_{\mu\nu}\, h \quad (18)$$

which may be referred to as "gravitational potentials". The Einstein field equation then becomes

$$\bar{h}_{\mu\nu,\alpha}{}^{\alpha} + \eta_{\mu\nu}\,\bar{h}_{\alpha\beta,}{}^{\alpha\beta} - \bar{h}_{\mu\alpha,}{}^{\alpha}{}_{\nu} - \bar{h}_{\nu\alpha,}{}^{\alpha}{}_{\mu} \;=\; -16\pi\, T_{\mu\nu}\,. \quad (19)$$

We now use a *gauge freedom* (an infinitesimal redefinition of coordinates, $(x^{\mu})' = x^{\mu} + \xi^{\mu}(\mathbf{x})$) to set $\bar{h}^{\mu\alpha}{}_{,\alpha} = 0$. (This gauge is analogous to the Lorentz gauge in electromagnetism: $A^{\alpha}_{,\alpha} = 0$). Eq. (19) then becomes

$$\bar{h}_{\mu\nu,\alpha}{}^{\alpha} \equiv \Box\bar{h}_{\mu\nu} = -16\pi\, T_{\mu\nu} \quad (20)$$

and in vacuum (where $T_{\mu\nu} = 0$):

$$\Box\bar{h}_{\mu\nu} = 0 \quad (21)$$

which implies that if $\bar{h}_{\mu\nu}$ changes with time, then the changes propagate at velocity c.

3. How do gravitational waves interact with matter?

Consider here a *plane-fronted* wave propagating in the $x^3 = z$ direction (using rectangular Cartesian coordinates). For this, we can write the Riemann tensor as

$$R_{\alpha\beta\gamma\delta} = R_{\alpha\beta\gamma\delta}(t - z)\,. \quad (22)$$

This satisfies the *Bianchi identities*:

$$R_{\alpha\beta[\gamma\delta;\epsilon]} = 0 \quad (23)$$

where the semi-colon denotes a *covariant derivative* and the square brackets [ ] denote anti-symmetrization. For a plane wave on a flat background, these give:

$$R_{\alpha\beta12,0} = 0 \Rightarrow R_{\alpha\beta12} = 0 \quad (24)$$

$$R_{\alpha\beta13,0} - R_{\alpha\beta10,3} = 0 \Rightarrow R_{\alpha\beta13} = -R_{\alpha\beta10} \quad (25)$$

$$R_{\alpha\beta23,0} - R_{\alpha\beta20,3} = 0 \Rightarrow R_{\alpha\beta23} = -R_{\alpha\beta20}\,. \quad (26)$$

Therefore, using the symmetries of the Riemann tensor:

$$R_{\mu\nu\alpha\beta} = R_{\alpha\beta\mu\nu} \tag{27}$$

$$R_{\alpha\beta\mu\nu} = -R_{\alpha\beta\nu\mu} \tag{28}$$

it follows that any pair of purely spatial indices $(12, 13, 23, etc.)$ *either* gives a vanishing component *or* can be converted to a spacetime pair 10 or 20. There are then six independent components:

$$\begin{aligned} &R_{1010}\ R_{1020}\ R_{1030} \\ &R_{2020}\ R_{2030}\ R_{3030}\,. \end{aligned} \tag{29}$$

However, Einstein's equation for vacuum $R_{\mu\nu} = R^{\alpha}_{\mu\alpha\nu} = 0$ then reduces this to two:

$$\begin{aligned} R_{x0x0} &= -R_{y0y0} = -\tfrac{1}{4}(\bar{h}_{xx} - \bar{h}_{yy})_{,00} \\ R_{x0y0} &= \ R_{y0x0} \ \ = -\tfrac{1}{2}\,\bar{h}_{xy,00}\,. \end{aligned} \tag{30}$$

The only relevant components of $\bar{h}_{\mu\nu}$ are purely *transverse* (i.e. only x and y are involved) and a further gauge freedom can be used to make it *traceless*. This is called the *Transverse Traceless* (TT) gauge:

$$\begin{aligned} h^{TT}_{xx} &= \tfrac{1}{2}\,(\bar{h}_{xx} - \bar{h}_{yy}) = -h^{TT}_{yy} \\ h^{TT}_{xy} &= \bar{h}_{xy} \end{aligned} \tag{31}$$

with all of the other components being zero. Then

$$R_{j0k0}\,(t-z) = -\tfrac{1}{2}h^{TT}_{jk,00} \qquad (\text{for } j,k = 1 \text{ or } 2). \tag{32}$$

We can write h^{TT}_{jk} in terms of two *polarizations*, introducing the polarization tensors $\mathbf{e}^{+}$ and $\mathbf{e}^{\times}$ such that

$$\begin{aligned} e^{+}_{xx} &= -e^{+}_{yy} = 1 \\ e^{\times}_{xy} &= e^{\times}_{yx} \quad = 1 \end{aligned} \tag{33}$$

with all other components being zero. Then

$$h^{TT}_{jk} = h_{+}\,e^{+}_{jk} + h_{\times}\,e^{\times}_{jk}\,. \tag{34}$$

We are now ready to calculate the effect of the wave on matter.

Consider two adjacent particles (with separation vector, ξ^{j}) hit by a gravitational wave. Their relative acceleration can be calculated using the *equation of geodesic deviation*:

$$\frac{d^2\xi^j}{dt^2} = -R_{j0k0}(t)\,\xi^k \tag{35}$$

$$= \tfrac{1}{2}\,h^{TT}_{jk,tt}\,\xi^k \tag{36}$$

and the change produced in their separation is then

$$\delta\xi^j = \tfrac{1}{2}\,h^{TT}_{jk}\,\xi^k \tag{37}$$

giving an overall fractional change of

$$\frac{\delta\xi}{\xi} \sim h \tag{38}$$

where h is the amplitude of the metric perturbation.

We now consider the effect produced by a periodic gravitational wave on a ring of test particles oriented perpendicular to the direction of propagation of the wave, looking separately at the effects of the two polarization modes $\mathbf{e}^+$ and $\mathbf{e}^\times$ (see the figure below). In both cases, the originally circular ring first becomes elliptical, then returns to being circular again, then becomes elliptical again with the semi-major axis being in a perpendicular direction to before and then becomes circular again. The cycle is

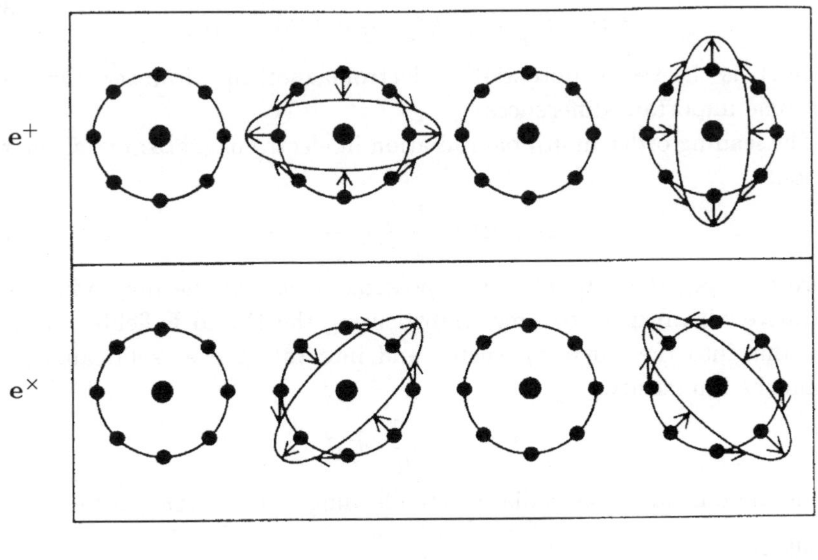

Figure 1. Deformation of a ring of test particles under the influence of different gravitational wave polarizations

then repeated. A bar of material placed perpendicular to the direction of propagation of the wave would experience oscillations in its length. Note that the behaviour shown in the figure indicates that the particle seen as mediating the gravitational interaction (the graviton) should have spin 2 since there is an invariance under rotation through π and one would expect invariance under rotation through $2\pi/S$, where S is the spin of the intermediary particle.

4. How is gravitational wave emission linked to changes in the source?

Changes in a spherically symmetric source which maintain its spherical symmetry, cause no change in the external field. This result is known as Birkhoff's theorem and holds both in Newtonian theory and also in GR. Since there is no change in the external field in this case, it follows that no gravitational waves are produced. For getting gravitational-wave emission, it is necessary to have *time-dependent*, *non-spherical* behaviour of the source of the gravitational field.

There are some notable similarities between gravity and electromagnetism which are relevant for what we are discussing here:

$$\begin{array}{llll} \text{em:} & A_0 = 0 & A_{i,i} = 0 & \Box A_i = 0 \\ \text{GR:} & h^{TT}_{0\mu} = 0 & h^{TT}_{jk,k} = 0 & \Box h^{TT}_{jk} = 0 \end{array} \tag{39}$$

where A_i is the vector potential of electromagnetism. However, there are also some important differences.

The leading order multipole radiation in electromagnetism is *dipole* radiation:

$$A_j\,(t, \mathbf{x}) = \frac{1}{cr}\,\dot{d}_j\,(t - r/c) \tag{40}$$

where $r \equiv |\mathbf{x}|$, $\mathbf{d}$ is the electric dipole moment and the dot indicates a derivative with respect to time. Substituting the $\mathbf{B}$ and $\mathbf{E}$ fields obtained from this into the Poynting vector and integrating over solid angle, we obtain the luminosity:

$$L_{em} \;=\; \frac{2}{3c^3}\,\ddot{d}_j\,\ddot{d}_j \tag{41}$$

In GR there is no dipole radiation; the leading order is *quadrupole*.

Recall that

$$\Box\bar{h}_{\mu\nu} = -\frac{16\pi G}{c^4}\,T_{\mu\nu}\,. \tag{42}$$

Integrating this, one gets after some manipulation

$$h_{jk}^{TT}(t, \mathbf{x}) = \frac{2}{r}\frac{G}{c^4}\ddot{\mathcal{I}}_{jk}^{TT}(t - r/c) \tag{43}$$

where $\mathcal{I}_{jk}$ is the mass quadrupole moment given by:

$$\mathcal{I}_{jk} = \sum_A m_A\left[x_j^A x_k^A - \tfrac{1}{3}\delta_{jk}(x^A)^2\right] \tag{44}$$

(following the definition of Misner, Thorne & Wheeler [2]).

The energy flux is given by

$$T_{0r} = \frac{c^4}{32\pi G}\left\langle h_{jk,0}^{TT}\, h_{jk,r}^{TT}\right\rangle \tag{45}$$

where $\langle\ \rangle$ indicates the average over several cycles.

Inserting the expression for h_{jk}^{TT} into this and integrating over the solid angle, we get the luminosity:

$$L_{GW} \equiv -\frac{\mathrm{d}E}{\mathrm{d}t} \tag{46}$$

$$= \frac{1}{5}\frac{G}{c^5}\left\langle \dddot{\mathcal{I}}_{jk}\,\dddot{\mathcal{I}}_{jk}\right\rangle . \tag{47}$$

This is known as the *quadrupole formula* for gravitational radiation. Although, as presented here, it applies for just the weak-field regime, it does, in fact, have a wider range of validity if $\mathcal{I}_{jk}$ is suitably defined.

If the source is *non-axisymmetric*, gravitational waves can also carry away angular momentum:

$$\frac{\mathrm{d}J_i}{\mathrm{d}t} = -\frac{2}{5}\frac{G}{c^5}\varepsilon_{ijk}\left\langle \ddot{\mathcal{I}}_{jm}\,\dddot{\mathcal{I}}_{km}\right\rangle \tag{48}$$

where ε_{ijk} is the permutation tensor.

5. Order of magnitude estimates

In this section, we make some order of magnitude estimates of gravitational wave emission to show how detectability of gravitational waves is linked to the characteristics of the source. The third time derivative of $\mathcal{I}$ can be very roughly approximated by

$$\dddot{\mathcal{I}}_{jk} \sim \frac{MR^2}{T^3} \sim \frac{Mv^3}{R} \tag{49}$$

where M, R and T are the characteristic mass, size and timescale of the source and v is a characteristic velocity.

From Eq. (47) we then get

$$L_{GW} \sim \frac{G}{c^5}\left(\frac{M}{R}\right)^2 v^6 \tag{50}$$

$$\sim L_0 \left(\frac{R_s}{R}\right)^2 \left(\frac{v}{c}\right)^6 \tag{51}$$

where $L_0 \equiv c^5/G \rightarrow 3.6 \times 10^{59}\, erg/s$ and R_s $(= 2GM/c^2)$ is the Schwarzschild radius of the source. It follows that the most powerful sources will be *compact* (with $R \sim R_s$) and *fast moving* (with $v \sim c$.)

For detectors, the important quantity is h since it is this which indicates the relative strain produced by an incident gravitational wave. From Eq. (43):

$$h_{jk}^{TT} = \frac{2}{r}\frac{G}{c^4}\ddot{I}_{jk}^{TT} \tag{52}$$

we have that

$$h \sim \left(\frac{R_s}{R}\right)\left(\frac{v}{c}\right)^2 \frac{R}{r}. \tag{53}$$

The behaviour $h \propto 1/r$ is a general feature of gravitational waves and indicates that distant sources are more easily observable than one would expect on the basis of a normal $(1/r^2)$ fall-off.

There is a connection between compactness of the source and typical velocities of motion since it is often the case that the *kinetic energy* of a system is of the same order as its *gravitational potential energy*, i.e.

$$\tfrac{1}{2}v^2 \sim \frac{GM}{R} \tag{54}$$

$$\Rightarrow \frac{v^2}{c^2} \sim \frac{R_s}{R}. \tag{55}$$

Inserting this into Eq. (53) gives

$$h \sim \left(\frac{R_s}{R}\right)^2 \frac{R}{r}. \tag{56}$$

6. Summary of astronomical sources of gravitational waves

In this section, we give a list of the main predicted astronomical sources of gravitational waves, focusing particularly on those which can be good candidates for detection by the new generation laser-interferometric detectors such as LIGO, VIRGO and GEO600 which are most sensitive to frequencies in the range from ten Hz to a few hundred Hz. We group the sources into three classes: *burst sources*, for which there is a sharp pulse of gravitational radiation emitted; *periodic sources*, where gravitational waves are emitted over a very large number of similar cycles; and *stochastic sources*, where the signals from many objects mix to form a "noise" background. Different detection strategies will be used for these different types of source and hence the detection thresholds are very different (see the article by Thorne [3] for more details). For a long-lived periodic source, there is the possibility of integrating over very many cycles in order to extract the signal from detector noise and this can give as much as six orders of magnitude enhancement in sensitivity as compared with burst sources. For stochastic sources, which give a background whose overall features change only very slowly with time, it is again possible to make use of time integration to enhance sensitivity and here there can be a gain of up to three orders of magnitude in sensitivity with respect to burst sources. In the lists that follow, rough sensitivity thresholds are given for each type of source, appropriate for the Advanced LIGO detector. Note that, in fact, these thresholds are dependent on frequency of the waves but the values given are typical ones for our frequency range from ten to a few hundred Hz.

Burst sources ($h \gtrsim 10^{-22}$ for detection by Advanced LIGO)

- Gravitational collapse to form stellar mass black holes and neutron stars (associated with supernovae)
- Coalescence of neutron star and black hole binaries
- Infall of a star into a large black hole (lower frequency than the LIGO range)

Periodic Sources ($h \gtrsim 10^{-28}$)

- Rotating neutron stars
  - Young neutron stars in supernovae which are non-axisymmetric as a result of the growth of unstable modes
  - Neutron stars which are non-axisymmetric as a result of misaligned strong magnetic fields

- Binary stars (lower frequency than the LIGO range)

Stochastic Sources ($h \gtrsim 10^{-25}$)

- Supernovae
- Binary stars (lower frequency than the LIGO range)
- Early universe, cosmic strings and phase transitions (mostly lower frequency than the LIGO range)
- Population III stars (mostly lower frequency than the LIGO range)

Note that the sensitivity thresholds (the minimum induced fractional strains measurable by the detector) are rather impressive numbers! The best possibilities for early detection by laser interferometers seem to be

- Coalescing neutron star binaries
- Rotating non-axisymmetric neutron stars
- Coalescing black hole binaries

For all of these, it is extremely important to produce *templates* of the expected wave signals to aid the extraction of signals from detector noise. This is an area of physics where experiment and theory need to proceed very closely together.

For us as astrophysicists, the greatest excitement in the search for gravitational waves concerns the possibility of opening a new window on the universe to enable us to get information about phenomena of relativistic astrophysics which are largely hidden from us for as long as we are constrained to make observations only by means of electromagnetic radiation. However, in addition to the interest for astronomers, there is also the aspect that gravitational wave observations are likely to produce output of great interest for *basic physics* as well. In particular, we can mention:

- Confirmation of the existence of black holes
- Better understanding of gravity
- Better understanding of neutron stars (information about the physics of high density matter)
- Better understanding of the early universe (information about ultra-high energy physics)

7. Gravitational waves and the binary pulsar PSR 1913+16

In conclusion, we turn to a brief discussion of this famous object which has presented the strongest observational evidence so far for gravitational waves actually being emitted. It was discovered by Hulse and Taylor in 1974. They saw a single pulsar (a rotating neutron star) with a period of 59 ms and, from Doppler shifts in the frequency, they were able to infer that it was in orbital motion around an unseen companion with an orbital period of ~ 8 hours. The orbital velocity was measured at $\sim 300\,km/s$, giving $v/c \sim 10^{-3}$. Now, after many years of observations, all of the parameters of the system are known to high accuracy. In particular, the orbital period is seen to be decreasing at a rate $\dot{P} = -2.425 \times 10^{-12}\,s\,s^{-1}$ ($\sim 0.1\,ms/year$). For a relativistic binary system such as this, GR predicts that gravitational radiation will carry away orbital angular momentum and cause the two components of the binary to spiral towards each other with a progressively shortening orbital period, exactly as observed for PSR 1913+16. The observed period change is in excellent quantitative agreement with the theoretical prediction using GR, giving strong circumstantial evidence that this object is indeed emitting gravitational waves. In 1991, a second rather similar object was discovered (called PSR 1534+12). This again shows evidence for orbital angular momentum being carried away by gravitational waves.

These results are very encouraging. However, the really exciting moment will come when there is the first direct evidence for gravitational waves from an astronomical source actually being picked up by a detector.

Further reading

[1] S.L. Shapiro and S.A. Teukolsky, *Black holes, white dwarfs and neutron stars: the physics of compact objects* (Wiley, New York, 1983); pp. 466 – 498.

[2] C.W. Misner, K.S. Thorne and J.A. Wheeler, *Gravitation* (Freeman, New York, 1973); pp. 941 – 1044.

[3] K.S. Thorne, in *Three hundred years of gravitation* (eds. S.W. Hawking and W. Israel - Cambridge University Press, 1987); pp. 330 – 458.

GRAVITATIONAL WAVES AND THEIR DETECTION

EUGENIO COCCIA

Dipartimento di Fisica, Università di Roma "Tor Vergata" and INFN Sezione di Roma II, via della Ricerca Scientifica, Roma, Italy,
INFN, Laboratori Nazionali del Gran Sasso, I-67010 Assergi, L'Aquila, Italy

Direct detection of gravitational waves is one of the great challenges of contemporary experimental physics. Gravitational waves constitute a powerful testing ground for theories of gravity and provide unique information about the coherent bulk motion of the matter in the Universe. After more then 40 years from the beginning of the experimental search for cosmic gravitational waves, several detectors are today monitoring the strongest potential sources in our Galaxy and in the local group. We report here the principles of operation and the main techinques adopted by resonant-mass and interferometric detectors.

1. Introduction

Einstein's theory of gravitation, General Relativity (GR), predicts the existence of gravitational waves (GW). The quantitative analysis of experimental data obtained from the observations of binary star systems, in particular binary pulsars such as the one discovered by Hulse and Taylor, gives strong evidence that these systems indeed loose energy through gravitational radiation.

The detection of GW as such provides one of the most fundamental tests of Einstein's theory. However GR is not the only metric theory of gravity. The distinguishing characteristics of such waves, i.e. propagation speed, polarization states and multipolar structure, vary from one theory to another, and so direct measurement of the effect of a GW is potentially a powerful tool to select a candidate theory.

Quite independently of which is the correct theory of the gravitational field, the analysis of GW will open a new window for the observation of the universe, thereby founding a new Astronomy. Astrophysical objects and forms of matter which do not emit electromagnetic radiation are normally invisible although some information also comes to us in the form of neutrinos. As gravity is the only force known in the universe to which no object or form of matter can escape, gravitational waves can in principle carry information about all kinds of systems, those which are detectable otherwise and those which are not. GW are almost unaffected by matter so they travel across the Universe and arrive to the Earth in their original condition. In GR their effect is to cause a quadrupole strain $h = \delta L/L$ in space-time, perpendicular to the direction of propagation and thus in all objects imbedded in it. GW detectors aim at measuring such strain against a background of perturbing influences.

The experimental search for cosmic GW was initiated by Joseph Weber in the early 60, at a time when almost nothing was known about possible cosmic sources [1]. His detector consisted of a massive metallic bar with a fundamental longitudinal frequency of about 1kHz and of a motion sensor converting the vibration of the bar into an electric signal.

The excitement following the Weber announcement in 1969 of simultaneous signals in two detectors - one near Washington DC, the other near Chicago - produced the birth of other research groups around the world. Even if their results gave no evidence that GW were being seen, the enormous potential pay off that could follow the successful detection of GW (made clear by many theoretical efforts) stimulated new generations of resonant bars, involving the use of cryogenics and superconducting techniques for noise reduction and the development of new detectors based on the laser interferometry between widely spaced bodies. Prototype interferometric detector development has proceeded to a stage where proposed long baseline ($3 \div 4$ Km) facilities have now started to take data. Resonant mass detectors development brought to a 10^4 fold improvement in energy sensitivity over Weber's original antennae. Several such detectors are now in the continuous observational mode with sensitivity h$\simeq 6 \times 10^{-19}$, or, in spectral units, 10^{-21} Hz$^{-1/2}$ over bandwidth of a few Hz. The strongest potential sources of GW bursts in our Galaxy and in the local group are today monitored by such instruments. In parallel with the observations, experimental development work is very active. Resonant-mass detectors have enormous potential for improvement.

We review here the principles of operation, the present status and the future prospects of gravitational wave detectors.

2. Gravitational Waves and Detectable Sources

General relativistic GW are ripples in the curvature of space-time that propagates with the speed of light. If we call $\mathrm{R}^{gw}_{\alpha\beta\gamma\delta}$ the contribution to the Riemann curvature tensor due to a GW having wavelength very short compared to the length scale on which the background curvature varies, then a freely falling (or slowly accelerating) observer measuring in its proper reference frame the position of a test particle momentarily at rest at location x^j, will experience tiny oscillatory changes δx^j in the position of the test particle, such that

$$\frac{d^2\delta x^j}{dt^2} = -R^{gw}_{joko} = \frac{1}{2}\frac{\partial^2 h^{TT}_{jk}}{\partial t^2}x^k \quad (1)$$

This equation of motion is known as equation of geodesic deviation and is the basis for all the experimental attempts to detect GW. The dimensionless amplitude h^{TT}_{jk} is the primary entity for describing a GW because of its simple relation to the displacement produced by the wave. In fact since in all realistic wave the oscillatory change is much smaller than the distance of the particle from the origin, x^k can be considered constant in the eq. (1) and the result of its integration is

$$\delta x^j = \frac{1}{2}h^{TT}_{jk}x^k \quad (2)$$

h^{TT}_{jk} plays the role of strain of space; this change in the separation of points in space provides the observable effect of the passage of a GW. Put in other words, there is a change in the proper time taken by light to pass to and fro between two fixed points in space.

The superscript TT means that, according to general relativity, the GW field is transverse and traceless. Orienting the z axes of a set of cartesian coordinate along the wave propagation direction, the trasversality of the wave means that the only non-zero components of the wave field are h^{TT}_{xx}, $h^{TT}_{xy} = h^{TT}_{yx}$ and h^{TT}_{yy}. The trace free property means that $h^{TT}_{xx} = -h^{TT}_{yy}$. Thus GW have only two independent components, i.e. two polarization states.

Because of their TT nature, GW produce a quadrupolar, divergence free force field. The two components of this field correspond to the two polarization states of the wave. They are indicated by the quantity $h_+ \equiv h^{TT}_{xx} = -h^{TT}_{yy}$ which produces a force field with the orientation of a '+' sign while $h_\times \equiv h^{TT}_{xy} = h^{TT}_{yx}$ produces one with the orientation of a '$\times$' sign. h_+ and $h_\times$ are called the 'plus' and 'cross' GW amplitudes.

The quadrupolar symmetry tells us that GW must be associated with quanta of spin two ('gravitons'). Also tells us that the convenient technique for computing wave generation is the 'quadrupole formalism' which writes the GW field generated by a source having weak internal gravity, small internal stresses and in slow motion as

$$h_{ij}^{TT} = \frac{2G}{rc^4} \ddot{I}_{ij}^{TT} (t - \frac{r}{c}) \tag{3}$$

Here r is the distance to the source and t the proper time as measured by an observer at rest respect to the source. I_{ij}^{TT} is the TT part of the quadrupole moment of the source:

$$I_{ij}^{TT} = \int \rho(t)[x^j x^k - \frac{1}{3} r^2 \delta_{ik}] d^3x \tag{4}$$

The GW carry energy, which is used to deform space-time. The GW energy flows per unit area and unit time is

$$F = \frac{c^3}{16\pi G} < \dot{h}_+^2 + \dot{h}_\times^2 >= \frac{G}{8\pi r^2 c^5} < \dddot{I}_{ij}^{TT} \dddot{I}_{ij}^{TT} > \tag{5}$$

The amplitude of a GW is so small, that no conceivable laboratory source can produce a measurable response in any practical detector. It is therefore mandatory to look for astronomical systems, some of which possess masses and rates of change of their quadrupole moment large enough to produce GW of considerable intensity.

A review of the possible sources with estimates of the GW intensity on Earth are given in [2,3]. We list in table 1 various sources and the most mature methods of detection.

Sources of GW can be divided into at least three groups according to the spectral character of the signal produced:

- burst sources: e.g. collisions between black holes and between black holes and neutron stars; coalescence of a compact binary system; collapsing and bouncing supernovae cores: births of black holes; starquakes in neutron stars
- continuous sources: e.g., binary star systems; rotating deformed neutron stars and white dwarfs; pulsations from white dwarfs after novae outbursts,
- stochastic sources: e.g. coalescence radiation from black holes formed from Population III stars; fluctuations in the density of the early universe, radiation from speculative objects such as cosmic strings.

Gravitational collapses, like supernovae, have been the primary goal of GW detector development. We know little about the precise wave form to expect, but traditional theoretical arguments suggest that the burst will

Frequency	Sources	Detection method and projects
10^{-16} Hz	Primordial	Anisotropy of CBR Background Radiation
10^{-9} Hz	Primordial Cosmic Strings	Timing of ms pulsars
10^{-4} to 10^{-1} Hz	Binary stars Supermassive BH(10^3-10^7 $M_\odot$)	Doppler tracking of Spacecraft Laser interferometers in Space LISA
$10 - 10^3$ Hz	Inspiral of NS and BH binaries (1 - 10^3 $M_\odot$) Supernovae Pulsars	Laser interferometers on Earth LIGO, VIRGO, GEO600, TAMA
10^3 Hz	Coalescence of NS and BH binaries Supernovae ms Pulsars	Cryogenic resonant detectors ALLEGRO, AURIGA, EXPLORER, NAUTILUS, MiniGRAIL, Mario Schenberg

have a central frequency of about 1000 Hz and will last for one or two cycles.

Consider for instance a GW burst of duration τ_g described as a sinusoidal wave of amplitude h_0 and angular frequency ω_0 for $|t| < \tau_g/2$ and zero for $|t| > \tau_g/2$.

The strain amplitude of a wave burst at a distance R from the source which radiates an energy $M_{GW}c^2$ in a time τ_g (typically a few milliseconds) is:

$$h = \sqrt{\frac{16GM_{GW}c^2}{c^3R2\omega_0^2\tau_g}} = 1.38 \times 10^{-17}\frac{1000Hz}{f}\frac{1000pc}{R}\sqrt{\frac{M_{GW}}{10^{-3}M_\odot}\frac{10^{-3}s}{\tau_g}} \quad (6)$$

where $M_\odot$ is one solar mass. For a supernova in the center of our Galaxy (10 kpc) with an energy release of $10^{-3}M_\odot$ this gives $h = 1.38 \times 10^{-18}$. Note that this amplitude corresponds to an energy flux of 0.5 MJ$/m^2$ on

Earth or a power of 500 MW in 1ms!

The best present-day GW antennae have strain sensitivities around 2×10^{-19} quite enough to detect supernovae events in our Galaxy or in the nearby ones. The big problem is the scarcity of such events in our surroundings galaxy. We must look at a distance $\sim$ 1000 times larger to find a cluster like Virgo, with about 2000 galaxies, to increase the probability of observation to several events per year. From (6) we see that the required strain sensitivity for the Virgo cluster (R$\cong 10^7$ parsec) in the same condition as the preceding example is of order 10^{-21} which is a step of a factor 100 in present-day achieved strain amplitude.

3. Resonant Mass Detectors

3.1. *The resonant body*

Any vibrational mode of a resonant body that has a mass quadrupole moment, such as the fundamental longitudinal mode of a cylindrical antenna, can be excited by a GW with nonzero energy spectral density at the mode eigenfrequency.

The size of a resonant antenna is determined by the frequency and the velocity of sound ν_s in the material used. Since ν_s is always orders of magnitude less than the speed of propagation of gravitational radiation, resonant-mass antennas are always much smaller than the wavelength of the radiation. Typical antennas are thin cylindrical bars made of aluminium with a fundamental longitudinal resonance in the frequency band around 1kHZ.

Because of the forces responsible for the antenna's elasticity, the GW does work and, in the case of a cylindrical antenna, deposits energy only in the odd-order longitudinal modes. Because of the quadrupole nature of the radiation, the even-order modes are not excited.

The mechanical oscillation induced in the antenna by interaction with the GW is transformed into an electrical signal by a motion or strain transducer and then amplified by an electrical amplifier. Unavoidably, Brownian motion noise associated with dissipation in the antenna and the transducer, and electronic noise from the amplifier, limit the sensitivity of the detector. Typically, the detector output is filtered by a suitable linear filter that is designed to optimize the signal-to-noise ratio.

The study of the interaction of one mode of the resonant body with a GW is simplified if we consider the measured displacement of the detector as the displacement of an equivalent oscillator of mass m and length l, having the same angular frequency ω_k of the given mode.

If we monitor the displacements of a cylindrical bar end face, it is easy to demonstrate, by energy considerations, that $m = M/2$ and $l = \frac{4}{\pi^2}L$, where M and L are the mass and length of the cylindrical bar. If we monitor the radial displacements of the lowest quadrupole mode of a sphere of mass M_s and radius R it is: $m = M_s$ and $l = 0.6R$.

The unidimensional displacement x(t) of the oscillator mass m follows the equation

$$\ddot{x}\,(t) + 2\beta_1\,\dot{x}\,(t) + \omega_0^2 x(t) = \frac{l}{2}\,\ddot{h}\,(t) \tag{7}$$

where h indicates h_+ or $h_\times$. The quantity $2\beta_1$ expresses the oscillator losses and is related to the quality factor Q and to the decay time τ by $2\ \beta_1 = \omega_0/Q = \tau^{-1}$.

It is conventional to characterize the energy absorbed by the resonant body in the frequency domain in term of a cross section $\sigma(\omega)$. The energy deposited by a burst of gravitational radiation in an antenna initially at rest is then

$$E = \int F(\omega)\sigma(\omega)d\omega \simeq F(\omega_n)\int \sigma(\omega)d\omega = F(\omega_n)\sigma_{tot} \tag{8}$$

where $F(\omega)$ is the energy spectral density of the signal pulse, ω_n is the antenna mode eigenfrequency and σ_{tot} is the total integrated cross section. The approximate equality holds if we consider a burst signal of duration less than Q times the antenna oscillation period, so that the spectrum of the incident radiation can be considered uniform and much wider than the antenna resonance. The cross section is sharply peaked at $\omega = \omega_n$. For a cylindrical antenna this cross section (unilateral) is given by

$$\sigma(\omega) = \frac{1}{n^2}\frac{8}{\pi}\frac{G}{c^3}Mv_s^2\frac{1}{\pi}\frac{\tau_n^{-1}}{(\omega-\omega_n)^2+(2\tau_n)^{-2}}sin^4\theta cos^2 2\varphi \tag{9}$$

where θ is the angle between the direction of propagation of the GW and the cylinder longitudinal axis, φ the angle of the wave's polarization ellipse.

The cross section depends on the antenna material through the combination Mv_s^2. Writing $M = \rho V$ (density ρ times volume V) and the speed of sound as $v_s = (Y/\rho)^{1/2}$, as for the longitudinal vibrations of an isotropic and homogeneous material having Young modulus Y, we get $Mv_s^2{=}VY$. In accordance with the geometrodynamical interpretation of a GW, the energy absorbed by an extended oscillator, deformed by a tidal force, is proportional to the deformed volume V and to the material stiffness, represented by the Young modulus.

Near a given mode, the cross section $\sigma(\omega)$ is a narrow-band resonant function centered at the mode eigenfrequency with a full width at half maximum of $(2\omega_o/Q)$, where Q is the mechanical quality factor of the antenna. This behaviour of the antenna cross section seems to imply that a high-Q antenna will have a very narrow bandwidth. This is not the case. Both the strain signal spectrum and the thermal Brownian motion noise in the antenna exhibit the same resonant response near the mode eigenfrequency. Thus the signal-to-noise ratio is not bandwidth limited by the antenna thermal noise. A more significant sensitivity and bandwidth limitation comes from the transducer-amplifier readout. In the past, resonant-mass antennas were narrow-band devices $(\Delta f/f_0 \ll 1)$, but in the last few years, the efforts of researchers managed to increase the bandwidth to several tens of Hz [4,5].

3.2. *The motion sensor*

After a signal has been picked up by the antenna, it must be amplified and recorded for analysis. The only viable way to achieve this is to transform the signal into electromagnetic energy, and then use state-of-the-art techniques for electromagnetic amplification and read-out.

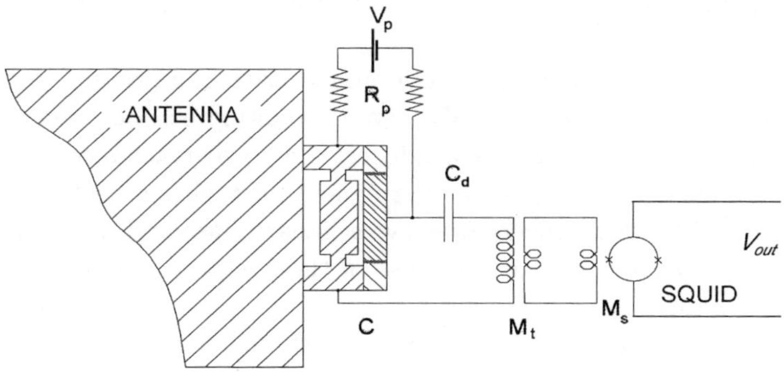

Figure 1. The read-out system of the Rome gravitational wave detectors.

The electromechanical transducer can be represented with the components of the Z_{ij} matrix which connects the input variables (force f(t) acting on the transducer and velocity $\dot{x}$(t) of the transducer mechanical parts) with the output variables (voltage v(t) and current i(t)):

$$f(t) = Z_{11}\,\dot{x}\,(t) + Z_{12}i(t) \tag{10}$$

$$v(t) = Z_{21}\,\dot{x}\,(t) + Z_{22}i(t) \tag{11}$$

In important cases the Z_{ij} components satisfy the relationships $Z_{11}Z_{22} = Z_{12}Z_{21}$ and $Z_{12} = Z_{21}$.

An important parameter is the ratio β of the electrical energy in the transducer to the total energy in the resonant body:

$$\beta = \frac{1}{m\omega}\frac{|Z_{21}|^2}{Z_{22}} \tag{12}$$

The principle of all transducers is to store electromagnetic energy in a very small volume, usually a narrow gap, one of the walls of which is part of the antenna. The motions of this wall, arising from vibrations in the antenna, induce a modulation of this energy which is detected and amplified as an electrical signal. Transducers of the sensitivity required to detect the extremely small signals of gravitational radiation are not available commercially. Development of transducers and components of the read-out system are therefore an important part of the R&D efforts of all gravitational wave experiments.

Strain transducers are classified as belonging to the following three categories: capacitive (electrostatic), inductive (magnetostatic), optical (electromagnetic).

Piezo-electric, magnetostrictive and μ-wave cavities are respective examples of such transducers. We can also distinguish between passive and active transducers. Passive transducers are linear transducers in which the source of energy in the gap is a permanent field, either electric or magnetic or both. These transducers preserve a linear phase and amplitude relation between input and output. Because of their relatively simple construction they are widely used.

In active transducer the gap is fed with an oscillating bias field at high frequency ω_p. The mechanical vibration of frequency ω_0 modulates the phase of this oscillating field and produces side-bands which contain the information of the mechanical signal. Good performance of this type of transducer is expected because of the gain ω_p/ω_0 resulting from the conversion of the pump frequency ω_p into the antenna frequency ω_0.

An important breakthrough in increasing the sensitivity of resonant-mass detectors was achieved when resonant transducers (which can be active or passive) were introduced [6]. In a resonant mass transducer an oscillator with a small effective mass m_t is coupled in resonance with the antenna, which has a large effective mass m. The maximum amplitude of the motion of the small oscillator will then be increased by a factor $\sqrt{m/m_t}$ relative to the amplitude of the antenna. In the expression of β the antenna mass is thus replaced by the transducer mass.

Several groups have analyzed resonant transducers and have developed their own designs [6,7,8,9,10]. The transducer is connected to an electrical amplifier whose noise can be characterized by two parameters. The two parameters are, usually, the power spectra of the voltage and current noise, V_n^2 and I_n^2, or their following combinations:

$$T_n = \frac{\sqrt{V_n^2 I_n^2}}{K} \tag{13}$$

$$R_n = \sqrt{\frac{V_n^2}{I_n^2}} \tag{14}$$

T_n is called the amplifier noise temperature and R_n the amplifier noise resistance. Another parameter, useful to express the matching between transducer and amplifier, is

$$\lambda = \frac{R_n}{|Z_{22}|} \tag{15}$$

A large fraction of the technological complexity of resonant mass detectors results from the optimization of the quantities β, T_n and R_n.

3.3. *Noise*

Two classes of noise sources have to be considered in the sensitivity analysis of a GW antenna:

- the intrinsic noise sources such as the thermal and electronic ones, which have Gaussian statistics and can be accurately modeled, and
- the external noise sources such as seismic noise and disturbances from cosmic rays, which are more difficult to characterize because they are non-Gaussian and often also non-stationary.

The depth of the peaks in Figure (2) is directly proportional to the thermodinamical temperature of the bar. So cooling the detector to cryogenic and ultracryogenic temperatures is, first of all, an effective way to reduce

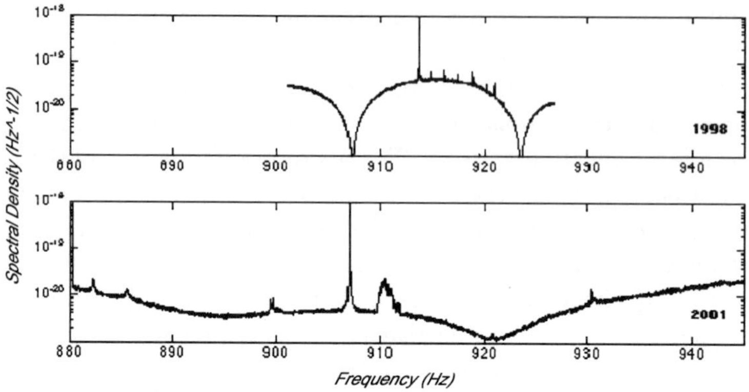

Figure 2. The strain sensitivity curve for the EXPLORER detector.

thermal noise and to improve the sensitivity of the antenna. This involves the use of low temperature techniques [11].

The useful bandwidth of the detector is, roughly speaking, given by the ratio β/T_n. This means that the effort of the groups must be focused on better transducers and low noise amplifiers. The device that, in the kHz region, has the best performances in terms of T_n is the d.c. SQUID, which is capable of getting very close to the quantum limit [12,13].

The last noise source, which the experimentalist can take care of, is seismic noise. Through FEM analysis, suspension systems with attenuation factors of more then 250 dB can be designed and used on real detectors.

The effects of other external unmodeled noises can be efficiently eliminated by coincidence measurements with several detectors located far away from each other

Let us give an explicit expression to the total detector noise. The thermal, or Brownian, noise is due to chaotic motion of the detector atoms in the thermal bath at the temperature T. The power spectrum of the stochastic force acting on the oscillator is

$$S_F(\omega) = 4\beta_1 mkT \tag{16}$$

The electronic noise has two terms: one is the back action stochastic force exerted by the current noise generator. This acts on the oscillator like the Brownian force, with a power spectrum

$$S_f = |Z_{12}|^2 I_n^2 = m\beta\omega_0 |Z_{22}| I_n^2 \tag{17}$$

The effect of the back action noise can be seen as an increment to the oscillator temperature T. The sum of Brownian noise of the oscillator at temperature T and back action noise can be attributed to the Brownian noise of an oscillator at temperature $T_e > T$ where

$$T_e = T(1 + \frac{\beta Q T_n}{2\lambda T}) \tag{18}$$

The other term is an additive noise due to amplifier; its power spectrum is

$$S_o = V_n^2 + I_n^2 |Z_{22}|^2 = kT_n |Z_{22}| (\lambda + 1/\lambda) \tag{19}$$

If the Z_{22} impedance is a smooth function of the frequency, the power spectrum S_0 can be considered white in the antenna bandwidth.

Another convenient dimensionless parameter is

$$\Gamma = \frac{T_n(\lambda + 1/\lambda)}{2\beta Q T_e} \tag{20}$$

that gives the ratio of the wide band noise in the resonance band width to the narrow band noise (in practice $\Gamma \ll 1$).

The sum at the output of the contributions given by the Brownian noise (at temperature T_e) and by the wide band electronic noise, gives the total detector noise. This can be referred to the input of the detector (as if it were a GW spectral density) and is usually indicated as $S_h(f)$:

$$S_h(f) = \frac{1}{\pi^3} \frac{kT_e f_0}{ml^2 Q f^4} \{1 + \Gamma[Q^2(1 - \frac{f^2}{f_0^2})^2 + \frac{f^2}{f_0^2}]\} \tag{21}$$

$S_h(f)$ represents the input GW spectrum that would produce a signal equal to the noise spectrum actually observed at the output of the antenna instrumentation. This function is independent of any assumption about the signal waveform. The half height width of this function gives the bandwidth of a resonant detector:

$$\Delta f = \frac{f_0}{Q} \Gamma^{-1/2} \tag{22}$$

This is much larger than the pure resonance linewidth f_0/Q. A natural way to express $S_h(f_0)$ is by means of the total integrated cross section:

$$S_h(f_0) = \frac{G}{c^3} \frac{4kT_e}{\sigma_{tot} Q f_0} \tag{23}$$

The two last relations characterize completely the sensitivity of a resonant-mass detector.

At present there are four resonant-mass detectors [14], operating with a noise level for broad-band gravity-waves bursts of $h \simeq 2 \times 10^{-19}$.

3.4. *The future of resonant-mass detectors*

The next generation of resonant-mass GW detectors will have spherical shape. At present, two of these detectors are being built, MiniGRAIL [15] (Leiden, The Netherlands) and Mario Schenberg [16] (Sao Paulo, Brasil). Both are spheres made of an alloy of copper and aluminum with a diameter of 65 cm and a mass of 1150 kg and a resonant frequency of about 3 kHz. The goal sensitivity of these detectors is $h \simeq 10^{-20}$.

Spheres have several advantages with respect to bars. First of all, a spherical detector is omni-directional, it is equally sensitive to a wave from any direction and it can also measure the polarization state of the wave. A single sphere is able to determine the source direction [17]. Besides, as will be shown below, the energy cross-section of a sphere is about 70 times larger than a typical bar at the same resonant frequency. Furthermore, the spherical geometry can, in principle, help in discriminating between different metric theories of gravity [18].

The total cross section of a sphere for each quadrupolar mode is given by [17]

$$\sigma_n = F_n \frac{G}{c^3} M_s v_s^2 \tag{24}$$

where n is the order of the quadrupole mode, M_s is the sphere mass, v_s the sound velocity and F_n is an dimensionless coefficient characteristic of each quadrupole mode. It is interesting to note that the cross section of the second-order quadrupolar mode is only a factor 2.61 lower than that of the first-order quadrupole mode. This means this detector can potentially be used at two frequencies.

It has been known for some time that a sphere has a gravitational cross section larger than that of an equivalent bar at the same frequency [17] for a single component of the gravitational tensor, a factor 18 if referred to the present bars. Moreover a sphere can detect all 5 indipendent components of the gravitational strain tensor, compared to only one for the bar. For a bar detector, it is well known that averaging over source direction and polarization [17] leads to a loss of energy resolution, compared to the optimum, by a factor $15/4 = 3.7$. Thus the net result is that the angle-averaged energy resolution of a sphere is $3.7 \times 18 = 67$ times better than the present bar detector (or about 8 times in h).

A "dual" resonator GW detector is formed of two mechanical massive resonators both sensitive to GW and whose relative vibrations are measured by non-resonant readouts [19]. Configuration as two nested spheres, an inner solid one and a hollow outer one, or two nested cylinders have been considered. The frequency of the first quadrupolar mode of the external resonator is lower than that of the inner one, thus, the two resonators are differently driven by the GW.

"Dual" resonators promise, both high sensitivity and wide bandwidth, of the order of one kHz, at about 2 kHz. A research and development activity is started.

4. Interferometric detectors

Interferometric detectors presently in operation are LIGO [20] in the US, GEO600 [22] in Germany and TAMA300 [23] in Japan, while VIRGO [21] in Italy is in the commissioning phase.

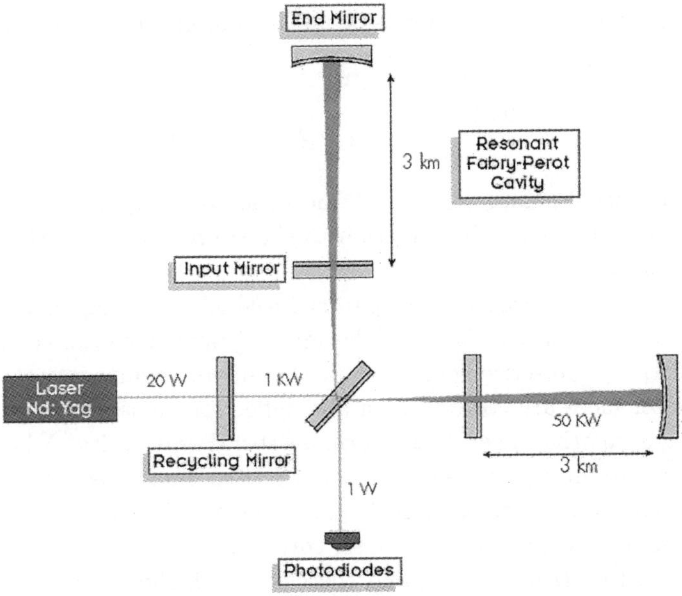

Figure 3. A schematic diagram of a laser interferometer gravitational wave detector.

The basic idea behind this kind of GW detector is the Michelson interferometer. The simplest design uses light that passes up and down each arm once. Real detectors are designed to store the light in each arm for longer than just one reflection: the optimum storage time of the light is half of the period of the GW (Ex. 200 Hz wave, $t_{stor} \sim 3$ ms $\Rightarrow$ L=1000

Km). The impracticality of very long arms has led to the development of schemes for folding a long optical path into a shorter length: delay lines and Fabry-Perot cavities.

A laser interferometer gravitational detector consists of masses that hang from vibration-isolated supports (shown in figure (3) is the optical system to monitor the separation between the masses). Two masses are close to each other, at the corner of an "L", and one mass is at the end of each of L's long arms. The arms lengths are nearly equal, $L_1 \approx L_2 = L$. When a GW, of amplitude $h(t)$, with frequencies higher than the masses' pendulum frequency (≈ 1 Hz), passes through the detector, it changes each arm length by a quantity equal to $\frac{1}{2}h(t)L_{1,2}$, thereby changing the arm-length difference, $\Delta L \equiv L_1 - L_2$. That change is monitored by laser interferometry in such a way that the variations in the output of the photodiode (the interferometers output) are directly proportional to $\Delta L(t)$.

The interferometer's output is a linear combination of the two wave fields h_+ and $h_\times$:

$$\frac{\Delta L(t)}{L} = F_+ h_+ + F_\times h_\times \equiv h(t) \tag{25}$$

The coefficients F_+ and $F_\times$ are of the order of unity and depend in a quadrupolar manner on the direction to the source and the orientation of the detector [2].

Test masses are made of transparent fused silica, though other materials might be used in the future. The masses' inner face are covered with high-reflectivity dielectric coating to fit the mirror requirements, while the masses outer faces are covered with anti-reflection coatings. The two mirrors facing each other on each arm form a Fabry-Perot cavity. A beam splitter splits a carefully prepared laser beam in two, and directs the resulting beams down the two arms. Each beam penetrates through the antireflection coating of its arm's corner mass, through the mass, and through the dielectric coating (the mirror); and thereby - with the length of the arm's Fabry-Perot cavity adjusted to be nearly an integral number of half wavelengths of light - the beam gets trapped in the cavity. Since the cavity's end mirror has much higher reflectivity than its corner mirror, most of the light impinging on the cavity is reflected backwards, and then hits the beam splitter where it recombines with light from the other arm. The output of the interferometer is kept on the dark fringe, so most of the recombined light would go back toward the laser but it is returned to the interferometer by a "light-recycling mirror".

The large magnitude of the low frequency seismic noise makes a "passive" interferometer design unworkable. The key is to use a *feedback* to keep the interferometer fixed at a chosen operating point (the dark fringe). The feedback is made of

- a sensor, producing an error signal measuring how far you are from the desired operating point;
- an actuator, a device that takes the error signal as input and that supplies the feedback influence to bring the interferometer toward this point.

When a GW hits the detector, changing the lengths L_1 and L_2 of the two cavities, it shifts each cavity's resonant frequency slightly relative to the laser frequency, and thereby changes the phase of the light in the cavity and the phase of the light that is reflected by the cavity toward the beam splitter. Corrispondingly, the relative phase of the two beams returning to the splitter is altered by an amount $\Delta\Phi \propto \Delta L$, and this relative phase shift causes a change in the intensity of the recombined light at the photodiode, $\Delta I_{pd} \propto \Delta\Phi \propto \Delta L \propto h(t)$. Thus, the change of photodiode output current is directly proportional to the GW amplitude $h(t)$. This method of monitoring $h(t)$ is capable of very high sensitivity.

5. Interferometers' noises

The significance of the various noise sources for the final design of a fully optimized interferometer to broad band GW is shown in Figure (4). At low frequencies, the sensitivity will be limited by seismic noise, at intermediate frequencies by the thermal noise of the mirror suspension, and at high frequencies by photon shot noise.

5.1. *Seismic noise*

At the frequency of 1 kHz, the displacement due to seismic ground motion is about 10^7 times larger than any possible signal due to GW. Thus, an effective way to isolate the test masses is required. Some very promising work on passive isolation has been done in the VIRGO Project. The Superattenuator [24] has been designed on the working principle of a multistage pendulum. With this solution a very good reduction of the seismic noise transmission to the test masses in all degrees of freedom has been obtained, extending the detection band in the low-frequency region down to a few hertz [25]. The system consists of an inverted pendulum, the seismic

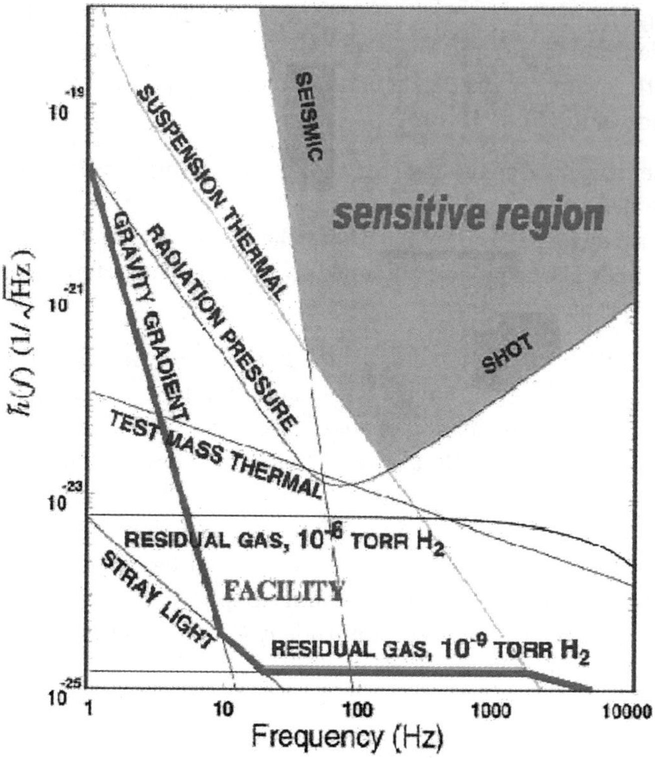

Figure 4. Noise sources limiting the sensitive region of earth-based interferometers.

filters (six and two in the long and short SA chains, respectively) connected to each other by metallic suspension wires and the last stage or payload. A more detailed description of each element including the short suspension system can be found in [24].

To the last stage of the SA chain is hung a mechanical filter; it consists of a marionette, a reference mass and a mirror. The marionette has been designed with four wings on which four small permanent magnets are attached. In front of these magnets four coils are placed. They are attached to the end of four aluminium pipes fixed on the bottom part of the previous filter. In this way the magnet-coil system and the marionette allow the control of the interferometer optical component in four degrees of freedom: the displacement along the beam direction (z) and the horizontal direction perpendicular to the beam (x), the rotation around the vertical axis (θ_y)

and the rotation around the horizontal axis perpendicular to the beam (θ_x).

From the marionette two pairs of thin wires start. The first pair supports the mirror and the second one supports the reference mass forming the last stage with a pendulum length of 0.7 m.

5.2. *Thermal noise*

Thermal noise is mostly important on the last stage of the suspension system. This is usually a simple wire pendulum made of a sling supporting the mirror. The resonance frequency is about 1 Hz and the frequency window of observation is usually above this frequency. The spectral density of apparent strain noise due to this effect is given by

$$\tilde{h} \approx \sqrt{\frac{16kT\omega_0}{mQ_S\omega^4 l^2}} \tag{26}$$

where m is the mirror mass, ω_0 the resonant frequency and Q_S the mechanical quality factor of the suspension pendulum. This Q can be much higher than the internal Q of the wire material because most of the energy of the pendulum is in the form of potential and kinetic energy of the swinging bob and not in the elastic energy of a bent wire. But it is extremely important that the wire support points be properly designed to avoid friction. Pendulum Q_S as high as 10^7 have been experimentally observed.

5.3. *Shot noise*

The sensitivity of a simple Michelson interferometer in the high frequency region is limited by photon shot noise to [26]

$$h \approx 2.4 \times 10^{-21} \left[\frac{\epsilon I_0}{50\,\mathrm{W}}\right]^{-1/2} \left[\frac{f}{1\,\mathrm{kHz}}\right]^{3/2} \tag{27}$$

where ϵ is the quantum efficiency of the detector, I_0 is the laser output power and f is the center frequency of the burst. Green light and a bandwidth of half the center frequency have been assumed in Equation (27). The first problem to be solved is thus the construction of a laser with sufficient output power in a stable sigle transverse and longitudinal mode.

This problem has been overcome in two ways: developing high power solid-state Nd:YAG lasers and recycling, which can be power and/or signal recycling.

Power recycling makes use of the fact that the interferometer output is held on a dark fringe by a feedback loop and almost all the light goes back towards the input. By placing a mirror in the input of the interferometer, a resonant optical cavity can be formed that uses the whole locked interferometer as an end mirror. So the circulating light power will be higher than the laser power by the inverse of the losses in the interferometer.

Signal recycling works similarly, except that it leads to a resonant enhancement of the signal instead of the light. By placing a mirror in the output port of the interferometer, a resonant cavity for the signal is formed. Depending on the reflectivity of this mirror, the detector can be made to operate narrow-band or wide-band and by changing the position of the mirror, the interferometer can be tuned to specific frequencies.

A combination of the above techniques is also possible, leading to the so called Dual Recycling. In this case, under the same assumptions as for Equation (27), the shot-noise limited sensitivity becomes [26]

$$h \approx 10^{-22} \left[\frac{\epsilon I_0}{50\,\mathrm{W}}\right]^{-1/2} \left[\frac{f}{1\,\mathrm{kHz}}\right] \left[\frac{1-R}{5\times 10^{-5}}\right]^{1/2} \left[\frac{l}{3\,\mathrm{km}}\right]^{-1/2} \tag{28}$$

where R is the reflectivity of the signal mirror. It is clear that, in order to make these detectors work, mirrors with extremely small losses are needed. This requires substrates with a microroughness of the order of an Ångström and reflective coatings with very small scatter and absorption. The effort of the last few years has focused on the problem of building highly polished surfaces and high reflecivity coatings. Nowadays, mirrors with reflection losses of much less than 50 parts per million are available.

6. Conclusions

Various technological aspects are pushed at the limit in the search for GW. Techniques based on resonant-mass and interferometer configurations are mature to explore a new astronomical window.

We can confidently say that today the strongest sources in our Galaxy will not pass unnoticed to earth-based GW detectors. This fact is extremely important as the search for GW is based on the technique of coincidences among two or more detectors.

Large interferometers are ambitious and promise a good rate of events in the next years. The future competitivity of resonant-mass detectors with advanced interferometers is connected to the possibility of overcoming the narrow band barrier ("dual" concept) and the detector directionality (omnidirectional spheres).

References

1. J. Weber, Phys. Rev. **117**, 306 (1960).
2. K.S. Thorne, in S.W. Hawking and W. Israel, editors, *Three Hundred Years of Gravitation*, pages 330-458. Cambridge University Press (1987)
3. B. F. Schutz, Class. Quantum Grav. **16**, A131-A156 (1999).
4. P. Astone *et al.*, Phys. Rev. Lett. **91**, 11 (2003).
5. J.P. Zendri *et al.*, Class. Quantum Grav. **19**, 1991-1996 (2002)
6. H. J. Paik, J. of Appl. Phys. **47**, 1168 (1976).
7. J-P. Richard, Rev. Sci. Instrum. **47**, 423 (1976).
8. P. Rapagnani, Nuovo Cimento **5C**, 385 (1982).
9. J-P. Richard, Phys. Rev. Lett. **521**, 165 (1984).
10. M. Bassan, Y. Minenkov, R. Simonetti, "*Proceedings of the VIRGO Conference*", F. Fidecaro and I. Ciufolini editors (World Scientific, Singapore, 1997).
11. P. Astone *et al.*, Astropart. Phys. **7**, 231-243 (1997).
12. P. Carelli *et al.*, Appl. Phys. Lett. **72**, (1998).
13. P. Falferi *et al.*, Appl. Phys. Lett. **82**, 931 (2003).
14. P. Astone *et al.*, Phys. Rev. D **68**, 022001 (2003)
15. A. de Waard *et al.*, Class. Quantum Grav. **20**, S143-S151 (2003).
16. O.D. Aguiar *et al.*, Braz. J. Phys. **32**, 866-868 (2002).
17. S. M. Merkowitz, W. Johnson, Phys. Rev. D **51**, 2546 (1995).
18. M. Bianchi *et al.*, Class. Quantum Grav. **13**, 2865 (1996).
19. M. Cerdonio *et al.*, Phys. Rev. Lett. **87**, 031101 (2001).
20. R. E. Vogt, *The U.S. LIGO Project*, in "International Report LIGO", pagg. 91-97, Caltech (USA) 1991.
21. F. Acernese *et al.*, Class. Quantum Grav. **20**, S609-S616 (2003).
22. M. Hewitson *et al.*, Class. Quantum Grav. **20**, S581-S591 (2003).
23. M. Ando *et al.*, Class. Quantum Grav. **19**, 1409-1412 (2002).
24. The VIRGO Collaboration, *VIRGO Final Design*, ETS Pisa, Italy (1995)
25. G. Ballardin *et al.*, Rev. Sci. Instrum. **72**, 3643-3652 (2001).
26. K. Danzmann, "*Proceedings of the Thirteenth International Conference On General Relativity and Gravitation*", R. J. Gleiser, C. N. Kozameh and O. M. Moreschi editors (Institute of Physics Publishing, London, 1993).

THE LEP LEGACY

G. GIACOMELLI & R. GIACOMELLI
Physics Dept., University of Bologna, INFN, Sezione di Bologna

In this lecture we shall summarize the scientific legacy of LEP, in particular in connection with the Standard Model of Particle Physics; we shall also discuss some historical and sociological aspects of the experimentation at LEP.

1. Introduction

In order to better understand the LEP contribution to particle physics it is appropriate to briefly recall the main features of the Standard Model (SM) of the Electroweak (EW) and Strong (SI) Interactions [1]. In this theory the fundamental constituents of matter are the quarks and the leptons, which may be considered as pointlike. The quarks and leptons appear in 3 families (generations), each made of 2 quarks and 2 leptons, one neutral and one negatively charged. The first family consists of the quarks u, d and the leptons ν_e , e^-. The second family is composed of the quarks charm (c) and strange (s) and by the leptons ν_μ and μ^- ; the third family is composed of the quark top (t) and bottom (b) and by the leptons τ^- and ν_τ. According to the Strong Interaction theory (QCD) each quark comes in three colours, green, red and blu. The only difference between families is the mass which becomes progressively larger when going from the first to the second and third family [and the lifetimes become smaller]. The SM does not explain why there are 3 families. In the SM to each quark and lepton corresponds an antiquark and an antilepton.

Quarks and leptons are subject to the EW Interaction mediated by the photon and the weak intermediate bosons W^+, W^-, Z^0. The SI between coloured quarks is mediated by 8 gluons (the leptons are not subject to the SI).

The formal structure of the SM is based on the gauge symmetry, which requires zero masses for quarks and leptons. In order to explain the observed masses, we introduce at least one scalar Higgs boson, which is needed for the spontaneous breaking of the symmetry and the generation of masses: the observed masses are due to the interaction of the Higgs boson with quarks and leptons. The Higgs also accounts for the large masses of the intermediate vector bosons. The coupling of the Higgs boson is predicted by the SM, but not its mass.

LEP yielded a very large number of important experimental results (see the Particle Data Books) and has placed the SM on a solid experimental ground.

The Large Electron Positron collider (LEP) was housed in a 27 km tunnel at ~100 m underground [2]. In 4 locations were placed the experiments ALEPH, DELPHI, L3 and OPAL. LEP took data from 1989 till 1995 at c.m. energies of ~91 GeV (LEP1) and from 1995 to 2000 at energies of 130-209 GeV (LEP2).

High energy colliders allow to study particle collisions at the highest energies, since the c.m. energy grows linearly with beam energy E_b, $E_{cm}=2E_b$. e^+e^- collisions allow the best study of the fundamental particles and of their interactions.

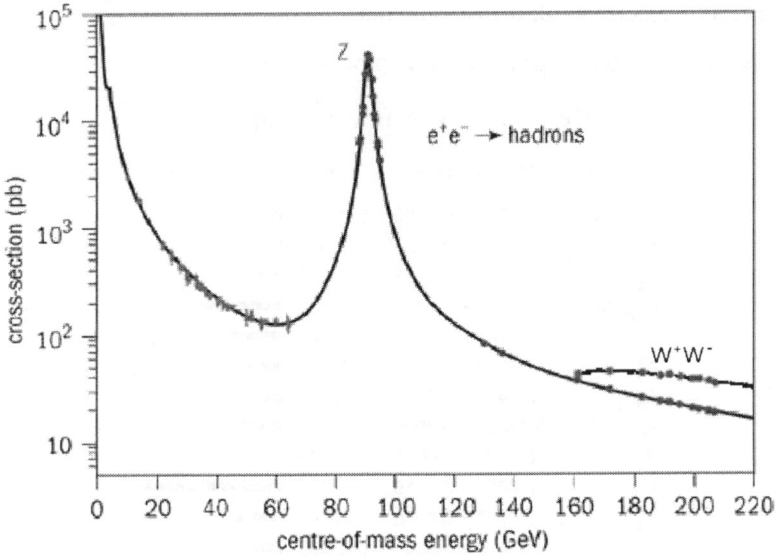

Fig.1. Hadron production cross section for $e^+e^- \rightarrow q\bar{q} \rightarrow$ hadrons vs c.m. energy.

Besides energy, another important parameter of a collider is its luminosity L, defined as that number which multiplied by a cross-section σ gives the collision rate N: $N=L\sigma$. LEP had luminosities of 10^{31}-10^{32} $cm^{-2}s^{-1}$ which yielded collision rates of ~1 event/s at LEP1 and ~0.01 event/s at LEP2. Recent lower energy e^+e^- factories have much larger luminosities.

Each of the 4 LEP experiments was a nearly 4π general purpose detector, made of many subdetectors. Their combined role was to measure the energy, direction, charge, and type of each produced particle. Apart from neutrinos and neutralinos, no particle was able to escape without leaving some sign of its passage. Each subdetector had a cylindrical structure with a "barrel" and two "end-caps". Tracking was performed by a central detector; electron and photon energy measurements were carried out by electromagnetic calorimeters; the magnet iron yoke was instrumented as a hadron calorimeter and was followed by a muon detector [3]. A forward detector completed the e.m. coverage and was used as a luminosity monitor. The quality of the detectors and the relatively low event rate allowed to study in detail each event.

Fig.1 shows a compilation of data on $e^+e^- \rightarrow$hadrons up to the highest LEP energies: up to 60 GeV the cross section decreases smoothly, then it is dominated by the Z^0; at higher energies it decreases, and above 160 GeV there is a structure connected with the opening up of the $e^+e^- \rightarrow W^+W^-$ channel.

2. Precision electroweak measurements

At energies around the Z^0 peak the basic processes are

$$e^+e^- \rightarrow Z^0,\ \gamma \rightarrow f\bar{f}\ ,\ \ f\bar{f} = q\bar{q}\ ,\ l\bar{l}\,. \tag{1}$$

	Measurement	Fit
$\Delta\alpha^{(5)}_{had}(m_Z)$	0.02761 ± 0.00036	0.02770
m_Z [GeV]	91.1875 ± 0.0021	91.1874
Γ_Z [GeV]	2.4952 ± 0.0023	2.4965
σ^0_{had} [nb]	41.540 ± 0.037	41.481
R_l	20.767 ± 0.025	20.739
$A^{0,l}_{fb}$	0.01714 ± 0.00095	0.01642
$A_l(P_\tau)$	0.1465 ± 0.0032	0.1480
R_b	0.21630 ± 0.00066	0.21562
R_c	0.1723 ± 0.0031	0.1723
$A^{0,b}_{fb}$	0.0992 ± 0.0016	0.1037
$A^{0,c}_{fb}$	0.0707 ± 0.0035	0.0742
A_b	0.923 ± 0.020	0.935
A_c	0.670 ± 0.027	0.668
A_l(SLD)	0.1513 ± 0.0021	0.1480
$\sin^2\theta^{lept}_{eff}(Q_{fb})$	0.2324 ± 0.0012	0.2314
m_W [GeV]	80.425 ± 0.034	80.390
Γ_W [GeV]	2.133 ± 0.069	2.093
m_t [GeV]	178.0 ± 4.3	178.4

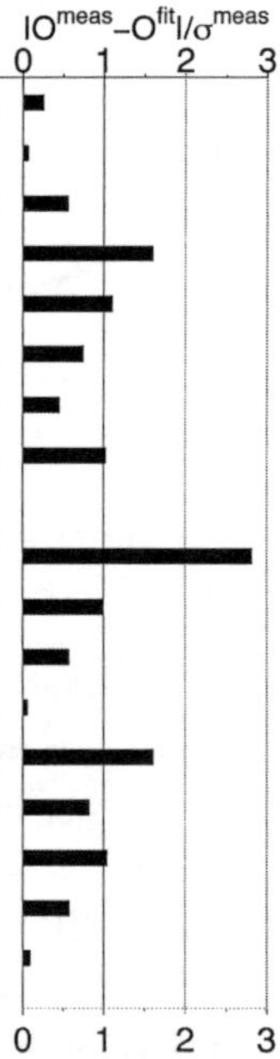

Table 1. Precision measurements of the electroweak parameters obtained by the global fit of all the data from the four LEP experiments [4].

The $q\bar{q}$ pairs are $u\bar{u}$, $d\bar{d}$, $s\bar{s}$, $c\bar{c}$, $b\bar{b}$. Each q or $\bar{q}$ hadronizes (fragments) into a jet of hadrons. The $l\bar{l}$ pairs are charged (e^+e^-, $\mu^+\mu^-$, $\tau^+\tau^-$) or neutral ($\nu_e\bar{\nu}_e$, $\nu_\mu\bar{\nu}_\mu$, $\nu_\tau\bar{\nu}_\tau$).

The behaviour of the cross-section around the Z^0 peak is typical of a resonant state with J=1, and is well described by a relativistic Breit-Wigner formula plus electromagnetic and interference terms. The formula has to be convoluted with initial state radiation. Around the Z^0, the last 2 terms are small corrections to the main Z^0 Breit-Wigner term. The formula depends on m_Z and on the partial width for the Z^0 decay into a fermion-antifermion pair. The total width Γ_Z is given by

$$\Gamma_Z = \Gamma_h + \Gamma_e + \Gamma_\mu + \Gamma_\tau + N_\nu \Gamma_\nu = \Gamma_{vis} + \Gamma_{inv} \tag{2}$$

where Γ_h is the hadronic width and Γ_e , Γ_μ , Γ_τ , Γ_ν are the leptonic widths.

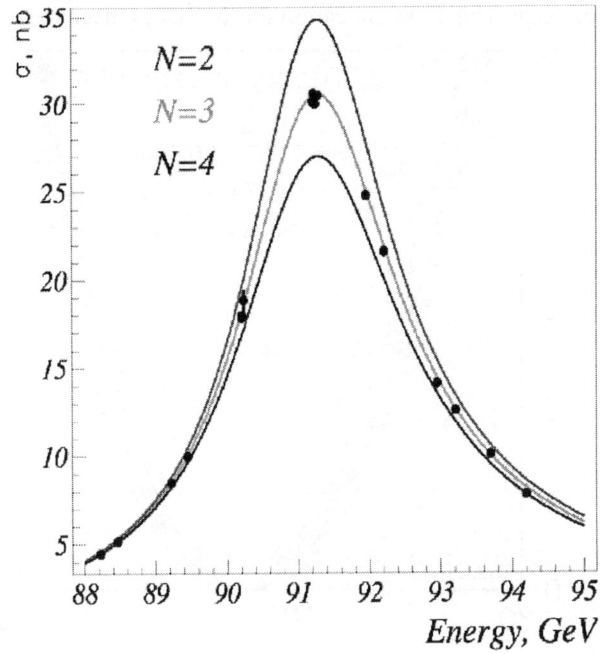

Fig.2. The shape of the Z^0 resonance yields information on the number of light neutrino types [three and only three].

At each energy around the Z^0 peak, measurements of the cross-sections were made for $Z^0 \to$hadrons, $\to e^+e^-$, $\to \mu^+\mu^-$, $\to \tau^+\tau^-$, the forward-backward lepton asymmetries A^e_{FB} , A^μ_{FB} , A^τ_{FB} , the τ polarization P_τ, the $b\bar{b}$ and $c\bar{c}$ partial widths and forward-backward asymmetries, and the $q\bar{q}$ charge asymmetry.

To combine results from the 4 LEP experiments, each experiment provided a set of optimized parameters (at beginning 9 parameters, 5 if lepton universality is assumed). Later more parameters were added. The latest ones are m_W, Γ_W.

Many parameters are expressed in terms of the effective EW mixing angle

$$\sin^2\theta^l_{eff} = \frac{1}{4}(1-\frac{g_v}{g_a}) \,. \tag{3}$$

The present best values of this and other parameters are given in Table 1.

The properties of the Z^0 have been studied with great precisions by the 4 experiments [4]: the Z^0 mass is now known with a precision of 2 parts in 100000 and the lifetime to 0.1 %. The measurements of the couplings of the Z^0 to quarks and leptons are tests of the SM to 0.1 %. Results were also obtained assuming lepton universality, which seems to be well established at LEP. The interactions of the Z^0 are those predicted by the gauge symmetry, while the masses do not reflect the symmetry. This is an indication for the Higgs mechanism.

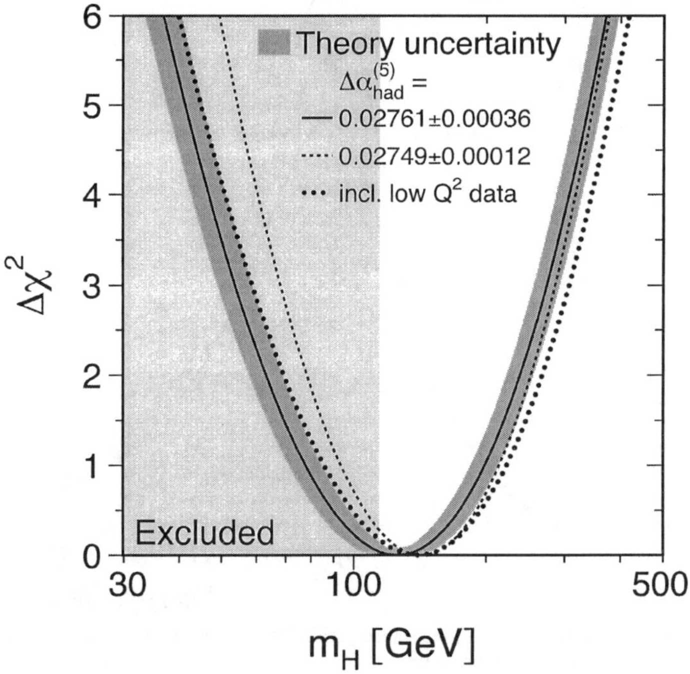

Fig.3. Fit probability vs m_H. The solid line is the result of the fit using all EW data; the band is an estimate of the theoretical error due to missing higher order corrections. The vertical line at m_H=115 GeV is the 95% C.L. limit from direct searches.

The number of neutrino types. The Z^0 decays "democratically" into any possible channel. Thus the width of the Z^0 increases with the number of generations (the number of neutrino types), Fig.2 [the lifetime decreases]. The combined measurement yields N=2.9841±0.0083: the number of neutrino types with masses lower than $m_Z/2$ is three and only three. This is one of the main results of LEP and SLC.

Determination below threshold of the top quark mass. Virtual particles affect the masses and couplings of the EW gauge bosons. In 1993 assuming the validity of the SM it was possible to deduce the top quark mass even if the top was not directly observable because it is too heavy (it was observed at Fermilab in 1995). The precision LEP measurements together with precise theoretical calculations allowed to determine accurately the t quark mass: it was a *discovery below threshold* [today the t quark mass determined below threshold is 171±10 GeV, to be compared with 174±5 GeV measured at Fermilab].

$\mathbf{m_H}$**.** The method used below threshold for the t quark was also used for the Higgs boson. Unfortunately the quantum corrections introduced by the H^0 are logarithmic, thus not as sensitive as those from the t quark. One could only establish that the Higgs boson mass had to be smaller than 200 GeV [95% C.L.], see Fig.3. It is an important constraint since in the SM the Higgs mas was only confined to be between 1 GeV and 1 TeV. The comparison with the direct searches, also shown in Fig.3, will be discussed later [5].

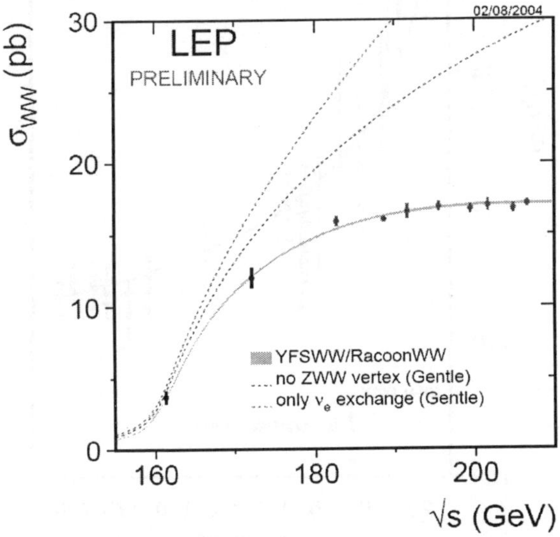

Fig.4. The total cross section for W-pair production at LEP2. The experimental data are compared with the SM prediction and also with the predictions assuming only ν_e exchange and no ZWW coupling (both excluded by the data).

Precision measurements at LEP2. The study of the reaction $e^+e^- \rightarrow W^+W^-$ allowed precision measurements of the W mass and the proof of the *existence of the triple bosonic vertex ZWW*, required by the SM, Fig.4 [6].

In the early LEP analyses one assumed lefthanded massless Dirac neutrinos and assumed separate conservations of the electron, muon and tau leptonic numbers. The presence of neutrino oscillations [7] forces to make changes: only the total lepton number $L=L_e+L_\mu+L_\tau$ seems now to be conserved, and one should include some right handed neutrinos. It is possible that these changes may be included in the SM, but it is also possible that the evidence for neutrino masses indicate physics beyond the SM [1].

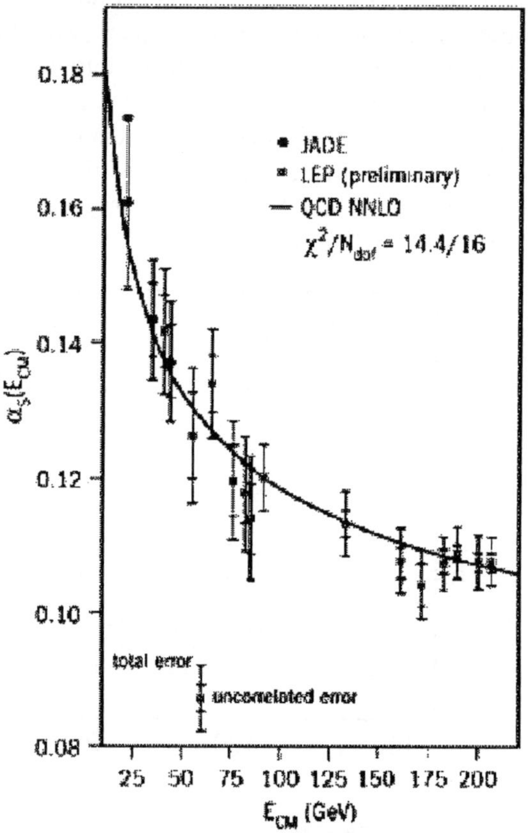

Fig.5. The running of the strong coupling constant.

3. QCD

Since the Z_0 decays predominantly into $q\bar{q}$ pairs, it yields a clean data sample with which to test quantum chromodynamics (QCD), the theory of the strong interaction. The $q\bar{q}$ pair is not observed directly, but it gives rise to two opposite jets of hadrons. Before fragmentation one of the quarks may radiate a gluon by a process similar to bremsstrahlung, yielding 3 jets of hadrons. The ratio of the number of 3-jets to the number of 2-jets is one way of measuring α_s, the strong coupling constant.

Multihadron production in e^+e^- annihilations proceeds via 4 phases [8].

1. The initial e^+e^- pair annihilates into a virtual Z^0/γ, which decays into a $q\overline{q}$ pair; a γ may be emitted by the initial e^+ or e^-. The production of the $q\overline{q}$ pair, described by the EW perturbative theory, occurs at distances of $\sim 10^{-17}$ cm.
2. In the second phase the q (or $\overline{q}$) radiates a gluon, which may then radiate another gluon (yielding a 3-gluon vertex), or may radiate a $q\overline{q}$ pair. This phase, described by perturbative QCD, occurs at distances of $\sim 10^{-15}$ cm.
3. Quarks and gluons hadronize (at distances of ~1 fm) into colourless hadrons.
4. In the 4th phase (described by models) the produced hadrons decay via strong or EM interactions; b-hadrons decay via WI with lifetimes of $\sim 10^{-12}$ s.

The Strong Coupling constant α_s. The coupling constant of the SI is a fundamental parameter which was precisely determined at LEP from many types of measurements. The experiments estabilished also the *flavour independence* of α_s and *the running of α_s*, that is its decrease with increasing energy, Fig.5 (Asymptotic freedom) [9]. At the Z_0 mass the value of the strong coupling constant is $\alpha_s(m_Z)=0.1176\pm0.009$: also this measurement is now a precision one!

Many other QCD studies have been made: the Colour Factors, the Physics of Heavy Flavours, the difference between hadron jets originated from quarks and from gluons [10]. Among the many phenomenological studies we may single out the study of Bose-Einstein and Fermi-Dirac correlations and the establishment of the dimensions and shapes of the hadron emission regions [11].

4. New particle searches

The SM has intrinsic inconsistencies and too many parameters. Many searches for new physics beyond the SM have been performed [12].

SUSY particles. In supersymmetric models, each particle has a SUSY partner whose spin differs by half a unit. A new multiplicative quantum number, R-parity is +1 for SM particles, -1 for SUSY partners. If R is conserved, sparticles are produced in pairs and decay to the lightest sparticle (LSP), which is the lowest mass neutralino. In the Minimal Supersymmetric Standard Model (MSSM), sparticle masses and experimental limits depend on 5 parameters.

Higgs bosons. In the MSSM one has 5 Higgs bosons, h^0, H^0, A^0, H^+, H^-; the neutral ones are searched for with methods as for the H^0_{SM} and limits are at the same level.
Charginos. The SUSY partners of the $W^\pm$ and of the $H^\pm$ mix to form 2 mass eigenstates for each sign, the charginos. Present limits are $m_{\tilde{\chi}^\pm} > 103.5$ GeV.
Charged sleptons. Each lepton has 2 scalar partners, the right and left-handed sleptons. They could be pair produced through s-channel γ/Z^0 exchange or t-channel neutralino exchange. Mass limits: $m_{\tilde{e}} > 99.4$, $m_{\tilde{\mu}} > 96.4$, $m_{\tilde{\tau}} > 87.1$ GeV.
Scalar quarks. The decay modes $\tilde{t} \rightarrow c + \tilde{\chi}_1^0$ yield $m_{\tilde{t}} > 95$ GeV.
Neutralinos. The $\tilde{\gamma}, \tilde{Z}^0, \tilde{h}^0, \tilde{H}^0$ mix to form 4 mass eigenstates, the neutralinos. They may be pair produced through s-channel Z exchange or t-channel electron exchange. The lowest mass neutralino is unobservable at LEP; one looks for $\tilde{\chi}_1^0$ production in $e^+e^- \rightarrow \tilde{\chi}_1^0 \tilde{\chi}_1^0 \gamma$ or $\rightarrow \tilde{\chi}_2^0 \tilde{\chi}_1^0$ with $\tilde{\chi}_2^0 \rightarrow \tilde{\chi}_1^0 l^+ l^-$. The limit for the neutralino mass is ~40 GeV. Better limits exist for specific values of the SUSY parameters. The lowest mass neutralino may be a component of the *Dark Matter*.
R-Parity violation. If R is violated, sparticles may be produced singly; there are no constraints on the nature and stability of the LSP (if it has a large lifetime it crosses the whole detector). Limits are given in the context of specific models.
Excited fermions. Compositeness. Composite models predict the existence of excited fermions, F^*, with the same EW couplings to the vector bosons as the fermions. They may be produced in pairs or singly. For photonic decays the final states involve two leptons and two photons; for neutrinos, the final states involve 2γ plus missing energy/momentum. Present limits for singly produced excited fermions are ~102 GeV.
Leptoquarks. Leptoquarks (LQs) are predicted in models which explain formally the symmetry between quarks and leptons; they may be produced in pairs and each LQ decays into lepton + quark. At LEP present mass limits are >100 GeV.
Heavy charged and neutral leptons. Searches for long-lived charged (neutral) heavy leptons, $e^+e^- \rightarrow L^+L^-$ ($e^+e^- \rightarrow N_l\overline{N}_l \rightarrow lW\bar{l}W$) used the central detectors and dE/dx measurements. Some searches for L^0's we made assuming $e^+e^- \rightarrow L^+L^-$, $L^\pm \rightarrow L^0 W^\pm$.
Fast heavily ionizing Dirac Magnetic Monopoles have been searched for directly, $e^+e^- \rightarrow M\overline{M}$, using nuclear track detectors or central detectors [12].

5. Historical and sociological aspects

In the last 50 years there were great changes in the organization and structure of particle physics experiments. In the 1950's the standard experiment was small and was performed by a small group of physicists, students and technicians from a single University. Experimentation at higher energies forced the concentration of experiments in large national and international labs, where accelerators were available. In the 1960's the bubble chamber experiments started to be performed by collaborations of few groups from different Institutions. In the 1970's the same trend was present in counter experiments. The experiments at LEP required another step, with tens of groups and hundreds of physicists and engineers, with interconnections at the national and regional levels (future experiments at the LHC require hundreds of groups and thousands of physicists, with interconnections at the world level). In the following we discuss some aspects of the sociology of the LEP experiments and their changes with time [13].

Preliminary workshops. The approval of the experiments. LEP was approved in '81. Before and after approval many physics workshops were held at CERN and in different countries. After the General Meeting in Villars, Switzerland, the collaborations started to form and prepare Letters of Intents, which were presented in 1982 to the CERN Director General and to the newly formed LEP Committee (LEPC). Later followed the Proposals, which were approved by the LEPC and the CERN Research Board. Every group of each collaboration had then to obtain the financial support from the National Research Institutions. In '83 the construction of LEP and of the 4 detectors started. This first period was evaluated to be positive and stimulating by a survey ECFA (European Committee for Future Accelerators).

Experiment construction. Each experiment had a 4π general purpose detector, with many subdetectors and hundreds of thousand electronic channels. While many young physicists were happy to construct equipment, others feared the lack of physics papers during the long construction period.

The first physics results. In august '89 the first beam became available: there was a strong competition among the experiments to observe the first event. In the subsequent runs, the shape of the Z^0 resonance was measured and this lead to one of the most outstanding results: there are 3 and only 3 types of neutrinos. The groups were busy completing and commissioning their detectors. During this period the number of young physicists at CERN increased considerably, and everybody was very active. It was a very exciting period.

Detector improvements. The available forward detectors allowed luminosity measurements (using the forward $e^+e^- \to e^+e^-$ cross section) to few %. In order to fully exploit the accelerator and the detectors, precision "Luminometers" were designed and built: they allowed measurements to better than 0.1%. It was also essential to compute the radiative corrections of forward Bhabha elastic scattering to ever increasing precision: there was a healthy collaboration between theorists and experimentalists to reach the desired goal. The final measurements were made with impressive precision, much better than expected.

Organizational structure of the collaborations. In the large LEP collaborations there was a need of an elaborate organization, which took a long time to set up. Each collaboration had a spokesman, physics coordinators, a governing body, a sort of parliament, project leaders, subdetector committees and a financial review committee. The governing body (Executive, Steering Committee, ...) consisted of a small number of physicists, with availability of experts when needed. The Collaboration Board with the group leader of each Institution, had the last word on most items.

From LEP1 to LEP2. From august '89 till the middle of '95, LEP was operated at energies around the Z^0 peak. All fields of research benefited from the increased luminosity. From 1994 all groups started improvements in order to be ready for higher energies. Among the improvements it is worth recalling the longer and more refined vertex detectors, designed to improve the performance of Higgs searches.

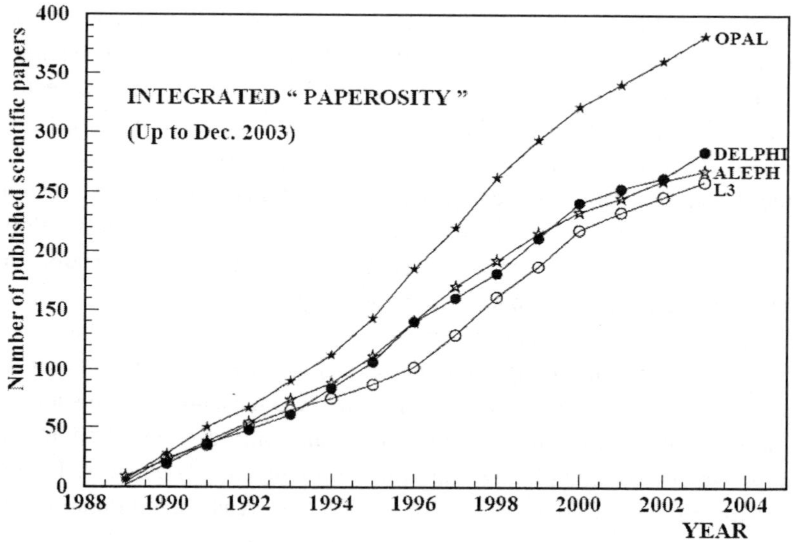

Fig.6. The integrated number of scientific papers published by each LEP collaboration from 1989 to 2003 (courtesy of Fabrizio Fabbri).

Paperosity. In the 11.5 years of LEP operation, 1020 scientific papers were published by the 4 collaborations, which totally included about 1730 authors. Fig.6 shows the integrated number of papers published until December 2003. Table 2 compares, for experiments at different colliders, the average number of authors and the ratio <R> = [number of papers/number of authors]. It is difficult to make a comparison since the duration of the experiments was different and the table does not include quality nor discoveries. But one can state that the LEP groups fare well in the comparison.

Collaboration	Average number of authors	<R>
LEP	330 – 550	0.5 – 1
CDF - D0	400 – 500	0.4 - 0.7
H1 – ZEUS	350 – 450	0.2 – 0.4
UA1 – UA2	65 – 150	0.33

Table 2: Various large high energy collaborations, their approximate average number of authors and the ratio <R> = [number of papers/number of authors] [11].

Visibility. For graduate students and young researchers it is important that their work be properly recognized. Visibility is not evident from papers with hundreds of authors. But each researcher may find his proper place inside a collaboration because of the fragmentation of responsabilities connected with the realization, operation, maintenance of complex equipments, and even more in physics analyses. Inside a collaboration there were presentations in working groups and in collaboration meetings, and also refereed and not refereed internal notes. Outside, there were presentations to conferences and invited papers. This may favour physicists who perform physics analyses, but there are also many technical workshops. From the results of the ECFA enquiry it seems that active young physicists can find a proper recognition, even inside large collaborations.

Scientific computing. There was on-line computing and off-line reconstruction of events, Monte Carlo studies and physics analyses. On-line computing was mostly done by clusters of Vax-stations; changes were few since it is difficult to do them in running experiments. Off-line computing had fast changes following the trend in the computing field. In 1983 the off line analyses were done by "large computers" of the type IBM 370/168 (referred to as "one unit"). At that time CERN promised to each experiment the availability at CERN of 3 computing units. At the end of LEP each experiment had a capacity of more than 1000 computing units! This change was done in steps, using new clusters of working stations. During the same time the memory space went from MegaBytes to GigaBytes, the interconnectivity improved dramatically and CERN made its most important invention: WWW, the World-Wide-Web.

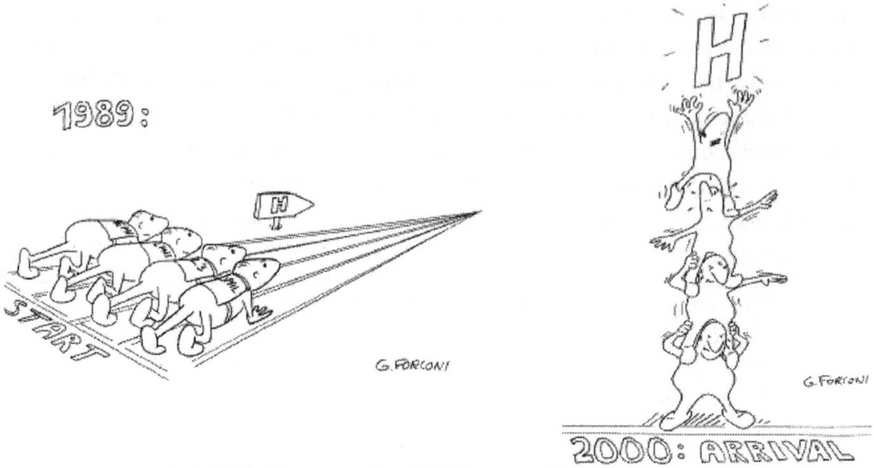

Fig.7. The Higgs search at LEP (courtesy of G. Forconi).

LEPC. It played an important role in all stages of the experiments. It came to an end at the end of year 2000: was it a glorious end? Did it investigate thoroughly the possibility of buying extra RF cavities, when it was still possible? An energy increase could have been important for the SM Higgs search.

Estimates of the Higgs boson mass came from precision EW data and from direct searches. While at the beginning of LEP, the 4 experiments raced one another in the search for the Higgs, at the end they combined their results as indicated in the cartoons in Fig.7. The combination of data from the 4 experiments became a standard procedure in most fields of research. It allowed to cross check data and obtain more precise results.

The secretariats. Each experiment had an efficient open door secretariat which provided scientific and bureaucratic information, and was called to solve every possible problem.

Sport. The 4 experiments participated with great enthusiasm in the sport life at CERN. Each experiment had several race teams for the annual CERN relay and road races of all categories (seniors, veterans, ladies, open, etc...). The team and sport spirit were at their best: people were happy also when they won the "random prize"!

Love affairs. In a large collaboration it is normal to have love affairs among collaborators, and even among members of different experiments! The sociology of "love affairs" followed the changing life pattern during the 15 years of the experiments. In the early 80's the word "fiancé" was a fine and used word, while later it almost disappeared and other terms were used, like partner, boyfriend, girlfriend, etc. There were "new experiences", course changes, encounters, new encounters and very few marriages. At the start of the new millennium there was an increase in the number of marriages, but "baby production" remains limited, much too low to compete with that in the developing countries!

6. Conclusions

The study of e^+e^- interactions at LEP provided many important results [14 15], like the 3 neutrino generations, precision determination of the electroweak parameters and of the strong interaction parameters, the running of the strong coupling constant, precise measurements of the lifetimes of short lived particles from b and τ decays, the existence of the triple bosonic vertex, precise measurement of the W mass, the determination below threshold of the mass of the top quark, possible indications of the Higgs mass, the physics of heavy flavours [16], (b, τ), etc. It may be worthwhile to stress that the precision reached in most measurements was much better than what anticipated, and that the Particle Data Book is full of LEP results. But no new surprice was found.

One should not neglect the very large number of Diploma, Laurea and PhD theses with data from LEP, and the strong impact of LEP on the public understanding of science.

It seems that most physicists involved in one of the 4 experiments considered the LEP experience an exciting experience, in particular when they were obtaining the most interesting physics results.

Acknowledgements. We thank many colleagues from LEP, the LEP experiments, theoreticians, the members of the Bologna team, the technical staff and B. Poli, F. Fabbri, Y. Becherini, A. Casoni for their collaboration.

References

1. A. Bartl, The SM and beyond, Lectures at this School.
2. E. Picasso, The LEP collider, Present and future collider Physics, Rome (1990).
3. See for example: K. Ahmet et al., Nucl. Instrum. Meth. A305(1991)275.
4. The LEP Collaborations and the LSD heavy flavour group, A combination of preliminary EW measurements and constraints on the SM, hep-ex/0312023 (2003).
   G. Abbiendi et al., Eur. Phys. J. C19(2001)587.
   G. Alexander et al., Z. Phys. C52(1991)175.
5. R. Barate et al., Phys. Lett. B565(2003)61.
   G. Abbiendi et al., Phys. Lett. B499(2001)38.
6. C. Sutton et al., The W and Z at CERN, CERN Courier 44, n. 4(2004)21.

7. Lectures on neutrinos at this School by S. Petcov, G. Giacomelli and D. Cowen. M. Ambrosio et al. Phys. Lett. B357(1995)481; Phys. Lett. B434(1998)451; Eur. Phys. J. C36(2004)323.
8. G. Giacomelli et al., Results from high energy colliders, Lectures at the 5th and 6th Particle Astrophysics Schools, DFUB 98/23; hep-ex/0202023.
9. D.J. Gross and F. Wilczek, Phys. Rev. D8(1973)3633.
   H.D. Politzer, Phys. Rev. Lett. 30(1973)1346.
   K. Ackerstaff et al., Eur. Phys. J. C7(1999)571.
10. G. Alexander et al., Z. Phys. C69(1996)543.
11. M. Cuffiani et al., Nuovo Saggiatore 18 (2002) n.1-2, pag. 46.
    G. Abbiendi et al., Eur. Phys. J. C16(2000)423.
12. S. Braibant et al., New Physics searches at LEP, ICHEP04, Beijing hep/ex.
    G. Giacomelli et al., Particle searches at LEP, Riv. del Nuovo Cimento 16 n.3(1993)1.
    K. Kinoshita et al., Phys. Rev. D46(1992)881.
13. F. Fabbri et al., Gli esperimenti al LEP del CERN: Risultati scientifici e aspetti storici e sociologici, Analysis n. 3/2002, 54.
14. G. Altarelli, Come LEP ha cambiato la fisica delle particelle, INFN-Notizie, dec. 2000, pag.12.
    C. Matteuzzi, L'eredità di LEP, INFN-La ricerca italiana in Fis. Subatomica, 2003.
15. J. Ellis, Conf. Summary, ICHEP 2004, Beijing, hep-ph/0409360.
    F. Teubert, Precision Electroweak interactions, ICHEP 2004, Beijing.
16. K. Ackerstaff et al., Eur. Phys. J. C5(1998)379; Z. Phys. C73(1997)401.

FUTURE ACCELERATORS, NEUTRINO FACTORIES, AND MUON COLLIDERS

R. A. CARRIGAN, JR.,
Fermi National Accelerator Laboratory[†], *Box 500*
Batavia, IL 60510, USA

This chapter reviews the current status of the accelerator field, accelerator facilities under development or discussion, and longer range possibilities. Some of the topics covered include the Tevatron, LHC, the new possibility of an International Linear Collider, new neutrino facilities including neutrino factories, and muon colliders. The possibilities of exotic accelerators like plasma wake field systems are also reviewed.

1. Introduction

Today particle physics faces several great questions; what are the fundamental symmetries of nature and their generators, the nature of space, the nature of the neutrino mass spectrum, the sources of missing matter and energy, and the origin of CP violation. In addition there may be other important questions yet to be identified. Accelerators are a key to answering many of these problems. It is not an overstatement to say that for some of these issues, the accelerator is almost the experiment. Indeed some of these questions require machines beyond our present capability.

As this volume attests, there are parts of the particle physics program that have been significantly advanced without the use of accelerators such as the subject of neutrino oscillations and many aspects of the particle-cosmology interface. At this stage in the development of physics both approaches are needed and important.

This chapter first reviews the status of the great accelerator facilities now in operation or coming on within the decade. Next, midrange possibilities are discussed including the International Linear Collider as well as others like neutrino factories, gamma-gamma colliders, muon colliders, and very large hadron colliders. Finally visionary possibilities are considered including plasma and laser accelerators.

[†] Operated by Universities Research Association, Inc. under contract No. DE-AC02-76CHO3000 with the United States Department of Energy

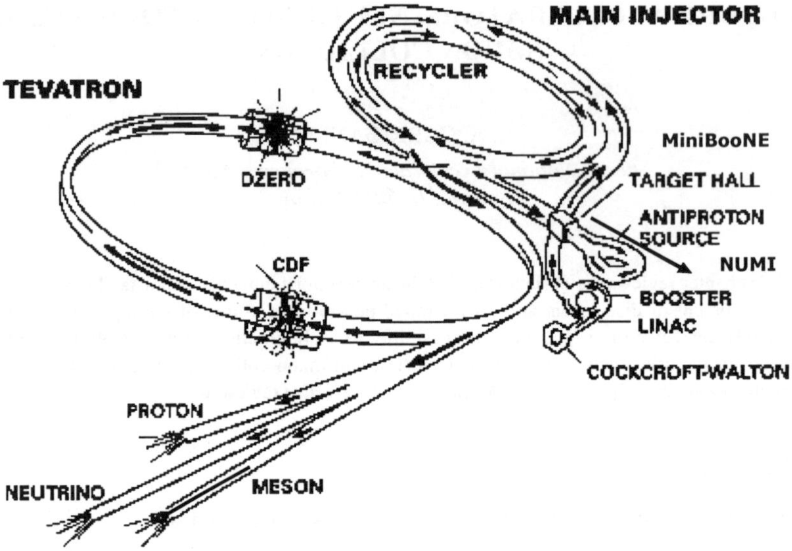

Figure 1: Tevatron schematic.

2. Accelerators from now to 2009

The main proton accelerator now in operation is the Tevatron. In 2007 it will be joined by LHC, the 800 pound gorilla of the accelerator world. Meanwhile accelerator-driven long base line neutrino facilities have come into operation. Asymmetric e^+e^- colliders (not discussed here) have made interesting progress on CP violation studies.

The upgraded Fermilab Tevatron started operation in the spring of 2001. The machine now operates with 36 colliding bunches rather than the 6 bunches used in the past. This has been made possible by the new Main Injector which takes the pre-Tevatron accelerator out of the Tevatron tunnel. This move also reduces backgrounds at the collider detectors. The Tevatron energy has been pushed closer to 1 TeV. Progress in 2004 has been spectacular with the luminosity doubling. The peak as of mid July, 2004 had passed $10^{32}/cm^2s^1$. With planned luminosity upgrades there is even some possibility of discovering supersymmetry and a low mass Higgs.

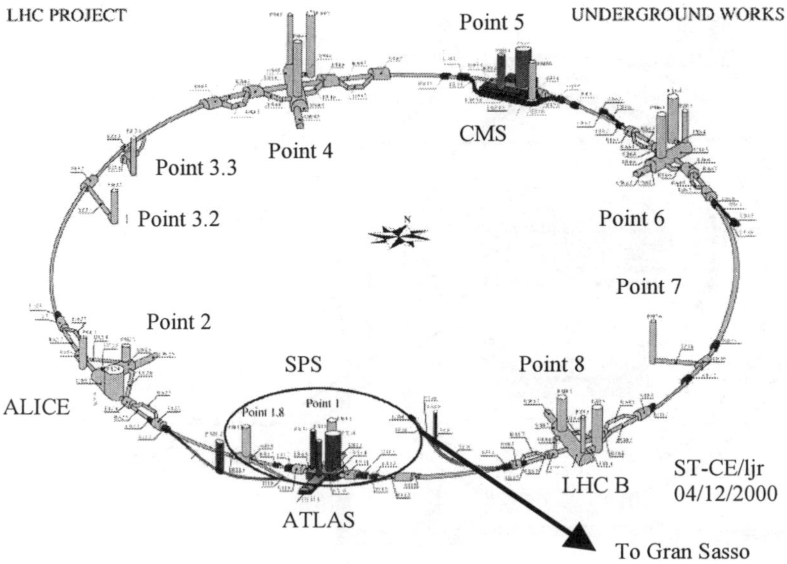

Figure 2. LHC schematic (from CERN).

The Fermilab Tevatron is shown schematically in Figure 1. The complex is a bestiary of different types of accelerators. The venerable Cockcroft-Walton accelerates protons (actually negative hydrogen ions) in a DC electrical field. The Linac uses a series of RF cavities to accelerate them up to 400 MeV. From there the hydrogen ions are stripped and the protons injected into the Booster. The Booster is an aging rapid cycling, combined function accelerator. From the Booster the protons travel to the new Main Injector where they are accelerated to 150 GeV. Main Injector protons are used to make antiprotons and feed protons and antiprotons to the Tevatron. The antiprotons are accumulated in three storage rings, first the debuncher, then the accumulator. Together these two rings are the antiproton source shown in the figure. Post store antiprotons can be saved in the Recycler in the Main Injector tunnel. The core parts of the physics program, the collider experiments, are located in the Tevatron tunnel. The collider detectors, CDF and D0, have been completely re-equipped to take care of the higher rates. The Main Injector protons are also used to support NuMI (Neutrinos at the Main Injector). In late 2004 the NuMI experiment will send a neutrino beam 735 km to a large, underground detector located at Soudan in Minnesota. This is discussed in more detail under the neutrino factory section. A second, lower energy neutrino experiment, MiniBooNE has been underway

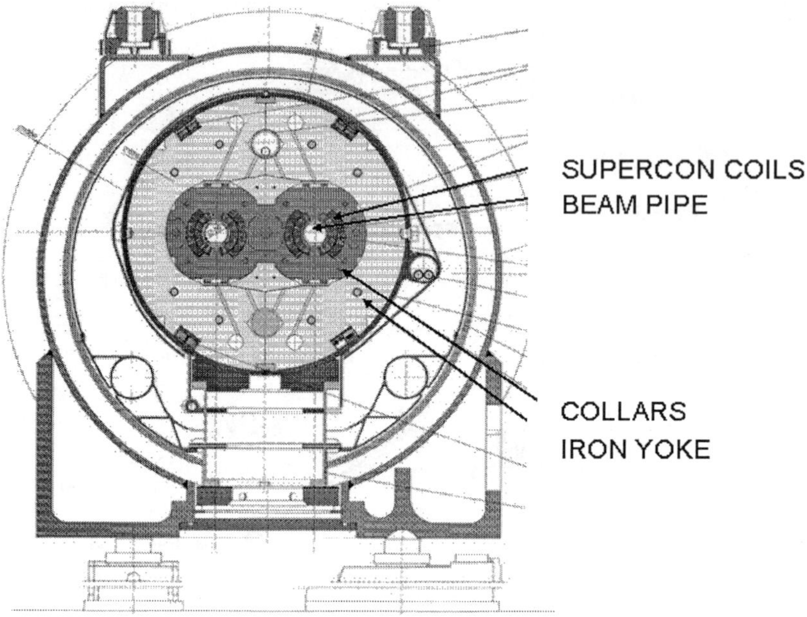

Figure 3: LHC magnet cross section (CERN).

for several years. The MiniBooNE neutrino beam is fed by protons from the Booster.

In 2007 the new Large Hadron Collider (LHC) will come into operation at CERN in the old LEP tunnel. Figure 2 shows the LHC layout. The big new detectors are ATLAS and CMS. Facilities have been provided in the complex to feed a neutrino beam to Gran Sasso. The LHC beam energy will be 7 TeV, seven times the Tevatron. The peak luminosity will be 10^{34} $cm^{-2}s^{-1}$ or 20 times the peak luminosity anticipated for the Tevatron. Many brands of SUSY and Higgs should be observable at LHC. The LHC is truly a marvelous accelerator. In fact, its existence raises the bar for future projects since they must be able to probe beyond the range of the LHC. This will be no easy challenge.

The key to any proton circular accelerator is the magnet system for deflecting the circulating beams. Figure 3 shows a cross section of the superconducting LHC dipole. The superconducting LHC magnet has a so-called two in one geometry to accommodate the two counter-rotating proton beams in the same iron yoke. Much of the field shaping comes from the superconducting coils. Note that the beam pipe is rather small to reduce the size of the magnet and the overall cost. The whole assembly is kept at the temperature of super fluid helium. These magnets are technology marvels. The sophisticated ensembles of machines at CERN and Fermilab represent the highest state of accelerator science and art.

Useful perspectives and mathematical tools for accelerator science and engineering can be found in the textbook by Edwards and Syphers [1] and the Tigner-Chao handbook [2]. The Tevatron and LHC rely on a number of individual components and systems. A breakthrough in any one of the areas could result in an important advance in the field. As an example particles are accelerated with radio frequency (RF) cavities. Linacs are just a series of RF cavities strung together. Conventional cavities (including superconducting models) can reach accelerating gradients of 50-100 MeV/m. A breakthrough here would be important but unexpected. Bending magnets are required for circular machines. These are now limited by the available superconductors to peak fields on the order of 10 Tesla. Focusing must be provided in addition to bending. Focusing is particularly important near colliding beam points. Again the technology has been pushed to the edge. Moving into future accelerator systems requires progress in these and other areas. In addition cost minimization is extremely important.

3. The next decade (2010-2019)

Looking beyond the LHC several interesting possibilities for accelerators have emerged. At the top of the list is the linear collider, colliding beams of electrons and positrons. A second possibility is the so-called muon collider. This is like an electron-positron collider in principle but the practical realization is completely different. Third is the possibility of high intensity neutrino facilities. Fourth is a logical extension of the LHC, the obviously-named, Very Large Hadron Collider or VLHC. The electron-positron collider can also be operated as a gamma-gamma collider. The muons needed for a muon collider might also be used to make a neutrino factory.

Each of these four possibilities has its physics advocates. Higgs signatures on the linear collider are clean. Indeed, CERN's LEP (an e^+e^- collider but not a linear collider) saw evidence of a Higgs signature that was later discounted [3]. On the other hand the cross section for Higgs production at a muon collider goes as $(m_\mu/m_e)^2$ so that all other things being equal the production rate would be 40,000 times larger on a muon collider. Neutrino facilities explore the very interesting topic of neutrino mass. Physics advocates for a very large hadron collider argue that it is the natural extension of the LHC.

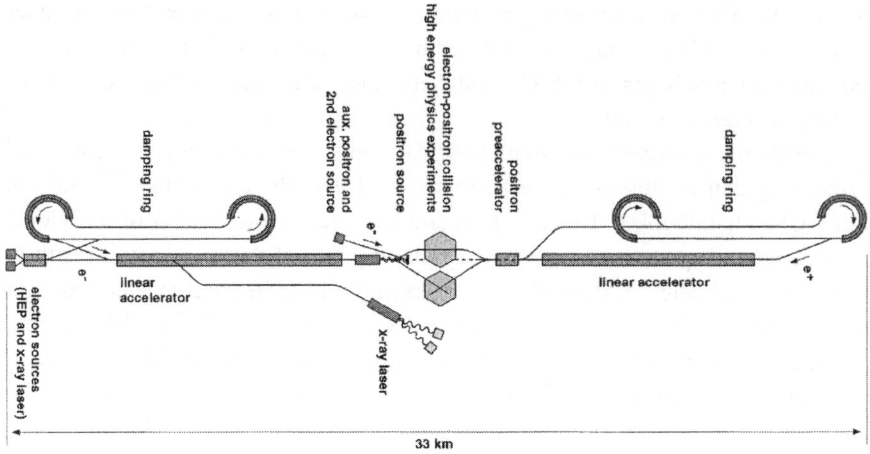

Figure 4-TESLA layout for neutrino production (DESY).

At present a linear collider is closest to realization. A muon collider is challenging. More likely is that a precursor muon factory will be tried, possibly for neutrino production. A VLHC is straightforward but very costly. There is also concern about starting a VLHC before early LHC results are in.

3.1. *Linear collider*

The ultimate requirement for an e^+e^- collider to be a linear machine rather than a circular one has been understood for many years [4]. This arises because synchrotron radiation in a circular machine rises rapidly with increasing energy. The first and only linear collider built so far was the Stanford Linear Collider (SLC) at SLAC. This 50 on 50 GeV collider came into full operation after LEP. It was a challenging machine to build because of the very small beam sizes required at the collision point to achieve useful luminosity. In one way SLC was better than the circular e^+e^- LEP at CERN because it could use polarized electrons. If there had been no LEP, SLC would have been considered a spectacular success.

To be useful for new physics the total energy of a linear collider must be in the 500-1000 GeV range and the luminosity must be O (10^{34} /cm^{2}s). With 300 fb^{-1} integrated luminosity a 250 fb cross section will give on the order of 75K Higgs in a year's running. Achieving this luminosity requires extremely high beam power and small beam cross sections. These are conflicting requirements which must be solved by the design.

In late 2004 an international consortium was created to design a superconducting linear collider, the International Linear Collider or ILC, paralleling the design for the DESY TESLA system. It would operate at a frequency of 1.3 GHz. A preliminary design report for TESLA was completed at DESY in the spring of 2001 [5].

The footprint for the original TESLA project is shown in Figure 4. The complex is 33 kM long. At present the RF gradient is about 25 MV/m. Future developments such as electro-polishing should increase the gradient. Damping rings are required to achieve the small emittance. The damping rings also require wigglers to cut the damping time. Recent development of flat beams at the Fermilab A0 test facility may moderate the requirements for the electron damping ring [6].

Another interesting possibility available with a linear collider is to construct a gamma-gamma collider. One physics channel this could study is the partial decay width of the Higgs into $\gamma\gamma$. High energy gamma rays are produced by Compton backscattering, that is by colliding a linac beam with an intense laser. The $e\gamma$ collision produces a spectrum of high energy gamma rays with a peak energy of 0.8 of the electron energy moving in a cone with an angle $1/\gamma$ in the direction of the initial electron. About 10^9 laser photons are needed to make one gamma ray. This factor is proportional to the square of the laser photon wavelength divided by the classical radius of the electron. A terawatt laser with an average power of tens of kilowatts is required. Note that separate systems are needed to handle the electron and positron beams. While the lasers are large, they are not impossible.

A linac can also be used to provide a powerful free electron laser based on self amplified spontaneous emission (SASE). SASE was first proposed at SLAC. The brilliance is typically 10^8 times present light sources. An FEL based on a linear collider will produce 1 Å x-rays with a pulse length of 100 fs. The TESLA Test Facility has already demonstrated SASE for 80-180 nm wavelengths A large material science and solid state community is interested in the development of these FELs. Typically an FEL would not use the full energy of a collider linac.

Finally a more ambitious linear collider project for the future has been discussed at CERN, the Compact Linear Collider or CLIC [7]. This is a so-called two-beam accelerator where an intense beam in one accelerator drives the 33 GHZ RF system in the second accelerator. This arrangement can produce an extremely high accelerating gradient but requires four linear accelerators rather than two. A prototype has been operated and has accelerated a probe beam by 60 MeV [8] but a working facility is far in the future.

3.2. *Muon colliders*

Muon storage rings have been around for a long time. Several rings have been built to study the anomalous magnetic moment of the muon. A recent experiment using a muon storage ring at Brookhaven has found a tantalizing difference between the measured anomalous muon magnetic moment and the QED prediction [9]. The problem with muons, in contrast to positrons and anti-protons is that they decay with a lifetime of 2.2 μs. However if long straight sections are added to a muon storage ring these decaying muons result in neutrinos beamed in the direction of the straight sections. There is a second problem in this process. Muons are produced by first making pions and kaons which decay into muons. The maximum transverse momentum of the decay muon is 30 MeV/c so that a 3 GeV muon could have an angular divergence up to 10 mrad. To overcome this, the muon beam must be cooled before it is injected into a storage ring in the same spirit that dampers are required for linear colliders. As will be seen, this a huge challenge.

Muon colliders probably would not have elicited much interest except for the fact that the Higgs cross section for $\mu^+\mu^-$ collisions is expected to be much larger than for the e^+e^- case. This is enough to cause one to investigate ways solve the big problems. In a typical design protons from a high intensity source are used to produce pions which decay into muons. The muons are captured by a very large, 20 T solenoid. The solenoid is followed by a so-called Siberian snake magnetic beam line which rotates the muon spin and provides momentum selection. Next follows an ionization cooling system where the muons are cycled 20 times through a system where they are slowed and lose both longitudinal and transverse energy and are then accelerated by a linac to regain the longitudinal energy. Because of the short muon lifetime all of this must occur very rapidly. From there they are transferred to re-circulating linacs where they are accelerated to the TeV regime. Typically, the beams collide in a smaller ring. The accelerating stages require large, rapid cycling magnets.

This is not an easy project! The ionization cooling is based on straightforward energy loss and multiple scattering. However cooling takes place in a six-dimension phase space. Every factor of two counts and the details of the tails of the distributions may be important. It is generally felt that a

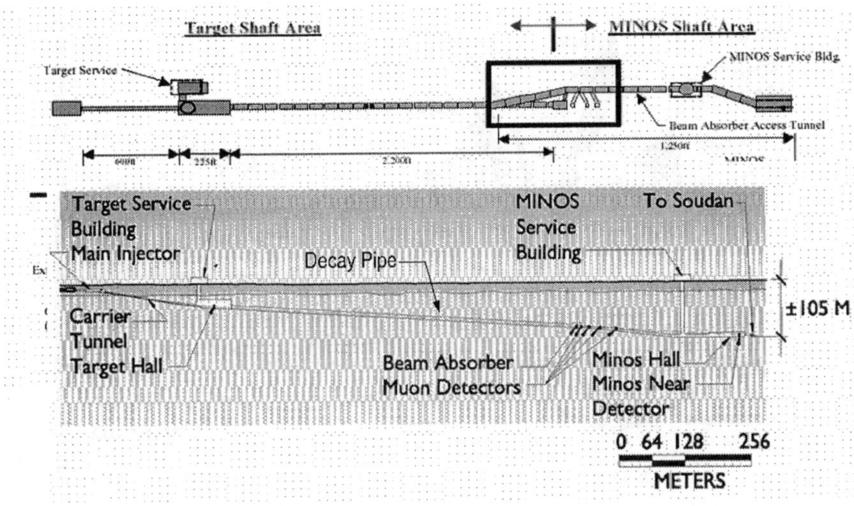

Figure 5. NuMI layout (Fermilab).

successful, full-scale test of ionization cooling will be necessary before any real design and building program could start.

3.3. *Super beams, neutrino factories, and beta beams.*

Intense neutrino facilities are another set of midrange possibilities. Some conventional beams in the works such as the NuMI facility at Fermilab are already important steps up in intensity. Beyond these developments are true neutrino factories. A non-accelerator possibility is to make improved reactor systems.

Figure 5 shows the layout of the Fermilab NuMI neutrino beam facility. The beam is driven by the 120 GeV Main Injector. The proton beam strikes the meson production target. The mesons that are produced are focused by horns and then decay in the 1 kM pipe. An impressive feature of this gigantic facility is the 57 mrad slope of the tunnel down to the Soudan detector 735 kM away.

NuMI produces a 10 microsecond pulse every 1.9 seconds. Initially the Main Injector will deliver $2.5*10^{13}$ protons per pulse. This will grow to $4*10^{13}$ later. The project will start commissioning in December, 2004. It will deliver 0.2 MW the first year rising to 0.4 W later.

Options beyond the current facilities include super beams, neutrino factories, and beta beams. Typically all of these facilities require a "proton driver", a very high current source of GeV protons. So-called super beams are

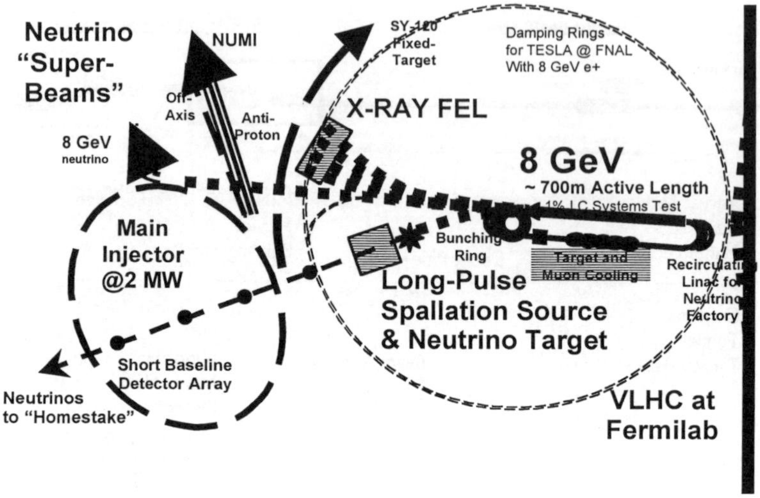

Figure 6. Design for a proton driver (Fermilab-W. Foster).

driven by proton beams with beam power above 2 MW. Storage rings for muon colliders also offer a possibility for producing intense neutrino beams. In fact, early investigations of muon colliders showed that some designs resulted in unacceptable off-site radiation from neutrinos. A neutrino beam based on stored muons is nearly as challenging as a muon collider. It is so daunting that the proponents have recently revisited conventional neutrino beam geometries using a proton driver alone. The estimated neutrino beams directly from a proton driver would be 0.1 as intense as a simple neutrino factory. A typical neutrino factory based on muon collider technology would be more flexible as far as unlike sign suppression. It could also produce electron neutrinos. The intriguing but challenging concept of beta beams was suggested by P. Zucchelli [10] several years ago. Zucchelli proposed that short-lived radioactive ions be boosted by a γ of O(100) and stored in a racetrack ring until they decay to produce pure neutrinos.

Moving to higher proton beam power for neutrino beams raises significant issues. These include the heat of ionization and the thermal stress on the target and horn. There are a number of radiation safety issues including groundwater protection, airborne activation, and prompt and residual activation of the neutrino beam area.

The next challenge is how to provide the increased proton beam power with some sort of proton driver. Fermilab is now looking at an eight GeV superconducting linac incorporating concepts both from TESLA and the Spallation Neutron Source at Oak Ridge. In this plan the SNS linac copy accelerates negative ions to 1.3 GeV. TESLA cryomodules push the beam on to 8 GeV. This system could supply 2 MW of beam power at both 8 and 120 GeV. The linac would also deliver small emittances so that the losses in the Main Injector would be relatively small. A design sketch for the system exists [11]. The costs are comparable to the Main Injector. Figure 6 illustrates the many roles such a proton driver could play. In addition to supplying the two familiar neutrino beams MiniBooNE and NuMI it could feed a short baseline detector array and a possible long baseline beam to someplace like the Homestake mine. It could supply a spallation source and an FEL as well as serving as a test bed for several linear collider components. A branch for a very large hadron collider and a recirculating system for a neutrino factory have even been sketched in.

4. Visionary possibilities (2020 and beyond)

Is there a way to break out of these complicated accelerating schemes? The path most often suggested is to find ways to get higher accelerating gradients. One suggestion is to use lasers. Tremendous strides have been made in the laser field and the future continues to look promising. In particular so-called table top Terawatt lasers have been developed in the last decade by "chirping" lasers [12]. The fundamental problem with the electric field produced by a laser is that the field is oscillating very rapidly so that it is difficult to couple it to a particle beam. The potentials and problems of laser acceleration were covered in a classic review by Palmer [13] written before the invention of chirped lasers.

One way acceleration could be achieved is through the use of an inverse free electron laser (IFEL). This is the inverse of the Free Electron Laser where a beam of x-rays is produced by an electron beam. In an IFEL an electron beam is accelerated transversely and longitudinally by a laser. A step on the path toward a coupled IFEL has recently been taken at the STELLA [14] experiment at the Brookhaven Accelerator Test Facility (ATF). A second laser acceleration R&D program, LEAP, is also underway at Stanford/SLAC [15].

Another acceleration possibility is to use plasmas. Pretty good plasmas oscillations can be generated with both lasers and charged particle beams. It is here that the most progress has been made. Good introductions are contained in articles by Dawson [16] and Tajima and Dawson [17]. Esarey and his collaborators [18] have prepared a useful up-to-date review.

Figure 7. Wake field surfer metaphor for plasma wakefield acceleration (courtesy S. and L. Carrigan).

A metaphor for plasma wakefield acceleration is shown in Figure 7. The power boat in the lower left is moving through a fluid, the river in this case. It creates a moving wave behind it. Notice that no rope connects the surfer to the boat, he is riding the wave. If the boat accelerates, the surfer will accelerate meanwhile moving slightly down the wave. There are lots of ways the surfer can fall including the results of turbulence, drifting too far down the wake, and improper phasing, that is getting too far up toward the crest. In a plasma accelerator the fluid is typically ionized electrons in a gas. The plasma wave driver is a charged particle beam or a laser rather than a boat. The surfer is a witness beam of particles accelerated by waves in the charged particle plasma.

These plasmas can generate high accelerating gradients. The "non-relativistic wavebreaking" field for a plasma is G (V/cm) = $0.96\ (n_0)^{1/2}$ where n_0 is the electron density. For $n_0 = 10^{18}/\text{cm}^3$ (a possible plasma density in a gas) the accelerating gradient is G = 1 GV/cm. This should be compared to a good RF cavity gradient of 0.0005 GV/cm.

A recent example of a plasma wakefield accelerator is SLAC experiment E157 which has been set up in the SLAC Final Focus Test Beam [19]. They use a 30 GeV electron beam with $2*10^{10}$ e in a 0.65 mm bunch. The head of the bunch sets up a plasma wave in a Li gas plasma. The tail of the bunch can then be accelerated. The configuration used up until now produces both accelerated and decelerated particles in the beam. Barov and Rosenzweig [20] have seen similar results in a plasma at the Fermilab A0 photo-injector. One exciting possibility for the SLAC project is the prospect of using a plasma afterburner as an energy doubler for the SLAC linac [21].

A different path to plasma wakefield acceleration is to use a laser pulse some tens of femtoseconds long to produce a plasma in a millimeter thick gas stream without the requirement of an initiating electron beam system. Using this approach several teams [22] have recently produced small divergence, 100 MeV electron beams with energies spreads of only several percent.

While progress has been interesting, particularly since the development of chirped terawatt lasers a decade ago, there is still much to be accomplished. Finally, particle physics applications of plasma acceleration require high luminosity. This means emittance must be small and bunch charges must be large. Little work has been done on this so far.

As noted above the plasma acceleration gradient is proportional to the square root of plasma density. An interesting possibility is to use solids instead of gases as the plasma medium. The plasma density could be up to ten thousand times higher leading to gradients of 100 GV/cm. Concepts for solid state acceleration have been developed in some detail by Chen and Noble [23]. A particular feature of their approach is the use of particle channeling to enhance the process. Particle channeling [24] occurs when a charged particle moves close to a crystal plane or axis in a single crystal. For positive particles channeling can reduce energy loss and multiple scattering and possibly provide some focusing.

Clearly there are big problems associated with solid state acceleration. The intense laser or particle beams needed as drivers will destroy the material. However, for femto-second lasers the crystal lattice might survive long enough to accelerate the beam. Daunting as the large laser power densities are the extremely high gradients and the potential for radiative damping are interesting.

At the Fermilab-NICADD Photoinjector Laboratory at A0 [25] our Fermilab-Darmstadt collaboration [26] has looked at channeling radiation to see whether channeling itself would survive as the beam densities are raised toward those required for acceleration.. The experiment has demonstrated that the channeling radiation yield is constant over the bunch charge range from $2*10^{-4}$ to 10 nC. Past experiments have been at least a factor of 10^8 away from the destruction regime. Our experiment has extended this reach by a factor of 50-100. This is a step but further experiments are needed to reach the regime of solid state acceleration.

5. Summary

The next decades look bright for particle physics. The Tevatron will operate with a luminosity that has a chance of discovering supersymmetry and maybe even a Higgs. The LHC will come on in the latter part of the decade and should easily be able to find a Higgs. That facility will be a fine resource until well past 2020. Several interesting new long base-line neutrino facilities will come into operation. If there is strong international collaboration construction of a superconducting linear collider may be underway by 2010. That complex will almost surely include the possibility of an FEL and possibly a gamma-gamma collider. Somewhere in the next decades one or more existing or planned facilities may be retooled to function as neutrino factories. Beyond these machines the future is cloudy. A VLHC is extremely expensive. A muon collider is technically challenging and also expensive.

Can exotic devices like plasma accelerators enter the picture and offer a new path to the energy frontier? Some progress has been made but much more information is needed. They may turn out to be impractical or the rate of progress may be too slow. Only time will tell.

References

1 D. Edwards and M. Syphers, *An Introduction to the Physics of High Energy Accelerators*, (Wiley, New York , 1993).
2 *Handbook of Accelerator Physics and Engineering*, eds. A. Chao and M. Tigner, (World Scientific, Singapore , 2002).
3. LHWG Note/2001-03, Report No. CERN-EP/2001-055 (2001).
4. M. Tigner, Nuovo Cimento **37**, 1228 , (1965).
5. *TESLA Technical Design Report*, Eds. F. Richard, J. Schneider, D. Trines, and A. Wagner (DESY, Hamburg, 2001).
6. D. Edwards, *et al.*, *Status of Flat Beam Electron Production*, Fermilab-Conf.-01/218-E (2001).
7. *A 3 TeV e^+e^- Linear Collider Based on CLIC Technology*, ed. G. Guignard, CERN 2000-008, (Geneva, 2000).
8. H. Braun, CERN/PS 2001-008 (AE), (CERN, Geneva, 2001).
9. G. W. Bennett, *et. al.*, Phys. Rev. Lett. **89**, 101804 (2002).
10. P. Zucchelli, Phys. Lett. **B532**, 166 (2002).
11. W. Foster, http://www.fnal.gov/orgs/fermilab_users_org/ users_mtg/2004/foster.pdf.

12. See, for example, M. Perry and G. Mourou, Science **264**, 917 (1994).
13. R. Palmer, Particle Accelerators **11**, 81 (1980).
14. W. D. Kimura *et al.*, Phys. Rev. Lett. **92**, 054801 (2004).
15. C. D. Barnes, E. R. Colby, and T. Plettner, p. 294 in *Advanced Accelerator Concepts: Tenth Workshop*, eds. C. E. Clayton and P. Muggli, Amer. Inst. of Physics Press CP647, New York (2002).
16. J. Dawson, Scientific American, March, 1989 (p. 54).
17. T. Tajima and J. Dawson, Phys. Rev. Lett. **43**, 267 (1979).
18. E. Esarey, *et al.*, IEEE Trans. On Plasma Sci., **24**, 252 (1996).
19. M. Hogan, *et al.*, Phys. of Plasmas **7**, 2241 (2000).
20. N. Barov, K. Bishofberger, *et al.*, *Ultra high-gradient energy loss by a pulsed electron beam in a plasma*, Particle Accelerator Conference Proceedings (2001), to be published.
21. *An Energy Doubler for a Linear Collider*, S. Lee *et al.*, USC (2001).
22. S. Mangles, *et al.*, Nature **431**, 535 (2004), C. Geddes , *et al.* Nature **431**, 538 (2004), J. Faure, *et al.* Nature **431**, 541 (2004).
23. P. Chen and R. J. Noble , p. 517 in *Relativistic Channeling*, eds. R. A. Carrigan, Jr. and J. A. Ellison (Plenum, New York, 1987). P. Chen and R. Noble, p. 273 in *Advanced Accelerator Concepts*, eds. S. Chattopadhyay, J. McCullough, and P. Dahl (AIP Press C398, New York, 1997).
24. See, for example, *Relativistic Channeling*, eds. R. A. Carrigan, Jr. and J. A. Ellison (Plenum, New York, 1987). V. Biryukov, Y. Chesnokov, and V. Kotov, *Crystal Channeling and Its Applications at High-Energy Accelerators* (Springer, Berlin, 1996).
25. J.-P. Carneiro, et al, *Proceedings of the 1999 Particle Accelerator Conference*, A. Luccio and W. MacKay, Editors, p. 2027-2029 (IEEE Publishing, Piscataway, New Jersey, 1999).
26. R. A. Carrigan, Jr. *et al.*, Phys. Rev. **A68**, 062901 (2003).

DETECTORS AND DATA ACQUISITION

D.F. COWEN
Pennsylvania State University

The field of particle astrophysics features a wide array of detection techniques. The detectors vary in size from tabletop to cubic kilometers, yet they have a surprisingly long list of similarities. We present in this paper an overview of particle astrophysics detectors, highlighting the chief similarities and differences, with an emphasis on neutrino detectors.

1. Introduction

Particle astrophysics experiments exhibit tremendous variation. They range in size from centimeters to kilometers and are situated in locations as disparate as the deep ocean, satellites, balloons, deep underground labs, buried in thick ice caps and in plain old laboratories. They attempt to use as astronomical messengers neutrinos, protons, gravitons and WIMPs, at energies that span an enormous range, from ZeV down to keV.

Despite this impressive variation, particle astrophysics experiments also exhibit a remarkable set of common threads. We will first focus on these similarities, as understanding them forms a foundation we can build upon to understand the fundamental operating principles of many of these experiments. It is worth noting that these common threads provide experimental particle astrophysicists with an enviable amount of portability, i.e., the ability to move without significant re-tooling of knowledge between experiments with very different physics goals.

After discussing the similarities, we will then focus on the differences, of which there are many. As will be seen, the variety and striking contrasts among particle astrophysics experiments are indicative of the creativity and intellectual ferment driving this exciting and relatively new field of physics.

2. The Similarities

There are many commonalities shared by particle astrophysics detectors, both in their construction and in the requirements imposed upon them to accomplish their physics goals. These similarities include:

- the use of photomultiplier tubes (PMTs) to detect Cherenkov light,

- the use of custom application-specific integrated circuits (ASICs) in their data acquisition systems,
- low data rates compared to accelerator-based high energy physics (HEP) experiments,
- low expected signal rates with corresponding sensitivity to potential backgrounds,
- instrumentation of very large detection volumes, and
- the use of neutrinos as astronomical messengers.

2.1. *Photomultiplier Tubes*

Many particle astrophysics detectors, from low energy solar neutrino detectors to ultrahigh energy neutrino telescopes, use PMTs as their fundamental detector sub-element. PMTs are capable of detecting single photons and converting them into electrical pulses which can be processed by modern electronics and computers. According to Professor Francis Halzen, a theorist heading the large IceCube neutrino telescope experiment, a PMT is "...a light bulb run in reverse." As theorists like to say, this is only an approximation, and we will now endeavor to give a somewhat more comprehensive explanation of how PMTs function [1].

Photomultiplier tubes come in all shapes and sizes. One of the largest is the one used by SuperKamiokande, a PMT roughly 50 cm in diameter. A photo of such a PMT, along with a photo of an assortment of PMTs (made by Photonis) are shown in Figure 1. Photomultiplier tubes consist of a photocathode, a dynode structure to which high voltage is applied, and an anode that collects the resulting charge pulse.

Figure 1: Above: A sample of photomultiplier tubes made by Photonis [1]. Right: A single large Hamamatsu PMT, one of about 11,000 used by the SuperKamiokande collaboration [2].

When a photon hits the photocathode, it can cause an

electron to be released. This electron is accelerated to the first dynode by the electric field applied to the dynode structure. When it hits the first dynode, it causes several other electrons to be released. This process is repeated at each dynode, resulting in a large number of electrons hitting the anode, causing a detectable electronic pulse to be output by the PMT. Typical amplification factors are 10^6-10^9. The interior of a PMT is held at vacuum to allow the electrons to propagate through the dynode structure with minimal scattering. A diagram of a PMT is shown in Figure 2.

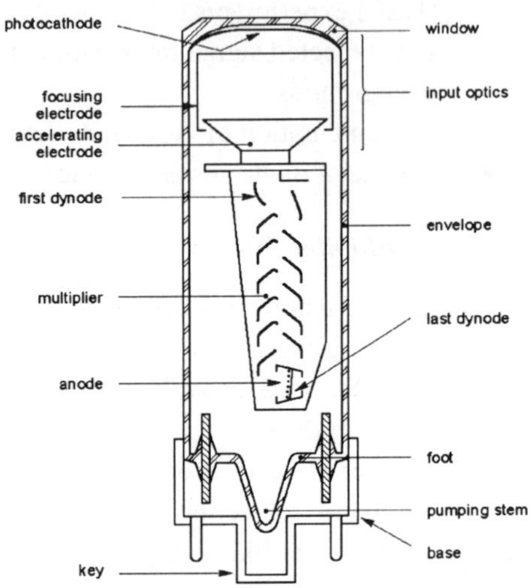

Figure 2: A diagram of a PMT (from Ref. [1]).

A variety of materials are used as photocathodes. Important parameters in the selection of photocathode material include spectral sensitivity and quantum efficiency. The spectral sensitivity is a measure of the PMT sensitivity to photons as a function of photon wavelength, and materials are chosen to optimize this sensitivity for a given application. The quantum efficiency is defined as the ratio of the number of emitted electrons to the number of incident photons, and generally does not exceed 30% for most photocathode materials. Other important characteristics of PMTs include:

- Noise rate: rate at which the PMT registers signals due to thermionic emission of electrons, leakage currents, and decaying radioactive contaminants in the materials used to construct the PMT.
- Gain: the ratio of anode current to cathode photocurrent.
- Response time and time resolution: The time for an output pulse to go from 10% to 90% of its full value, in response to an idealized delta function light pulse at the photocathode, and the spread in time between photon arrival and anode pulse output. For particle astrophysics experiments, the time resolution is typically more important, and is usually less than roughly 2 ns.

- Energy resolution: the spread in the difference between the flux of photons at the photocathode and the integrated charge produced at the anode.
- Effect of external magnetic fields: PMTs are sensitive to external magnetic fields because these fields are can deflect the electron trajectories in the dynode structure.

Photomultiplier tubes can be optimized not only by selection of photocathode material, but also by changing the dynode configuration, applied voltages, overall PMT geometry, and so on. The parameter space for this optimization is clearly quite large.

Due to statistical fluctuations and noise, the response of a PMT to a beam of monochromatic photons results in a spectrum of pulses at the anode. The extent to which single photons can be distinguished from noise hits from, say, thermionic emission, is an important measure of PMT quality, and is referred to as the "peak to valley ratio." The response of a PMT to multiple, time-correlated photons impinging on its photocathode is essentially a superposition of its response to single photons. However, the output signals get correspondingly more difficult to disentangle by feature extraction algorithms, and if too many time-correlated arrive in too short a period of time, it can saturate the PMT response.

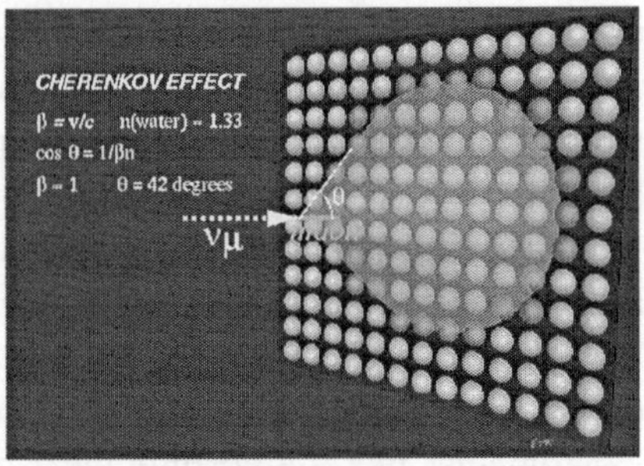

Figure 3: An array of PMTs can be used to image the Cherenkov light produced by a neutrino-induced event. In the figure shown, taken from the SuperKamiokande website, a muon neutrino has produced a muon, and the muon is emitting Cherenkov light at the angle Θ as described in the text.

Photomultiplier tubes are particularly well-suited to the detection of Cherenkov light. Cherenkov light is produced when charged particles move in a transparent medium at velocities that exceed the speed of light in that medium. When this occurs, electromagnetic radiation will be emitted by the charged particles in a shock wave with the characteristic angle $\cos\Theta = 1/\beta n$, where n is the index of refraction of the medium. Cherenkov light is produced at wavelengths that correspond to high spectral sensitivity in PMTs, and the light can be imaged by arrays of PMTs as shown in Figure 3 [2].

2.2. *Data Acquisition*

Data acquisition (DAQ) systems are another salient commonality held by many particle astrophysics experiments. The task of DAQ systems is to collect the data produced by one or more sub-detectors and render this information into a form such that it can be analyzed by physicists using standard computers. A DAQ system typically consists of custom electronics that collect the data from an analog source, such as a PMT, and then custom and/or off-the-shelf electronics and/or computers that form a "trigger," i.e., decide whether or not a particular chunk of data is of physics interest and, if so, build that data into an event and store it in a form amenable to later analysis.

Disentangling multiple photons from a complex waveform is made much easier with the advent of DAQ systems that use fast digitizing electronics. Being able to digitize such a waveform permits one to use digital signal processing algorithms to extract the information desired. Digitization is accomplished by the use of analog-to-digital converters (ADCs) that convert the voltage at the anode of a PMT to a number. This conversion is done at fixed time intervals (e.g., every few nanoseconds) to build up a full, digital picture of the analog waveform. Unfortunately, digitizers that are fast and flexible enough for particle astrophysics applications are either unavailable or too expensive, since there are no high-volume applications for this kind of technology (yet!). Hence, particle astrophysicists need to custom-design and build digitizers to do this, and given the typical speed and power consumption requirements, one often designs and builds an custom application specific integration circuit (ASIC) for this purpose.

These custom ASICs are the heart of many DAQ systems. Older particle astrophysics experiments, like the solar and atmospheric neutrino telescopes SNO and SuperKamiokande, use custom ASICs not to digitize waveforms but to extract key pieces of information from them, like pulse time and charge. Newer experiments, like the cosmological neutrino telescopes ANTARES, NESTOR and IceCube have a physics-driven need for full waveform digitization, and use their ASICs to extract that information.

2.3. *Other Similarities*

In contrast to many detectors in particle physics, nuclear physics and astrophysics, particle astrophysics detectors are typically subject to very low data rates. Neutrino detectors, for example, expect to detect O(1-100) neutrinos daily, while taking data at a trigger rate of O(100Hz). By contrast, at the Large Hadron Collider beam crossings occur at a rate of O(GHz) and acquire "useful" data at a rate of about several kHz.

Neutrino and other particle astrophysics detectors generally have a high sensitivity to background levels. This is especially true of detectors sensitive to low energy signals, such as solar neutrino and direct WIMP detection experiments. In these cases, ubiquitous natural radioactivity can mimic the expected signal. To mitigate against such backgrounds, physicists who build these experiments go to great lengths to shield their detector, typically by burying it deep underground to take advantage of the cosmic-ray filtering properties of the overburden, and by keeping the experimental device itself as free of contaminants as possible.

Detectors searching for higher energy signals have large fiducial volumes in common with one another. This is especially true of ultrahigh energy neutrino detectors, for which large volume translates to increased likelihood of a) a neutrino interacting in or near the instrumented volume and b) containing most or all of an event's interaction products when an interaction actually occurs. For example, a charged-current electron-neutrino interaction at E = 1 PeV will produce a region illuminated by Cherenkov light that is roughly 250 m in radius.

3. The Differences

To get a feeling for the striking differences among particle astrophysics detectors, we'll describe in some detail how a number of modern detectors function. We will begin with neutrino detectors, starting with the lower energy neutrino detectors SNO and KamLAND, and then move to higher energy detectors, highlighting AMANDA, IceCube, and ANITA.[a] We will then describe the Pierre Auger ultrahigh energy cosmic ray detector. Finally, we will describe the direct WIMP detector CDMS-II.

3.1. *SNO and KamLAND*

The Sudbury Neutrino Observatory (SNO) [3] and the Kamioka Liquid scintillator Anti-neutrino Detector (KamLAND) [4] are detectors designed to study neutrinos with energies in the range 1-10 MeV emitted from the sun and nuclear reactors, respectively. At these energies, backgrounds from residual radioactivity are a serious concern. Each detector has thousands of PMTs looking at kiloton-size active volumes. For SNO, one *gram* of dust from the mine in which it was being built would have caused a 10%-level background in its heavy water volume.

[a] For information on other high energy neutrino detectors, please see the writeup by J. Carr in this volume.

The SNO detector, which was used to solve the 30-year-old "solar neutrino problem," is located 6,800 feet underground in an active nickel mine in Sudbury, Ontario, Canada. It consists of a kiloton of ultra-pure heavy water (D_2O) surrounded by very pure light water and a spherical array of about 9,500 PMTs, as shown in Figure 4. A solar neutrino can interact with an electron or a deuteron in the D_2O, causing one or more electrons to be accelerated and emit Cherenkov light. As shown in Figure 3, this Cherenkov light can then be imaged by the PMT array, and events thereby reconstructed.

SNO was able to answer the solar neutrino problem by virtue of its ability to detect solar neutrinos via three distinct interactions: 1) elastic scattering (ES) $\nu_x + e^- \rightarrow \nu_x + e^-$, 2) charged current (CC) $\nu_e + d \rightarrow p + p + e^-$, and 3) neutral current (NC) $\nu_x + d \rightarrow \nu_x + p + n$. This ability depended in turn on the capability of SNO to reconstruct the direction, energy and vertex of each putative solar neutrino interaction. The ES interaction is highly directional and chiefly sensitive to ν_e, the CC interaction is somewhat directional and exclusively sensitive to ν_e, and the NC interaction (detected once the free neutron is captured on deuterium, releasing a gamma-ray that Compton-scatters one or more electrons to produce visible Cherenkov light) is non-directional but nearly mono-energetic and equally sensitive to all neutrino flavors. Fitting the data to a combination of ES, CC and NC interaction allowed SNO to determine that solar ν_e were oscillating into another flavor and thereby escaping detection in previous detectors.

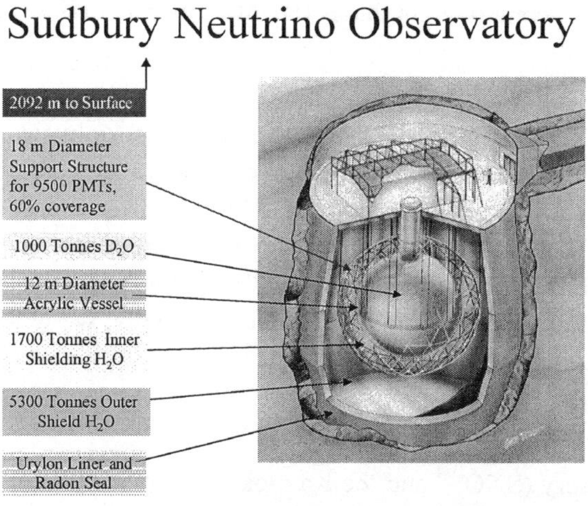

Figure 4: A schematic of the Sudbury Neutrino Observatory, showing its 1 kton of heavy water in an acrylic vessel surrounded by light water and an array of 9500 PMTs [3].

The sensitivity of SNO to NC interactions was enhanced by adding chlorine-rich salt. The cross section for capture of neutrons on chlorine is much larger than that on deuterium, and the energy of the subsequently emitted gamma-ray is about 25% higher. Both these factors make it easier to detect neutrons. Finally, SNO has just inserted an array of proportional counters filled with helium that will allow it to identify neutron captures on an event-by-event (not just statistical) basis. This will further improve SNO's neutral current sensitivity.

The KamLAND detector used electron anti-neutrinos produced in reactors in Japan and South Korea to confirm and refine the SNO solar neutrino result. KamLAND consists of 1 kton of liquid scintillator in a containment vessel, surrounded by approximately 1,900 PMTs and water. The advantage of liquid scintillator over D_2O is a lower threshold and a clean signal for electron anti-neutrinos due to the interaction anti-ν_e + p $\rightarrow$ n + e^+ that produces prompt Cherenkov light from the positron followed by a correlated burst of light roughly 200μs later, when the neutron capture causes more Cherenkov light to be emitted. Although this type of correlated signal is a powerful way to reduce contamination from misreconstructed backgrounds, the main challenge for KamLAND was to make the liquid scintillator extremely clean to reduce background from radioactive contaminants. KamLAND succeeded admirably in this effort, reducing, for example, ^{238}U to $3{\cdot}10^{-18}$ g/g.

3.2. *AMANDA and IceCube*

Ultrahigh energy neutrino telescopes are built to detect neutrinos of cosmological origin at energies exceeding roughly 10 TeV. To do this, a large, clear, "pre-fabricated" detection volume is needed. Large size is important to increase the likelihood of interaction and containment of interactions when they do occur. Also, a clear medium makes construction of a device affordable, since it allows one to use a low pixelization density to detect the Cherenkov light emitted by the daughters of neutrino interactions. At least two locales satisfy this requirement: deep water sites in various locations, especially the Meditteranean, and the deep ice at the South Pole. We focus here on the detectors built in deep ice sites, the Antarctic Muon and Neutrino Detector Array (AMANDA) and its enlarged successor, IceCube.

AMANDA [5] and IceCube [6] consist of arrays of PMTs placed in pressure vessels and inserted into holes melted into the ice at the South Pole. The holes then refreeze, permanently entombing the "optical modules" in the ice cap. The modules are read out by means of ~2 km long electrical and/or optical cables. AMANDA sends analog electrical or optical signals to the surface where these signals are digitized and recorded whenever time-correlated hits are detected. IceCube will digitize PMT output pulses in situ and then send up the digitized data via electrical cable to the surface, where like AMANDA it will then record data whenever time-correlated hits are detected.

AMANDA currently consists of 677 PMTs buried mostly between 1500 and 2000 m below the surface of the ice.

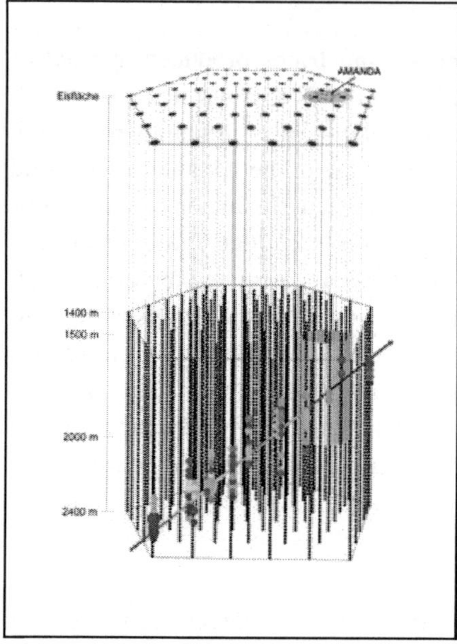

Figure 5: The IceCube detector will consist of 4800 digital optical modules (DOMs) deployed on 80 strings as shown. It will also include an air shower array on the surface [6].

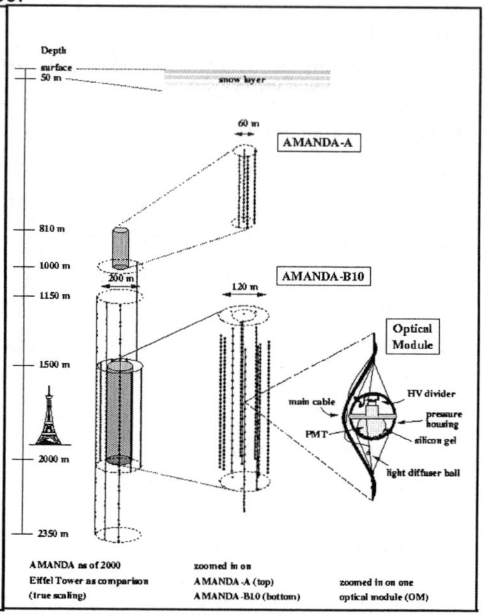

Figure 6: The AMANDA detector consists of 677 PMTs encased in glass pressure spheres, comprising an optical module. These OMs are arranged as shown [5].

The array is shaped like a narrow cylinder 200 m in diameter and 500 m in length. IceCube will consist of 4800 "digital optical modules" (DOMs) each housing a PMT, digitizing electronics, and remotely controllable light sources, all inside a pressure vessel. A diagram of the existing AMANDA detector is shown in Figure 6, and a simulated IceCube event, showing the anticipated detector layout, is depicted in Figure 5.

AMANDA, IceCube and other ultrahigh energy neutrino detectors operate by imaging the Cherenkov light produced by the daughters of neutrino interactions in the detector fiducial volume or surrounding medium. A muon neutrino can create a charged muon, resulting in a track crossing the detector volume, or an electron neutrino can create an electron, resulting in a "cascade" manifested as an expanding, roughly spherical ball of light fully or partially contained in the detector volume. A very energetic muon or cascade can produce complex multiphotoelectron waveforms in PMTs, and these waveforms in principle contain very rich information useful for reconstructing the energy, direction and vertex of the event. Taking full advantage of this information requires the use of digitizing electronics, a feature which AMANDA recently added and which IceCube will have as a designed-in feature from the start.

The DAQ system for IceCube is very similar to those of other ultrahigh neutrino experiments. Remarkably, IceCube and NESTOR (an ultrahigh neutrino experiment off the coast of Greece) use the same custom ASIC[b] to digitize PMT waveforms. Even more remarkably, the much lower energy KamLAND experiment also uses a variant of the exact same ASIC. As mentioned earlier, DAQ systems, and the physicists who can design them, are often very portable across very different particle astrophysics experiments.

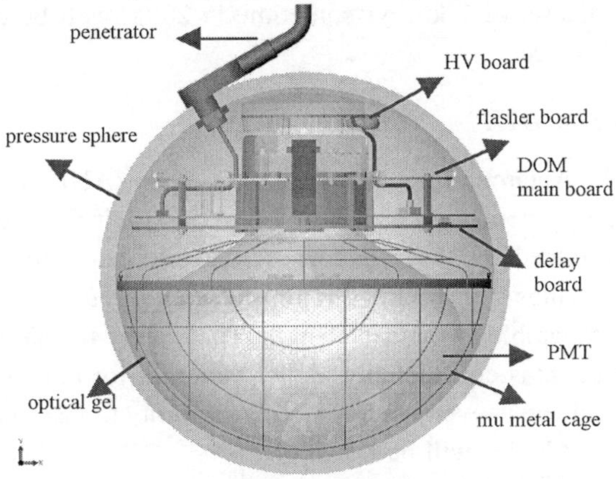

Figure 7: A schematic of an IceCube digital optical module (DOM). The DOM is the heart of the IceCube experiment and consists of a PMT, electronics to provide high voltage, digitization and communication, a light source, all encased in a glass pressure sphere [6].

The heart of the IceCube experiment is the digital optical module (DOM). Each DOM consists of a pressure sphere that contains a PMT, a high voltage source, digitizing and communication electronics, and a light source (see Figure 7). The digitization electronics is based on the custom ASIC called the ATWD that can digitize four independent channels' worth of data with a period of roughly 3 ns for a depth of 128 samplings, or a total time period of about 400 ns. To lengthen the digitization time and remain sensitive to late-arriving Cherenkov light expected from, for example, very high energy events or events with multiple cascades, a separate flash ADC is used that samples at a 25 ns period for a total time period of about 6 μs.

Given the large distances separating IceCube DOMs—over 1 km in many cases—and the requirement that each DOM maintain roughly a 7 ns time resolution so that event reconstructions can be performed accurately, one of the major challenges for the IceCube DAQ system is to calibrate the DOM-to-DOM

[b] The ASIC in question was designed at LBNL and is called the Analog Transient Waveform Digitizer, or ATWD.

relative timing. This is accomplished by a system which uses identical analog pulse generation electronics on the surface and in each DOM to send and receive timing pulses on a regular basis, e.g., at a rate of about 0.1 Hz. This system has been shown to meet the timing requirement, both using an AMANDA prototype DOM string deployed in 2000 and laboratory measurements made more recently on final IceCube DOMs.

IceCube is scheduled to deploy its first four strings in the austral summer 2004/2005 season. Over the next five-six years, the balance of the full 80-string detector will be deployed. As the detector grows, the DAQ system will permit IceCube to continually take data, and it is expected that IceCube will accumulate a dataset of 1 km^3·yr sometime in 2008, well before the full detector has been deployed.

3.3. *ANITA*

The Antarctic Impulsive Transient Array (ANITA) [7] is also looking for ultrahigh energy neutrinos, but it will try to exploit the radio rather than optical signal produced by these events. At several hundred MHz, cascades produced by neutrinos will produce radio pulses *coherently* since the size of the shower is comparable to the wavelength. The energy threshold of about $10^{17\text{-}18}$ eV for this process is considerably higher than that for optical Cherenkov devices, but this handicap is overcome by ANITA's ability to instrument an enormous volume.

ANITA will be flown on a balloon about 37 km high and it will take advantage of the circumpolar winds around Antarctica to fly one or more times around the continent. During the flight(s), ANITA will be sensitive to neutrino-induced cascades that occur in *gigatons* of ice below it. An overview of this scenario is shown in Figure 8.

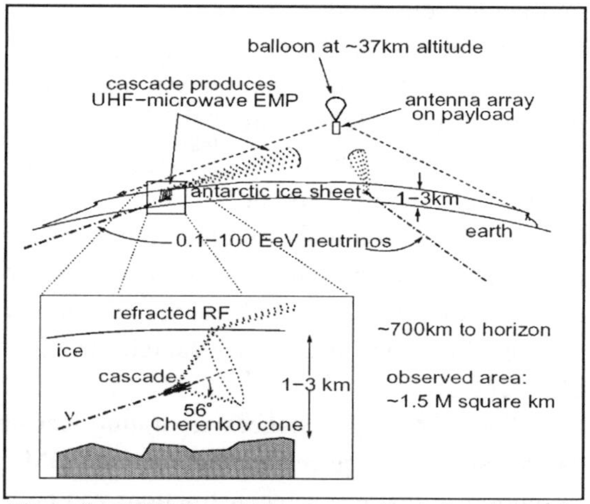

Figure 8: ANITA is sensitive to neutrino-induced cascades that occur in the ice 37 km below it. These cascades will produce radio pulses that refract out of the icecap that can be detected by ANITA's antenna array [7].

ANITA will consist of 40 horn antennas arranged as shown in Figure 9. The separation between the two antenna arrays helps to pinpoint the source of detected radio pulses. Photovoltaic panels provide power for the array.

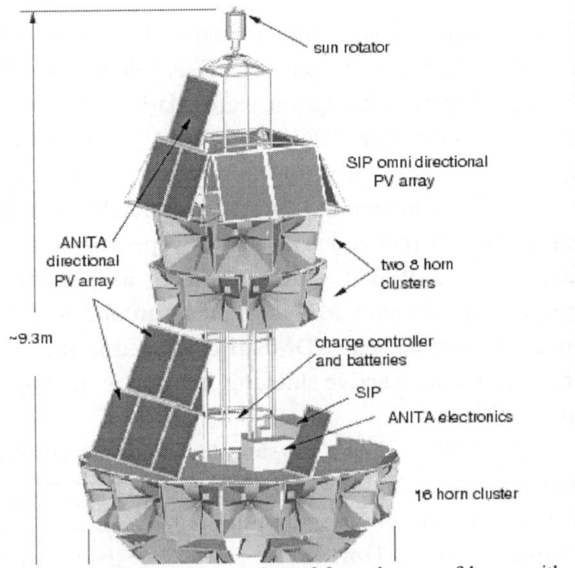

Figure 9: The ANITA device consists of four clusters of horns with a separation that helps pinpoint the source location. Photovoltaic panels provide power [7].

3.4. *Pierre Auger*

The Pierre Auger air shower array aims to study the flux of enigmatic ultrahigh energy cosmic rays at $E>10^{18}$eV. Previous air shower arrays have done this using one of two main techniques: 1) surface-based particle detectors and 2) fluorescence telescopes.

The Auger experiment will run both of these techniques simultaneously. This will reduce the systematic problems associated with these two techniques when they are run independently.

Each technique uses PMTs to detect either Cherenkov light produced by charged particles traversing tanks of water, or, on clear moonless nights, the fluorescence light produced by cosmic-ray induced showers in the atmosphere. The Auger array will comprise 1600 tanks over an area of about 3000 km^2, and four stations with 24 fluorescence detectors each. The resulting PMT signals will all be digitized.

One of the biggest challenges for Auger DAQ was how to provide power for each tank, and then how to gather up all the signals from all the tanks. Stringing cables between tanks was not practical due to the large distances involved. Power is provided by solar cells, and cell-phone technology is used to transmit data from each tank to various central locations for further processing.

3.5. *CDMS-II*

The Cryogenic Dark Matter Search II experiment [8] is one of a number of experiments trying to perform *direct* detection of WIMP dark matter. Here, "direct" implies a design in which dark matter interacts in, and is detected by, the experiment. For CDMS-II, a WIMP dark matter particle interacting in their tabletop-size detector would 1) rattle the crystal lattice of their detector, creating a detectable phonon and 2) create detectable ionization. Not surprisingly, it is paramount to reduce backgrounds, especially those that produce neutrons that can mimic a WIMP signal, as much as possible. This is accomplished by moving the detector as deeply underground as possible to reduce cosmic-ray induced neutron flux (CDMS-II is located in the Soudan mine), by surrounding the detector with active shielding, and by producing the detector itself as cleanly as possible.

The CDMS-II DAQ system looks for the correlated signals due to phonons and ionization, and also applies a risetime cut on the phonon signal to eliminate a troublesome background originating on the detector surfaces. Figure 10 shows a plot of the signals that various calibration sources produce in the detector [9]. The DAQ system is relatively simple due to the low event rates and consists of commercial ADCs, computers and LabView software. The main DAQ challenge has been to make the entire system remotely operable, so that people did not have to live in a mine in Soudan, Minnesota, USA, to run the experiment. This was accomplished using modern software techniques, in particular JAVA with RMI.

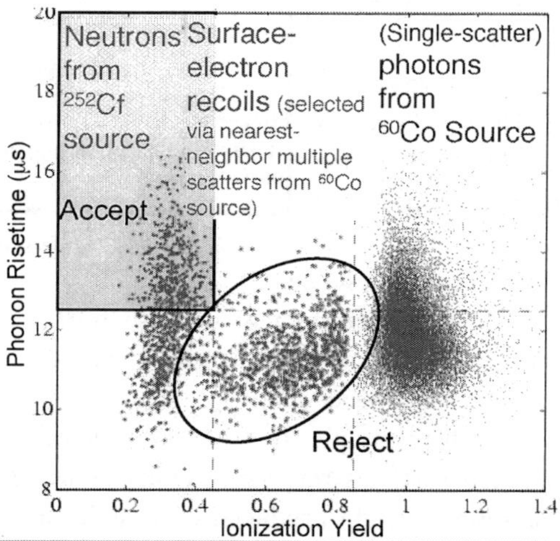

Figure 10: Phonon risetime vs. ionization yield for the CDMS-II detector for a variety of calibration sources. The neutron source produces events most closely like the expected WIMP signal. Electrons and photons are used to mimic backgrounds.

4. Conclusions

We have seen that particle astrophysics detectors exhibit striking variety in their design, locale, physical size and fundamental detector technology. In fact, this discussion has only covered a subset of these detectors, so the variety is even wider than glimpsed here. However, there are also salient similarities. The underlying technology in DAQ is very similar, and many, but by no means all, detectors use PMTs as their fundamental detector sub-element.

The field of particle astrophysics is successfully leveraging on the previous detector-building experience of its scientists, who hail from the fields of particle physics, nuclear physics, cosmic-ray astrophysics and high-energy gamma-ray astronomy, and is at the same time being extremely innovative. The challenges are great, but experimental particle astrophysics are rising to them, and having a lot of fun doing it!

References

1. Much of the information about PMTs in this and other sections can be found in E. Flyckt and C. Marmonier, *Photomultiplier Tubes, Principles and Applications*, Photonis, Brive, France, 2002. Also available online at http://www.photonis.com/en/product_range/photomultipliers/AppBook.htm.
2. Figure taken from the SuperKamiokande website, http://www-sk.icrr.u-tokyo.ac.jp/doc/sk/.
3. SNO home page, http://www.sno.phy.queensu.ca/
4. KamLAND home page, http://www.awa.tohoku.ac.jp/KamLAND/.
5. AMANDA home page, http://amanda.uci.edu/
6. IceCube home page, http://icecube.wisc.edu/
7. ANITA Proposal, http://www.ps.uci.edu/~barwick/anitaprop.pdf.
8. For information on the CDMS experiment, please see their web site at http://cdms.berkeley.edu/.
9. R. Schnee, "Results from CDMS," Neutrino 2004, http://neutrino2004.in2p3.fr/.

LARGE SCALE COMPUTING

PAOLO CAPILUPPI

Dipartimento di Fisica dell'Università di Bologna & INFN
I-40127 Bologna, Italy

Large Scale Computing is acquiring an important role in the field of data analysis and treatment for many Sciences and also for some Social activities. The present paper discusses the characteristics of Computing when it becomes "Large Scale" and the current state of the art for some particular application needing such a large distributed resources and organization. High Energy Particle Physics (HEP) Experiments are discussed in this respect; in particular the Large Hadron Collider (LHC) Experiments are analyzed. The Computing Models of LHC Experiments represent the current prototype implementation of Large Scale Computing and describe the level of maturity of the possible deployment solutions. Some of the most recent results on the measurements of the performances and functionalities of the LHC Experiments' testing are discussed.

1. Introduction

Computing requiring large quantities of resources is becoming usual for many current applications both in Science and in Social activities. This Computing becomes "Large Scale" when all or many of the involved components reach large numbers, like:

- Several Petabytes* of information data
- Data distributed over more than hundred sites
- Global computing power larger than tens of millions SpecInt2000†
- Number of System's users larger than five thousand
- Complexity of the needed algorithms requiring more than 500 thousand lines of code per computing program
- Access to the data in a chaotic way by the users, that's to say that the users access the data in thousand of independent patterns per day
- The resources have a global cost exceeding hundreds million Euros (or US dollars)
- The heterogeneity of the local systems to be used is of the order of tens/hundreds of different systems

Facing the above "numbers" in a Computing environment is a today challenge that has to be studied and approached without forgetting the winning

* A Petabyte is 10^{15} Bytes ~= 500.000 DVD movies

† A modern PC's CPU has the power of ~ 1000 SpecInt2000

argument: fostering the local, distributed ability of the actors with their capacity to give solutions. The real difficulty is to manage the distributed persons' contribution into a coherent design.

The above-mentioned problem is the well known problem of distributed computing, even if there are new factors: distributed collaborative work and distributed data access and control.

In the following of the paper a possible approach to a solution of the "challenge" is discussed, mostly for a single Science application, the High Energy Physics (HEP) one. Implications of HEP solution into other Sciences (and even Social environment) are of importance and non-HEP solutions could well continue to benefit of the HEP implementations.

2. Applications of Large Scale Computing

Applications needing a large number of resources already exist: Experimental Physics data analysis (and in particular Particle accelerator experiments), Earth observation activities, Weather forecasts, World stock market, prime elements availability and location, Web mining, etc.

Particle accelerator experiments have been since time consumers of large quantities of computing power and data storage; the being deployed new experiments at the particle colliders currently under construction will increase the already large necessity of resources.

Physics experiments that use cosmic radiation will also require Large Scale Computing to analyze the recorded data, including the astro-particle analyses.

However HEP experiments at particle colliders are the today's largest providers of applications needing large scale computing.

2.1. *LHC Experiments*

LHC [1] (Large Hadron Collider) is a particle accelerator being built at CERN [2] that will study proton-proton collisions at the center of mass energy of 14 TeV. Four experiments are being built to study different properties of the collisions, and in particular they will look for the discovery of the Higgs particle.

The four experimental setups are: Alice, Atlas, CMS, LHCb. Figure 1 shows a simplified design of the four experiments that will take data in the year 2007.

The expected rate of collisions at LHC is 40 MHz thus giving a rate of data of the order of PBytes per second. Each interesting collision is superimposed to many other uninteresting collisions (~20 per bunch crossing at design Luminosity of LHC). The interesting signal has to be extracted among the many others (see for example Figure 2) and the "trigger system" of the experiments therefore selects about 100 interesting events per second to be permanently stored. The data flux from the experiments is of the order of 100-1000 MBytes per second for a total accumulated data per year of the order of tens of PetaBytes. Billions of interesting events (proton-proton collisions) are expected per year of data taking.

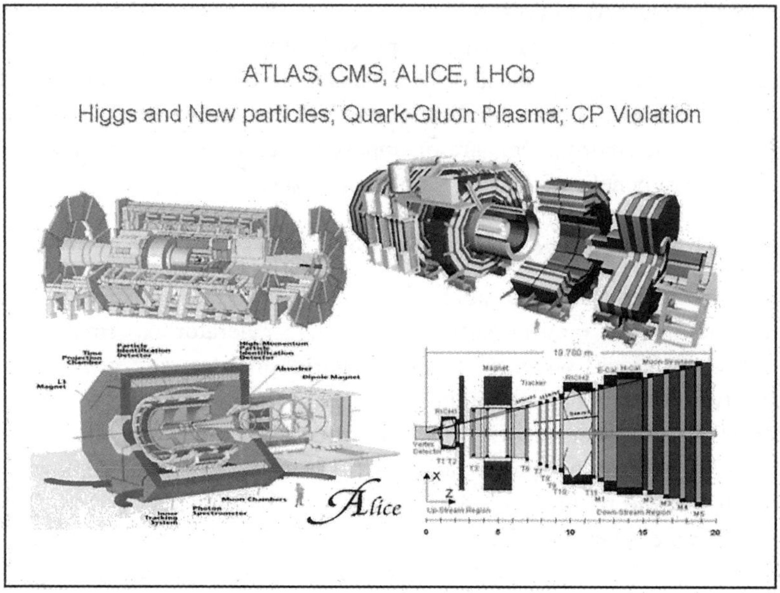

Figure 1. The four LHC Experiments.

More than 5000 Physicists and Engineers participate to the four LHC experiments Collaborations, from more than 300 Institutes of about 70 different Countries in the World. Therefore the actors of the Computing are dispersed and normally reside at their own Institution, where they run their applications. Moreover the resources are naturally distributed, as Funding Agencies would prefer to invest on local equipment and infrastructure.

The computing power necessary to investigate the data is globally larger than tens of thousands current CPUs, facing different environments and different applications (simulation, calibrations, different kind of analysis, etc.).

Because of the above mentioned scale of necessary resources the LHC Experiments well fit into the requirements of Large Scale Computing.

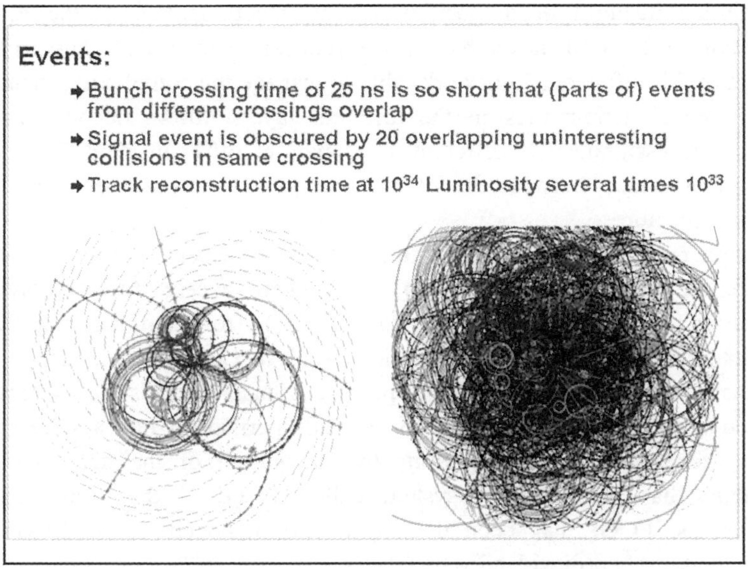

Figure 2. An example of the necessary selection power for the LHC events.

3. Complexities of Data Management

Having to manage a huge amount of data where the information is hidden and where there are correlations among the information, the complexity of data management and organization is of importance. The LHC experiments are of an unprecedented complexity (for HEP) as the detectors have millions of output channels, all giving a piece of information for each proton-proton collision.

The applications used to access the data follow the Object Oriented approach of programming and therefore the data have to be organized in objects.

Catalogs and databases are currently used in many computing applications to reduce the complexity of data access and location, and a mixture of these technologies is used in the design and implementation of the LHC Experiments Computing. Both relational and object-oriented databases are part of the design and the implementation.

Because of the dispersion of the actors around the World, the data have to be replicated into different places and thereby stored to guarantee equal opportunities to every analyzer. The consistency of the replicated data must be guaranteed, via databases synchronizations and catalogs automatic updates. Network connections of good quality and large bandwidth will have to be setup among the different sites of the Experiments' participating Institutions.

The architectural decomposition of the functionalities of the different applications to be used shows that a few layers can be identified, thus leading to a design of the software (framework) able to manage the complexity. Identifying the different functionalities in the different layers allows decomposing the problem and assignment to different developers of different identified task. Of course the interfaces between the different components have to be defined in the design architecture.

Another complexity aspect of the data management is the necessity to navigate the data following the objects links to access different layers of information: primary information (Raw data), reconstructed information (Reconstructed data), selected information (Analysis Object Data), etc.

Finally there is the need for a persistent data object store system. The recorded and created objects have to be made permanent into a store system, with an access catalog that permits the navigation of the data. The HEP community has developed a system called POOL [3] that matches those requirements. POOL has been developed and deployed in the LCG (LHC Computing Grid [4]) project leaded by CERN.

4. Distributed Computing and Data Access

Distributed Computing is a Computer Science well known issue. It has been studied since many years and possible solutions exist, for several types of applications. Large scale Computing differs from Distributed Computing as it has to solve the problem of distributed data access and availability. It's not only a problem of CPU coordination and distributed usage, the main point being indeed the data access.

Combining the distribution of resources (CPU power, data storage, infrastructure, network, services) into a coherent design is the challenge of Large Scale Computing.

HEP experiments applications are driving the effort to find a possible solution and can help to give hints for a more global solution. From the point of view of the design architecture, HEP applications can benefit of a couple of points:

- application executables are quite similar even if highly variable in time and scope;
- each running job access the event of particle collision with an intrinsic atomicity, the event itself.

The HEP community has developed a hierarchical model to better plan and deploy a system able to satisfy the large scale computing: the MONARC [5]

Model, called a *Tier model*. Figure 3 shows the main characteristics of the tiered model. It should be noted that the hierarchy is not based on amount of resources, but is based on the functional services to be provided to serve the data to the applications. Therefore the Tier1s have to provide much more services than the Tier2s, etc. The dimension of resources clearly depends from the complexity and amount of provided services so that as a consequence the Tier1s are much larger than the Tier2s, etc.

The model is somehow rigid and lacks of flexibility, which is necessary to allow the system to adapt to new and unforeseen requirements. In particular the LHC data analysis will be done by dispersed users organized in "analysis groups" looking to similar results in a collaborative way, and accessing the same replicated data. The Tier model has to be modified to incorporate this requirement. The *grid computing* paradigm is the best way we know today that is able to satisfy this need of flexibility.

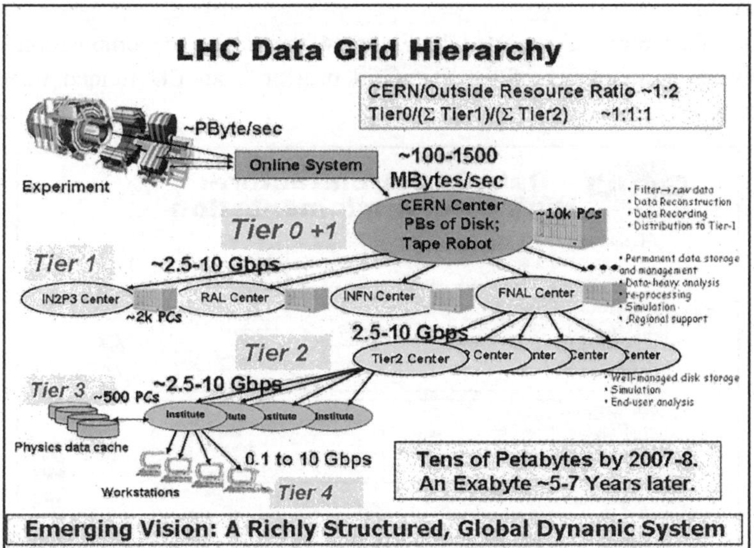

Figure 3. The MONARC Model of Tiers.

5. Grid Computing

The Grid Computing paradigm was proposed in the years 1998-1999 and introduced a layered architecture of components in the "middle" of basic services (mostly hardware related) and of high level services (mostly application software), hence the name of "middleware" for the Grid developed tools.

The word "grid" is taken from the electric power grid similitude, trying to build a standard from where all the applications get the necessary service and to which the providers (resources) offer computing and data services.

The "Grid" should be: dependable (can provide performance and functionality guarantees), consistent (uniform interfaces to a wide variety of resources) and pervasive (ability to "plug in" from anywhere).

INFN (the Italian funding agency for HEP) started one of the first projects about grid development in the year 2000 [6], in order to provide to the LHC experiments a middleware able to solve the Computing problem.

Many projects have been launched since then and the Grid solutions are being deployed. In particular the LHC community has launched a project (the LCG project) that is coordinating the development and implementation of a grid solution for the computing of the experiments. The middleware components of the LCG are based on the work done by previous and current projects, both in EU and US.

As an example is reported in Figure 4 the design decomposition of the middleware components of the DataGrid project [7], an EU funded three year project ended in 2003.

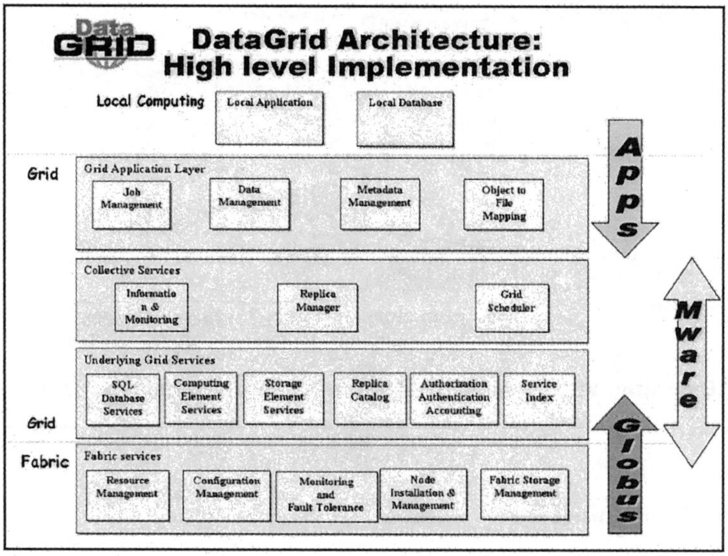

Figure 4. The DataGrid Project architecture.

Following the design of the many Grid project a number of service elements have been defined: the Computing Element (CE) that groups the services required for computation, the Storage Element (SE) that provides the services for storage, the Resource Broker (RB) that provides the services to match and balance the load of resources, the Information System that provides the services to gather the information about the grid available resources, the Authorization System, etc.

An example of how a job (application) can run on the grid and how it access the different services is sketched in Figure 5.

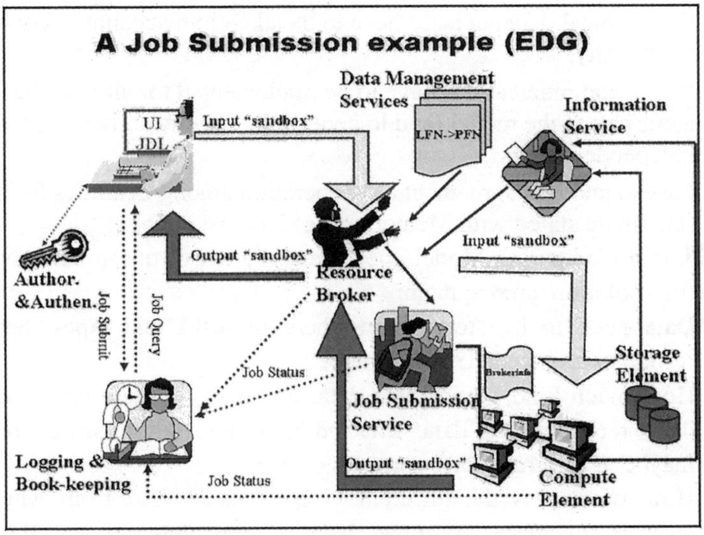

Figure 5. A job submission to the Grid.

6. Computing Models and basic components

A Computing Model is the set of rules that organize the access and elaboration of data and the way they are managed for getting information/results.

More in details a Computing Model has to organize the interaction and the coordination of:

- Infrastructure of resources,
- Middleware components,
- Implementation of Hardware resources,
- Applications software,
- User interfaces,
- Organization,
- Framework for applications,

- Data and application structure,
- And also System Architecture.

Every complex project accessing data at a large scale have to deal with a Computing Model, and all the HEP Experiments indeed have one (which progressively changes during time).

A Computing Model has to take into account some basic requirements:

- The model is a distributed effort by construction,
- Hierarchical dependencies have to be taken into account to control the complexity,
- Direct communications have to be implemented to mitigate the hierarchy of the model (and to control the chaotic activities of the independent users),
- Formal and real agreement of cooperation among Institutes for support (fair share stated with Memoranda of Understandings).

Building a Computing Model goes through a set of questions/answers that can drive to a solution, among them:

- Data need to be stored somewhere in disks and tapes: how many resources are needed? And where?
- How much kind of different data do we have? Raw data, calibration data, reconstructed data, selected data, etc., all requiring resources, maybe of different performances.
- How we access the different kind of data? And from where? The answers to these questions have implications on the distribution mechanism and catalogs. Who and where is going to provide those services? And how long?
- Which are the access methods to the data? The framework and the software have to give a solution here, providing also some simple user tool.
- Who is allowed to access the different kinds of data, and in which fraction? Policies, authorization, accounting and methods to set them must be provided.
- How much CPU power in needed for the different kind of activities? And where? Analysis is different from simulation or compilation and the requirements are much different. There are specific constraints on the resources to be provided, again in a distributed way.
- Where the analyses results will be stored? How can we provide share of access to the same data to the components of the same Group?

- Which will be the frequency of access to the different kinds of data? And from where? How important is the turnaround time for different activities?
- Which level of consistency do we need for replicated data? The answer has huge implications on the choice of replication methods, Databases, quality of connections, etc.
- How can we find the data we are interested in? There is the need of global worldwide catalogs or local catalogs are enough for the scope?
- Which kind of infrastructure and hierarchy can we setup? What are the constraints of the local provider of resources in the different parts of the World?
- Finally, which kind of organization is needed to guarantee the functionality of the Model? Who and where is going to guarantee the necessary service level agreements?

The definition of the Computing Model must account for different answers and moreover must be flexible enough to permit a smooth evolution to new performances and technologies.

It's a common understanding of the LHC Experiments that their Computing Models will have to be based on hierarchies of resources and services, managed via the common Grid Middleware. Preliminary designs for the Computing Models are under active study by all the LHC experiments and some indication about the required resources exists. Figure 6 shows preliminary estimates of the LCG necessary resources.

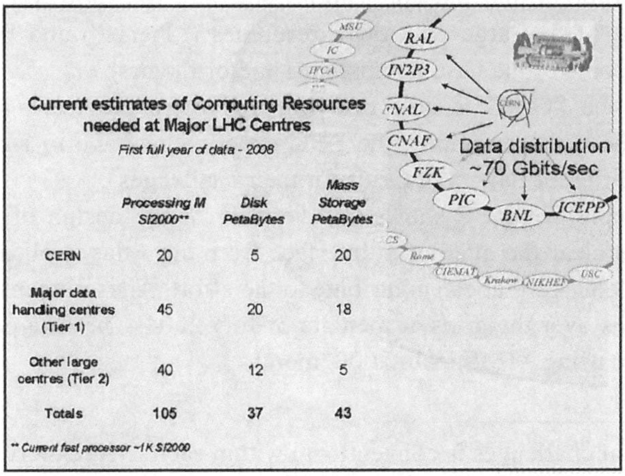

Figure 6. LCG planned resources for LHC Computing Models.

7. Measuring the Model's performances: Data Challenges

The Computing Model's components and their performance/scalability can only be tested using large scale resources. Some proof of concepts can certainly be demonstrated on limited amount of resources and sites; however the complexity of the provided components' solutions can be measured only with tests that map as closest as possible to real life.

These real life large scale tests are called "Data Challenges" for LHC Experiments. Indeed there is a subtle distinction between *Data* and *Physics* challenges: the first being mostly devoted to computing ability to provide solutions, the second being devoted to a similar scope but with use of the produced data for Detector performances studies.

The LHC machine is not still in operation, nor the experiments are built and running, therefore the only data that the LHC experiments can use for their challenges are simulated (Monte Carlo) data.

A number of Data/Physics challenges have been (and are being) performed by the LHC Experiments, progressively increasing complexity and scale. CMS [8] has recently ended its Data Challenge 04, Alice [9] is currently running its Physics Data Challenge 04, LHCb [10] is also running its Data Challenge 04 and Atlas [11] is underway for its Data Challenge 2[‡].

7.1. *First results from LHC Experiments' Data Challenges*

The LHC experiments have different preliminary Computing Models, as the data models are different and the applications software too; however they have in common most of the large provider's resources ("Tier1s") and the underlying Grid (indeed some of the Grid components and/or flavors).

Because the "Grid" is still evolving and there are many Grid projects looking for the better solutions, the LHC experiments have to face the current existence of different flavors of Grids for their challenges.

For example, Figure 7 shows an overview of the design of Atlas current challenge. It's clear the attempt to interface from the Atlas application software all the possible Grids that can contribute to the effort. Atlas ran tens of thousands simulation jobs over the grids at the date of July 2004. The data produced were tens of TBytes using ~1MSpecInt2000*months.

---

‡ Numbering of challenges is conventional within each LHC experiment; it's not simply the sequential numeration of challenges.

Alice, at the same date, ran ~50.000 different jobs producing tens of TBytes and using ~350 MSpecInt2000*hours. The data were produced using both Grid (LCG) and the specific Alice distributed system "Alien". The catalog ended to contain about 4 million entries, without measured degradation.

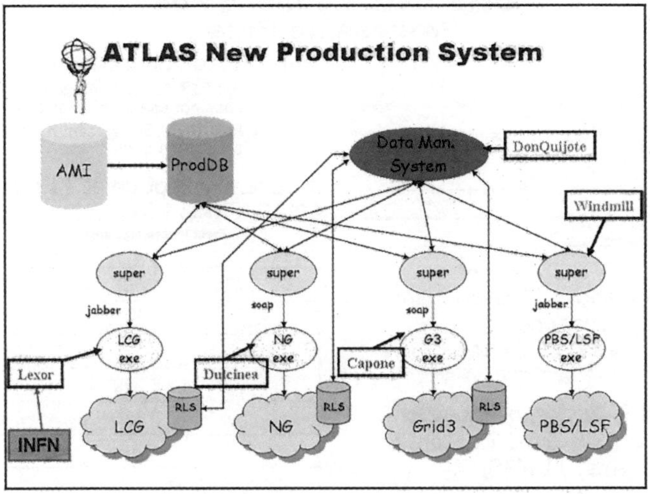

Figure 7. Atlas experiment design overview of Data Challenge 2.

CMS Data Challenge 04 had a slightly different scope if compared with the other LHC Experiments. There was a distributed production phase of simulated data (the pre-challenge production, similar to the previously described challenges of other experiments) that delivered ~80 million events of different characteristics, and a "true challenge" phase. The pre-production phase consisted of ~ 800.000 jobs over more than 30 distributed sites for a total of of about 100 TBytes of final data storage.

The "true challenge" phase lasted two months attempting to have a full chain demonstration in real time. Simulated data had to be reconstructed at the CERN Tier0, then sent to the Tier1s and finally thereby analyzed. The possibility to have also few Tier2s down in the chain was also approached. The final goal was to be able to sustain a data rate in the full processes of about 25 Hz (one quarter of the final nominal rate of real data taking of CMS in year 2007).

CMS ran during the "true challenge" phase more than 40.000 jobs, of which ~10.000 were of analysis. The system was able to manage the data along the full chain even if many problems were identified and promptly corrected. However there were down-times to apply the correction and to understand the problems: during the two months the target rate of 25 Hz was occasionally hit and only sustained for one complete day (Figure 8).

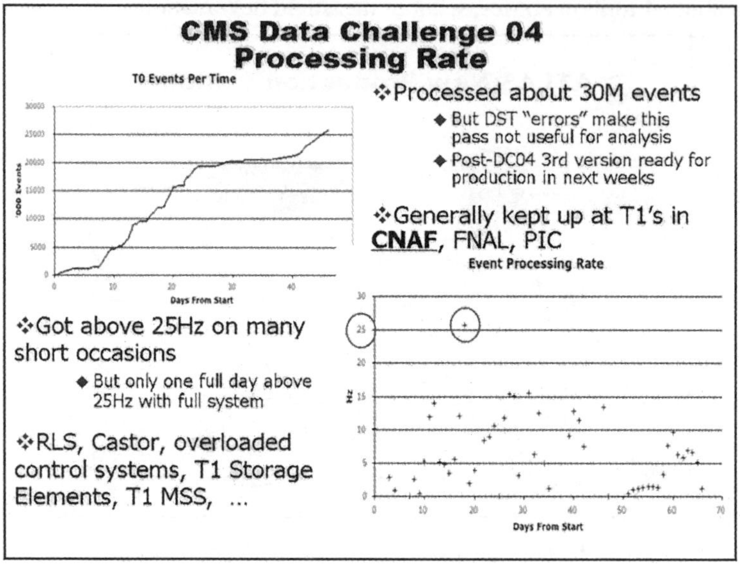

Figure 8. Results of the CMS experiment Data Challenge 04.

8. Perspectives on missing components

There are a number of missing components in the Computing Model and many details have to be still understood or implemented, the main worries being related to data models and data access/location. However the bigger missing component is the way the data analysis will be done. Analysis activity is a "chaotic" one, as the huge number of users cannot be coordinated in their day-by-day ability to extract scientific results. There is currently no measure of the level of load of such activity on the resources, nor have the proposed approaches been deeply investigated.

Some preliminary results have been obtained by Data Challenges (see for example Figure 9), however the results are biased by the predefined patter of access to the data and by the absence of really independent users. Moreover the basic questions of analysis have still to find a final answer; for example if it's better to move analysis jobs where the data reside or if it's better to move the data to the job (maybe a mixture of the two approaches is the correct one, and therefore the arbitration between the two procedures have to be designed and implemented).

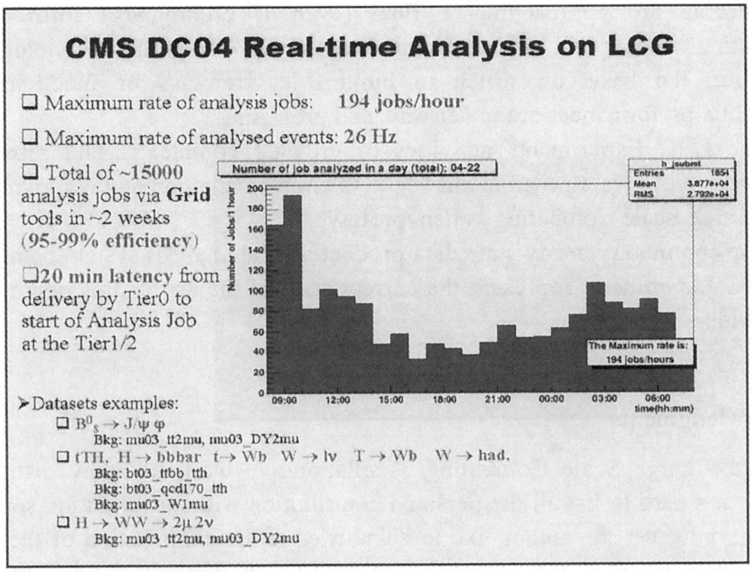

Figure 9. Real-time analysis results of the CMS experiment during Data Challenge 04.

Finally the scalability of the proposed solutions is far from having been demonstrated: there is the need of progressively increasing complexity and resources trials (Data Challenges), but the timescale of the LHC startup is becoming short and the urgency of working solutions could vain many of the efforts.

9. Conclusions

Large Scale Computing is growing in importance for many different applications requiring huge distributed resources and organization. There are many different applications existing today that require World coordination and access for data/information: from Social activities to Science.

Among the data applications that are fostering this kind of Computing there are the High Energy Experiments analyses, and in particular the being built LHC (Large Hadron Collider at CERN, Geneva) Experiments.

The challenge of this new worldwide Computing has some basic emerging technologies that can help to find a solution: the Grid Computing paradigm, the hardware evolution, the new software architectures and the infrastructure evolution (services and network).

There are missing components (with notably missing functionalities) to be still implemented and understood, however the Computing Models of LHC Experiments are approaching a final (even if preliminary) form. These Computing Models will well fit in the Large Scale Computing technology and will form the basis on which to build. Measurements of functionalities, scalability, performances are under way, and promising.

The LHC Experiments activities of growing complexity and resources' necessities will give a possible answer to the implementation and deployment of a first Large Scale Computing system prototype.

The continuous steady-state data production and analysis system being built for LHC Experiments represent the current state of the art for this new type of worldwide computing.

Acknowledgments

Being the Large Scale Computing a collaborative work of many distributed persons it's hard to list all the personal contribution without forgetting someone relevant, however the author like to acknowledge the contribution of the many colleagues working in the Computing of the LHC Experiments and in the Grid projects.

References

1. LHC Project: http://lhc.web.cern.ch/lhc/general/gen_info.htm
2. CERN (Geneva, CH): http://public.web.cern.ch/public/
3. POOL Project: http://lcgapp.cern.ch/project/persist/
4. LCG Project: http://lcg.cern.ch/
5. M. Aderholz et al.; MONARC: Models of Networked Analysis at Regional Centres for LHC Experiments (Phase 2 Report), CERN LCB-2000/001 (2000)
6. Grid.it Web Page: http://www.grid.it
7. EDG Project: http://eu-datagrid.web.cern.ch/eu-datagrid/
8. CMS Experiment: http://cmsdoc.cern.ch/cms.html
9. Alice Experiment: http://alice.web.cern.ch/Alice/AliceNew/collaboration/
10. LHCb Experiment: http://lhcb.web.cern.ch/lhcb/
11. Atlas Experiment: http://atlas.web.cern.ch/Atlas/internal/Welcome.html

SCIENCE, TECHNOLOGY AND SOCIETY

What can be done to make science more appealing and easier to understand

G. GIACOMELLI AND R. GIACOMELLI

Physics Department, University of Bologna and INFN, Sezione di Bologna

We shall discuss some aspects of science and technology, their increasing role in the society, the fast advances in modern science, the apparent decrease of interest of the young generation in basic sciences, the importance of proper science popularization for better public education and awareness in scientific fields.

1. Introduction

Science is interested in the laws of nature, while Technology applies scientific knowledge to make new things, new machinery and it may be used to "dominate" nature and to improve our life. The two aspects are deeply connected: without scientific research there is no technological progress and without technology we would not have new instruments for research. Usually the technological research improves and creates new instruments in known scientific fields, while most of the great technological revolutions are spin off of fundamental research. Just one example: WWW (World Wide Web), the key which opens every gate of Internet, the prefix most used by web navigators, was invented for improving communication in fundamental research in a large European Laboratory for fundamental physics, CERN in Geneva.

Fig. 1. Two colliding galaxies. They are not a galaxy and an antigalaxy because from the colliding region at the center of the picture we do not observe any drastic increase of luminosity, as it would be expected from particle-antiparticle annihilations.

Why do we perform research in general, and in particle physics in particular? For particle physics the standard answers are of the following type:

i) to understand the structure of matter and of what holds it together (Fig. 1), ii) to satisfy our curiosity, iii) because we enjoy doing it, iv) for technical spin-offs, v) for more modern teaching, others.

Can we justify the high costs of large scale research, like those in our field? It is now necessary to explain to the public what we do and how we spend the taxpayer money, besides assuring that we are not spoiling the environment. This is true in particular for research performed in the large international and national laboratories, like Brookhaven, CERN, etc. Scientific outreach (including science popularization, scientific awareness and appreciation of current research) has become an essential task for the research community and these activities must be made in a professional way [1, 2].

Technological development leads to economic progress, increased well-being, to new medical applications and more. The applications coming from modern physics have changed communications (TV, cellular phones, Internet) and make it possible to look inside the human body without opening it (x-rays, nuclear magnetic resonance, ultrasound, Positron Electron Tomography, new computer applications, etc.). Biotechnology is changing and will change biology, medicine and our lives even more. But we may have to face also ethical problems.

The recent scientific and technological progress has not been accompanied by appropriate penetration in modern society of the fundamental scientific concepts, and in the media often there is a considerable amount of "parascience" and even of antiscience [1, 2].

Fig. 2. **What is the World Made of?** (Particle Data Group). Why do so many things in this world share the same characteristics? People have come to realize that the matter of the world is made from a few fundamental building blocks of nature. By fundamental building blocks we mean objects that are simple and structureless -- not made of anything smaller. Since time immemorial people have been trying to understand what the Universe is made of. One of the earliest theories said that everything could be built from just four elements, Earth, Air, Fire and Water. This was a great scientific theory because it was simple. But it had one big drawback: it was wrong. The Greeks added also a fifth substance, quintessence, a term now used to describe the energy of the vacuum.

Science and technology play an increasing role in everyday life and progress in modern science and technology occurs quickly, both in specific subjects and because of the opening up of new fields and new interests. We should consider these changes in our school curricula, promote refresher courses, permanent education, and proper scientific outreach.

2. Science popularization

For researchers it is not easy to do science popularization and outreach in a simple and appealing manner, and one may remember what Gianni Puppi used to say in Bologna: "If you are not able to explain to your aunt in less than 5 minutes what you are doing in physics, then you have not really understood what you do", Fig. 2 [3].

Scientific knowledge update is needed to understand the great scientific and technological changes. But this must be done properly, stimulating the interest; the mass media often do not help since they may insist on aspects which are doubtful or incorrect or not scientific. An example: the press is full of articles about the "hydrogen economy", but they often forget to write that there are no "hydrogen mines"; they also write about getting hydrogen from water, forgetting that water is the result of combustion (it is like ash) and that to break water molecules into hydrogen and oxygen requires energy (which at the moment could only come from fossil fuels or nuclear fuels). A second example: vaccinations saved millions of people, but only the few unhappily lost lives become news in the press. Moreover the press and TV give great emphasis to magic, horoscopes, divinations and not enough to proper scientific information.

Most of the high school teachers followed university courses several years ago, when the teaching did not involve many of the present basic concepts, like subnuclear physics, the quark model, neutrinos, etc. It is clear that they would benefit from regular refresher courses.

While in the past the experimental research was done by small groups, now it is done by large collaborations involving many groups from different countries and large detectors which may be very costly [4, 5]. Furthermore the decision on which experiments to perform may involve not only scientists, but larger communities, including bureaucrats, politicians, citizens. This means that the science communication should involve a broader community, using all available media, written journals, conferences, TV, Internet, etc.

Both in developed and developing countries there has been a consistent decrease of university applications in basic scientific fields (physics, chemistry, mathematics, ...). It seems that in the young generations there is a decreasing interest in sciences, even if the students, the teachers and the people would like to know more about science.

The Universities worry about it and they ask their scientists to increase their efforts for scientific outreach, in newspapers, TV, conferences, debates and by innovative means.

Also for developing countries there are many initiatives, see [6]. It may be worth recalling the following statement by a Peruvian officer: "Peruvian society is not well informed about science and people think that it is something too difficult to be understood by lay people. We need to change this scenario in order to promote debate in science and technology issues, using as many different tools as possible" [6].

Fig. 3. My aunt after my explanation of particle physics (P. Waloscheck, DESY) [3].

3. Scientific outreach in Internet

The web site in Internet is one of the new methods of scientific outreach. All international and national large laboratories like CERN, NASA, ESA, INFN, have prepared nice outreach sites [7]. Now this is also done by smaller labs and by universities. Efforts are made to create simple and stimulating web sites, using interesting new approaches, with nice figures, often animated or interactive.

The multidisciplinary outreach web site of the University of Bologna is mainly addressed to last year high school students and university students, but it is also for a wider audience [8]. The purpose is to provide and promote scientific popularization and outreach in simple and appropriate ways, stimulating at the same time the curiosity of the younger generation. The site is "multidisciplinary": with topics such as The Brain, Antimatter, Molecular machines, Electrosmog, The radio window on the cosmos, Pollution of the environment, Artificial intelligence, Dark matter, Physics and fantasy, etc. Every page has a minimal text and uses photographs, figures, animated figures, and whatever can make the reading more inviting, without losing scientific correcteness. Particular notes are also devoted to scientists from Bologna: Luigi Galvani (Animal electricity), Guglielmo Marconi (Wireless telegraphy), etc.

Fig. 4. Astronomy picture of the day (23/7/04): Saturn's Rings. Astronomy, astrophysics and space science pictures provide an incredible amount of beautiful pictures, which can easily be used for science popularization.

In order to facilitate the comprehension of technical terms an interactive dictionary is readily available on line for each subject.

There are links to science popularization journals, to scientific web sites, with the purpose to help students to widen their interests. The site contains also "articoli di approfondimento".

An example of the content of the topic **Antimatter**, written in cooperation with CERN, is:

Antimatter: what is it? Something exotic and not real?
    Everything you wanted to know about antimatter
Short history of antimatter
    From the first revolutionary ideas to the present situation
Antimatter around us
    From antimatter in cosmic rays to antimatter use in the PET
Antimatter at the beginning of the Universe
    How much antimatter was there? How did it disappear?

Antimatter today in the Universe
  Recent searches for antimatter
Antimatter at the University of Bologna
Questions & answers
Glossary

Fig. 5. Artificial intelligence. (Credit: "*Roboethics Symposium*"). Artificial intelligence is an interdisciplinary scientific and technological field in which studies are made on intelligence, logic, robotics, learning, theoretical and applied informatics [8].

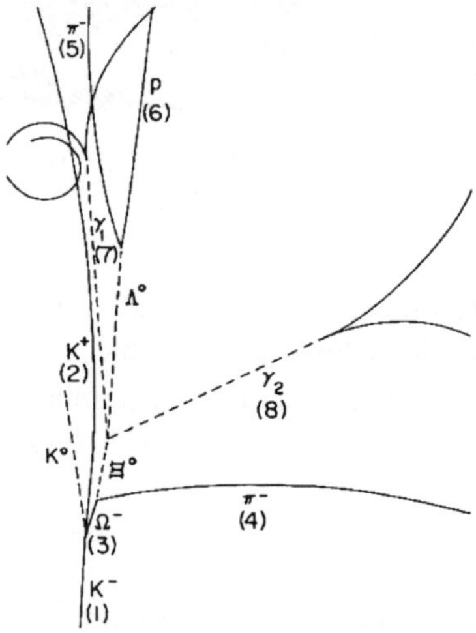

Fig. 6. The first observation of the Omega-minus particle in the Brookhaven hydrogen bubble chamber (1954). Selected bubble chamber photographs are very useful for the popularization of particle physics [3].

In the section "Antimatter around us", practical examples are given using cosmic rays, radioactive decays and their use in Positron Electron Tomography

(PET). Fig. 1 is a figure from the section on "Antimatter today in the Universe". Examples of other figures used in other topics discussed in the site are Figures 5, 6, 7, 8.

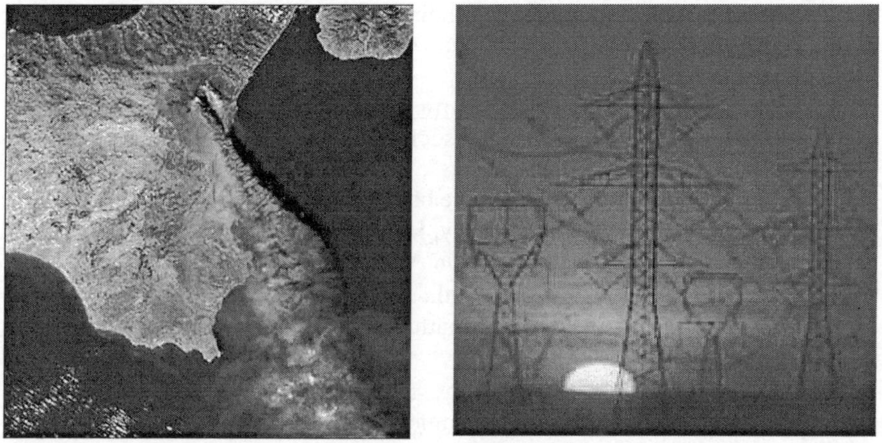

Fig.7 (a). Satellite photo of the Etna eruption in July 2002. (Credit: *Kathy Strabala and NASA*) [7, 8]; (b). Electrosmog (Electromagnetic pollution) is a term which generates wide discussions, often with exaggerated fears [8].

4. Conclusions

Science and technology play an increasing role in our lives, and progress in modern science and technology occur very quickly. Science and technology cannot give an answer to everything, but they lead to civic and economic evolutions improving the quality of our lives.

It is generally agreed that education and awareness in science have to be strengthened. Scientific outreach, improvements in teaching, proper scientific information are very important issues. Outreach should also be addressed to politicians and decision makers.

While for many researchers the main motivation for doing basic research remains scientific curiosity, for most of people the motivations involve also scientific progress, technological improvements, well being and the quality of everyday life, without spoiling the environment.

We should remember that the advanced techniques used in particle physics, can be applied in many fields, in particular in medicine.

We would like to thank A. Casoni and many colleagues for their cooperation.

References

1. E. Johansson, The importance of funding outreach, CERN COURIER.
2. T. Regge, Scienza e societa', Inaugurazione Anno Accademico 1995-1996, Politecnico di Torino. http://www1.polito.it/organizz/docuffic/inaugurazione96/Regge.html
3. P. Waloschek, Particle physics and society. Popularization of science, in "30 years of bubble chamber physics", http://www.bo.infn.it/~spurio/bubble.htm
4. G. Giacomelli et al., The LEP legacy, lecture at this School, hep-ex
5. K. Ahmet et al., Nucl. Instrum. Meth. A305 (1991) 275. S.P. Ahlen et al., Nucl. Instrum. Meth. A324 (1993) 337.
6. www.scidev.net: news and information about science, technology and the developing world.
7. NASA (www.nasa.gov) , ESA (www.esa.int/esa.cp/index.html), CERN (http://public.web.cern.ch/public/welcome.html), INFN (http://scienzapertutti.lnf.infn.it) Particle Data Group (http://pdg.ge.infn.it/particleadventure/)
8. University of Bologna (http://www.scienzagiovane.unibo.it/)

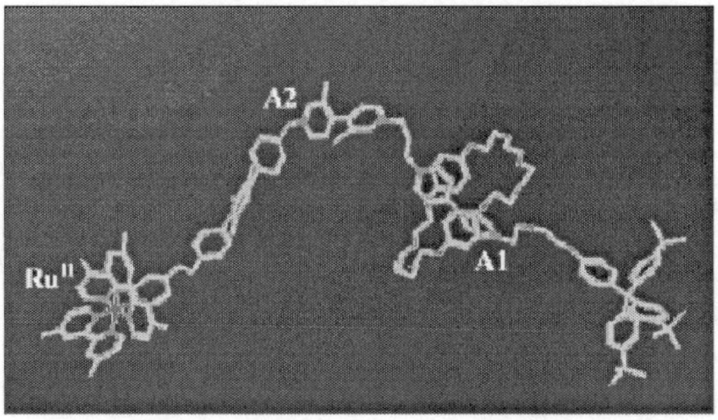

Fig. 8. Molecular machines. In order to reach extreme miniaturization, chemists start from molecules [bottom up approach]; assemblying them in appropriate ways they obtain "objects", "machines" (complicates molecules) of nanometer dimensions (10^{-9}m). The figure shows a "molecular shuttle" activated by luminous energy, see [8].

THE UNIVERSE TODAY, YESTERDAY AND TOMORROW

TULLIO REGGE

Politecnico di Torino and INFN, Sezione di Torino,
regge@isiosf.isi.it

After a short discussion of how we proceeded in discovering the structure of matter, a brief summary is made of the expanding universe, with some personal recollections.

1. Remembering Goedel

One of my most vivid memories of my years in Princeton is Kurt Goedel at noon time walking past my office at the IAS. He is certainly the most legendary figure I ever met in my life. He died in the late seventies and shortly after I left Princeton and came back to Italy.

Goedel's celebrated theorem had a shattering effect on mathematics of his time but not so much on physics. The ambitious project of Russell and Whitehead to condense all math in four thick volumes was discredited and even got the nickname of "diplococus".

I am not an expert of logics and of Goedel's theorem and I cannot go into details. The theorem states that any non trivial mathematicl formalism based on axioms contains propositions which are undecidable, loosely speaking there are questions which have no answer within the formalism. Given one such proposition you may decide whether it is true or false and once you do this the proposition becomes an axiom and you get a formal language which extends the one you started with. But then this new language itself will produce more undecidable propositions and more axioms and the procedure never ends. I've heard that assigning a truth value to a proposition is sometimes called "consulting an oracle". This unending procedure and the very idea of consulting an oracle always attracted me.

Physical laws are written in a mathematical language and I always wondered if Goedel's theorem had any relevance to physics and to the real world. I admit that I would be very disappointed and perplexed by a negative answer.

If a look back to the history of physics I see some interesting trends. At the time of Louis Philippe a very conservative french abbey had an argument with the king and fled to Turin, my home town. His name was Auguste Cauchy and he got a chair at the local university where he completed his celebrated theorem on complex variables. When he left the same chair was eventually given to Avogadro.

2. Avogadro and Mendeleyev

In 1811, Avogadro published an article in *Journal de physique* that clearly drew the distinction between the molecule and the atom. He pointed out that Dalton had confused the concepts of atoms and molecules. The "atoms" of nitrogen and oxygen are in reality "molecules" containing two atoms each. Thus two molecules of hydrogen can combine with one molecule of oxygen to produce two molecules of water. Avogadro suggested that "equal volumes of all gases at the same temperature and pressure contain the same number of molecules" which is now known as Avogadro's principle.

The principle provided a clear and unambigous definition of atomic weights which allowed Cannizzaro to derive a first list of weights and to present it at a chemistry meeting in Frankfurth in 1861. The list got eventually extended and got into the hands of Mendeleyev who produced the periodic system.

3. The beauty of Flaws and the infinite universe

The periodic system is the first table of elementary particles, the atoms. It was an exciting discovery but it has apparent flaws. In spite of the name the system was not exactly periodic, it had rare earths. Years later Niels Bohr came out with his atomic model and the flaws turned out to be essential in our understanding of the atomic structure.

Atoms ceased to be elementary particles and there appeared a new list of particles which included the electron, proton and neutron and kept on growing with the energy of accelerators. By the end of the 1930's the list was just as perplexing in number and in misterious flaws as the old periodic system.

We have now a standard model of elementary particles which includes three generations, quarks and leptons and describes the real world up to energies which are thousand billions of times higher than the ones of the time of Mendeleyev. The model works very well but is by no means simple and includes strange departures from symmetry which strongly remind me of the flaws of the early models.

A proposals has been already put forward by Harari in which quarks themselves are no longer elementary and are composed of new entities called preons. I am not an expert on this matter but I am ready to bet that preons or similar models, if succesful, will show interesting and unexpected flaws hinting at a even deeper level. An interesting hypothesis is that in fact there is no ultimate level, that experiment plays the role of Goedel's oracle and that the show will continue forever and stop only when sadly there will be no more government funds for accelerators.

4. The very small and the very large

The descent into the infinitely small is also relevant to cosmology and to our undestanding of the Big Bang. We need to know the behaviour of matter in extreme conditions in order to undestand early cosmology. Each step down has opened the door to a scientific revolution, Bohr's model has opened the way to quantum mechanics and quarks are themselves very queer particles.

If there is no ultimate model of elementary particles then the history of the universe up to the Big Bang will contain an infinity of eras , the flow of events should not be measured by our time but rather by the log of time and there will be no final theory and the Big Bang is infinitely remote in the past. I do not find this infinity distressing, it is a guarantee that the show will never end and that in the centuries to come there will be always something new, unexpected and I hope exciting.

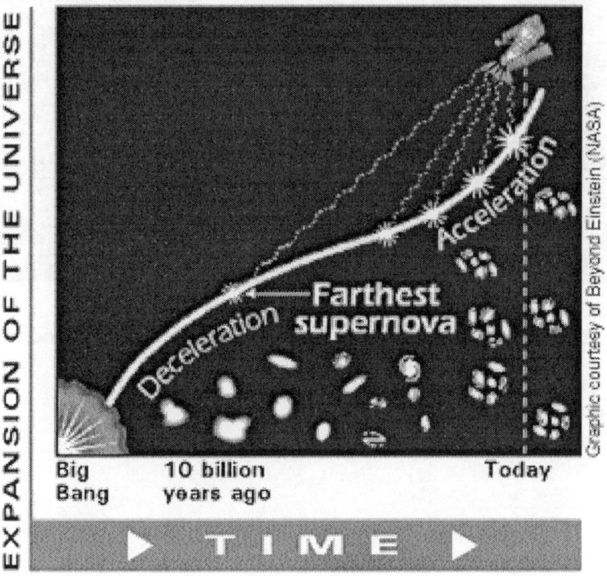

Figure 1. The Universe is expanding: every galaxy receeds from the others with a speed which increases with distance (Hubble's law). This was the situation until 1998; see Figure 2 for present understanding.

The current cosmological model of the universe is based on the theory of general relativity. When it was first conceived by Einstein the universe appeared as a 3-sphere of constant positive curvature and finite volume which kept on expanding. It appears now that the curvature is negative and that the universe is probably infinitely extended. Of this universe we see only a tiny expanding fragment which keeps on growing at an alarming slow rate. We have no idea of what lies beyond a few dozen billion light years away from earth but I would be very disappointed if it turned out to be a boring variation on what we already see from earth. If the universe is infinitely extended then any object in agreement with physical laws and no matter how improbable, insolent or ridicolous will exist somewhere. If we could gaze into infinity and have time to waste we would eventually see a giant 300 ft tall replica of the Venus of Milo in in pink marble. Why not?

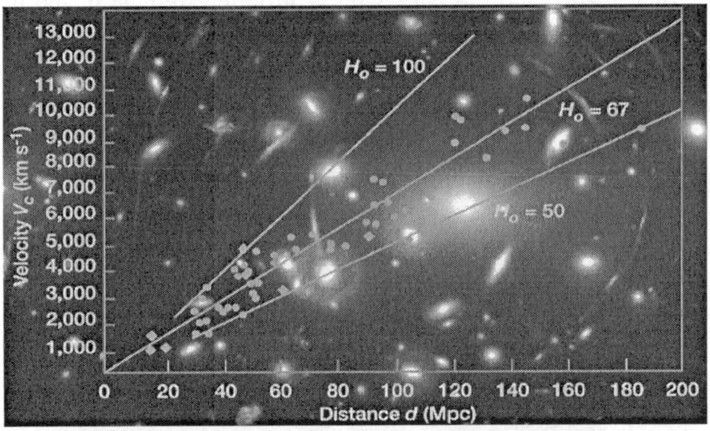

Figure 2. Schematic representation of the expansion of the Universe. The deceleration is due to gravity acting among celestial bodies; the acceleration is probably due to Dark Energy (Quintessence) pervading the vacuum of the Universe (it may be sumarized in a cosmological constant). (Courtesy of Beyond Einstein, NASA).

ONE HUNDRED YEARS OF SCIENCE

RICHARD A. CARRIGAN, JR.
Fermi National Accelerator Laboratory[†], *Box 500*
Batavia, IL 60510, USA

The last half-century has seen enormous strides in many scientific and technical areas. In particle physics previously unrelated fields like weak interactions and electromagnetism have been linked. Cosmology has been probed back to the Big Bang. Computing has moved from being a scientific tool to a subject that dominates the economy. The structure of DNA has been untangled. We may be on the threshold of understanding the origin of life and even discovering life elsewhere in the universe. Several of these diverse topics such as cosmology and fundamental physics are already profoundly coupled. Interestingly, ties exist between all of these subjects. These links are reviewed in light of opportunities ahead in the next decades.

1. Introduction

Books on astroparticle physics often cover the history of the universe in a chapter or two. Based on that, summarizing a hundred years of science in ten pages may not be so outrageous. A bird's eye view can uncover relationships that would be lost in more detailed perspectives. The key to this subject is the word science. Astroparticle physics is already a unification of astrophysics or cosmology and particle physics. This chapter searches for the implication of a larger unification, the interrelationship of all of science and technology.

I came to this subject along several paths. One influence has been the interesting cosmological diptych, *Astrophysics and the Ant*, created by my wife after decades of listening to astroparticle physicists[1]. In the left panel an astronomer asks "can I unweave the gauzy fabric of stars?" while in the right one the ant puzzles "can I untangle the matted carpet of earth?" There are two cosmologies here, the grand cosmology of this volume and the cosmology of the ant spanning biology, our planet, and our solar system. The second path has been my interest in SETI, the search for extraterrestrial intelligence. This subject involves many disciplines including astronomy, bioastronomy, computer science, and the nature and expression of knowledge.

[†] Operated by Universities Research Association, Inc. under contract No. DE-AC02-76CHO3000 with the United States Department of Energy.

The next question is what one hundred years to discuss. One of the great unifications in science was the merger of electricity and magnetism by Maxwell around 1870. Stepping forward from there would cover some elementary particle physics and modern biology as well as the beginnings of astroparticle physics. It would embrace Dawinian evolution and Medelian inheritance. Starting with Einstein and Planck about 1900 would bring us more or less to the present, but this gives no foothold on the future. In any case, that was the last century. Another choice with some future would start with the great wave mechanics developments of the twenties by Heisenberg and Schrödinger. Interestingly Schrödinger later turned to an influential investigation of the nature of life published at the height of World War II in neutral Ireland as *What is Life?* Shortly after that war there was a fantastic burst of new science from rocket technology, to the discovery of many mesons heralding the emergence of particle physics, to untangling DNA in biology, to the discovery of the Big Bang. Based on the fifties explosion I take my hundred years to be 1944 to 2043.

In what follows I will loosely call the area of particle physics and accelerator science "particles", the broad areas of space, astronomy, and planetary science "space", all of biology "DNA", and anything covered by computing, math, and electronics just "computing".

2. Particle Physics and Accelerator Science

This volume focuses on the subjects of particle physics and cosmology. In particle physics the work of many theorists along with experimenters using accelerator facilities and other resources has led us to a picture of a universe consisting of building blocks of quarks and leptons bound by force carriers such as photons, gluons, and the electroweak force carriers.

This picture relies on many developments including the discovery of the electron, antimatter, muons, strange particles, neutrinos, and even three generations of quarks as well as theories such as Fermi's model of beta decay and the standard model. Perhaps one quarter of these discoveries came before 1944. The discovery of the electron more than a century ago is an illustration of how fundamental science moves into the living room. The cathode ray tube used to investigate the electron became today's TV.

These discoveries are the past, what about the future? Recently Chris Quigg[2] has summarized a list of more than thirty challenging questions particle physics should address. To cite only a few, are quarks and leptons elementary? What is the relationship of quarks to leptons? Are there different kinds of matter? What do generations mean? And the granddaddy of them all, what is the grand unifying symmetry? With Quigg's challenge list in hand particle physics has many years of interesting work ahead.

Enrico Fermi, the developer of the theory of beta decay, was also one of the better accelerator engineers of our 100 years. Fermi was a key designer and builder of the University of Chicago synchrocyclotron. Fermi's cyclotron was among a dozen or so instruments built after the Second World War. These devices set the foundation for experimental particle physics. The tradition established by these pioneer accelerators has led to Fermilab in the US and will shortly give birth to the Large Hadron Collider in Europe. Without facilities like these progress in particle physics would have been slow at best. These facilities and others should throw wide the windows to Quigg's questions for the future.

These present facilities are extraordinary. The LHC should endure as a research facility to the end of our one hundred year period. The future beyond LHC is not so clear. One important possibility, an electron-positron linear collider, has challenging cost and technology problems. A full-scale effort is now underway world-wide to start construction on an international linear collider. Another path is to seek new accelerator technologies. The most-talked about possibility is plasma acceleration. While interesting progress has been made producing gradients 100 times higher than conventional RF cavities, there are colossal problems. In any case, the progress in the plasma acceleration field is too slow to serve the future of particle physics. Something needs to be done!

3. Cosmology

Shortly after the start of our hundred year period George Gamow proposed what came to be known as the "Big Bang" as an explanation for our universe. In this theory all of the visible universe originated from a tiny but energetic explosion thirteen billion years ago. This resulted in the expanding universe that we observe through red-shift measurements and the famous relic black body radiation. All of the elementary particles and atoms were formed in the early moments of that explosion. It was many years after Gamow's proposal before the relic black body radiation was found and even longer before the crucial interplay between the Big Bang and particle physics was appreciated.

The study of the Big Bang has gone from triumph to triumph. Some of these developments are mostly astronomical like studies of red shifts, the character of the relic three degree radiation, and the determination that most of the universe consists of "dark matter". But other studies have shed evidence on

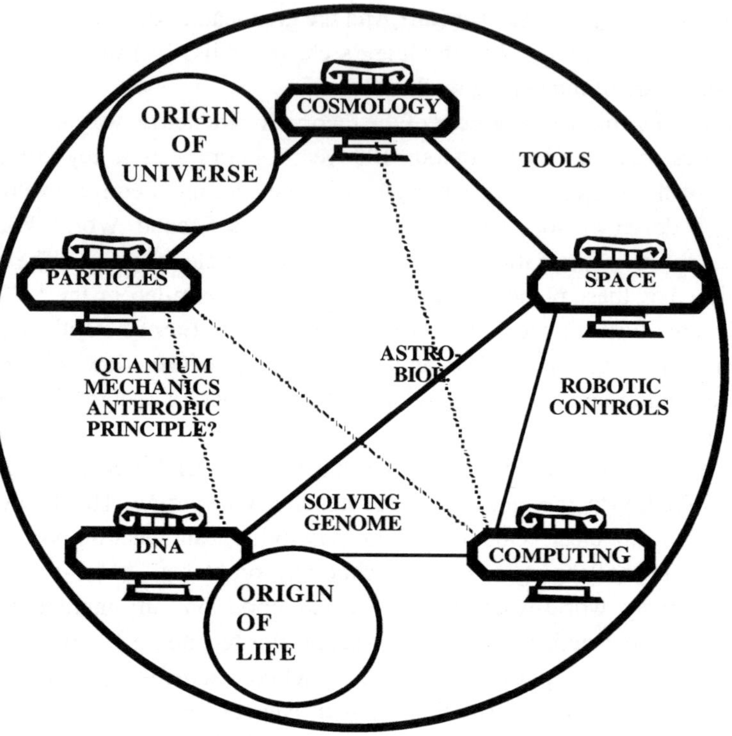

Figure 1.The five pillars of modern science.

the number of kinds of neutrinos and limits on exotic particles like magnetic monopoles.

In fact, as this volume attests, particle physics and cosmology are joined at the hip. Figure 1 shows these two great pillars of science linked together by insights on the origin of the universe. Cosmology explains the particles we have. Conversely, the rules for linking particles and the forces of nature set the path our universe took in this development. This emergence of the particle-cosmology linkage has been one of the great events of my life. I can remember the moment when I realized it was here to stay. I picked up Physical Review Letters one day in 1979 and saw John Preskill's article discussing the absence of magnetic monopoles in terms of inflation. I stopped searching for monopoles after that. The particle-cosmology link was explaining the origin of the universe and the absence of monopoles at the same time!

Nevertheless all the answers are not in and there are plenty of questions to fill out the remaining four decades. Even more precise observations of the relic radiation are needed. Dark matter and dark energy investigations must continue to get to the origins of these strange effects. More work is needed on gravity and gravity waves. Even the fabric of space is not understood. We don't know if there are more dimensions than the three we see. Particle physics may hold the key to some of these subjects but astronomical observations are also needed. In fact, most of these observations are not old fashioned astronomy but use techniques that have only come to flower during our period. These are the techniques of rockets and space science, the pillar lurking on the right hand side of Figure 1.

4. Space

Much of the recent progress in cosmology has come with satellite missions like the Wilkinson Microwave Anisotropy Probe or WMAP. Figure 2 highlights some events and technologies that have gone into the exploration of space. The V2 was the father of modern satellite technology. It was born at Penemunde before World War II and before our 100 years. The V2 in the upper left frame of Figure 2 is at White Sands Missile Range. It may have been the one outside my barracks during my short life as a rocket scientist at White Sands. V2 technology eventually led to manned systems like the Space Shuttle shown on a

Figure 2. Some highlights of the history of space.

service mission to Hubble in the upper right panel. These days it is fashionable to kick around man-in-space but in the context of the times the Hubble service work required astronauts. In the lower left panel the geologist Harrison Schmitt is shown on a visit to the moon in 1972. He was the last person and only scientist to go there. Recently robots have taken over from man. In 2004 two Mars Rovers have produced a travelogue of photos and measurements like the image on the lower right as they have cruised around the planet. The difference between 1973 and 2004 is that robotic intelligence has become greater and more compact so that a robot can now work much like a person.

Access to space has opened tremendous windows for science. In the last years WMAP has provided a beautiful picture of minute variations in the black body temperature over the universe. This map has done much to refine the understanding of cosmology. On a different front satellite pictures of Jupiter's Galilean satellites have shown four entirely different planet-like moons. This has opened the possibility of another abode for life outside of the planetary habitability zone imagined only a few years ago. Recent spectrograms from Hubble show the presence of oxygen and carbon, some of the ingredients of life, in the atmosphere of the famous extra-solar planet HD209458. Interestingly, none of these results depend directly on humans in space although people in space have been important in refurbishing Hubble.

The next several decades of science in space look promising. The NASA mission profile stretches to the end of our 100 years. Robots have become increasingly important. Humans have a role in space but for the distant future. Space telescopes are important for every phase of astronomy from the infrared to ultraviolet, to supernova searches, and for gravity wave observatories.

5. DNA

When I was in high school biology was a science entirely separate from physics. Biology involved dissections and tropical rain forests. These were fun in their way but they all seemed to be wet and sticky, unlike the physical sciences that were redolent with equations. Darwin's *The Origin of the Species* includes no mention of deoxyribonucleic acid or DNA. For that matter there is no citation to Mendel in the index of Darwin's book.

Around 1953 the picture changed forever and biology became a mathematical science. The fourth pillar of modern science, DNA, is shown in the lower left of Figure 1. Watson and Crick unfolded the wonderful structure of DNA with information from Franklin and Wilkins. These biologically oriented crystallographers used techniques that built on the insights of quantum mechanics and mathematical chemists. The DNA structure Crick and Watson found was like a computer tape made up of combinations of four chemical bases. The information content per DNA base pair is 1.44 bits. This is contained in a matrix with the mass of 600 atoms, incredible when compared to a silicon device. The human genome contains about 3 billion base pairs. Because of junk DNA the actual information content is more like 0.05 Gbytes,

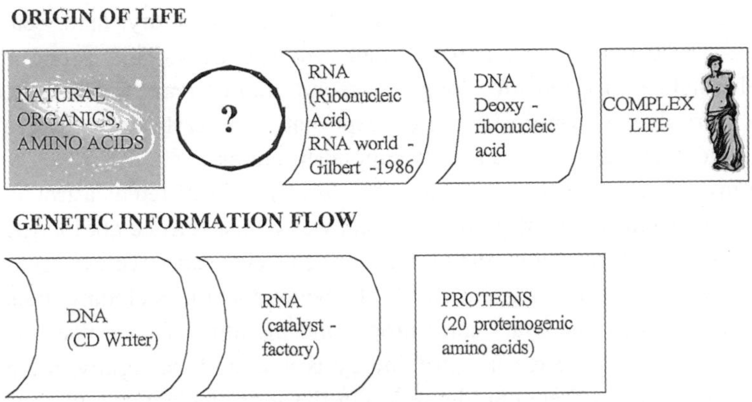

Figure 3. The Big Bang of Biology.

on the order of a Microsoft program. DNA plus Darwin's "survival of the fittest" plus the changing environment explains a great deal about biology on earth.

Like particle cosmology, biology has a big bang with associated parts similar in spirit to quarks and leptons. The Big Bang of Biology is illustrated in Figure 3. Life on the earth may have originated from a soup of natural organic molecules including amino acids. In one picture something happened and RNA formed. DNA evolved out of the RNA world. Complex life arose out of DNA. This process is the Big Bang part. The proteins formed by DNA are something like the parts of atoms. The linked action of DNA and RNA both propagates the genetic code quite well and also generates the proteins needed for life. It is as though one had a computer code capable of producing computers as needed.

As far as I know the origin of life is not understood. A good place to read about this is de Duve's 2002 book *Life Evolving* [3]. This is the intellectual soul mate of Weinberg's *The First Three Minutes* for the particle-astrophysics field. De Duve argues that getting RNA by chance is implausible. A complex chemical environment is needed. Perhaps there is a natural selection for molecules. One interesting clue may be chirality, handedness in molecules. The history of the first billion years on earth also offers clues.

A word of caution is in order here. This biology is not your daddy's biology. The tree of life I met in high school is only a small box on the current picture. A particularly important new portion is archaea which includes halophiles and thermophiles. These so-called extremophiles have been popping up in a number of unusual places such as hot deep-sea vents and very salty environments. Indeed nearly all the earth's surface layer, wet, dry, land, sea, and even ice is permeated with life. Life fits many environments! This is one of the aspects that helped to speed life along. A second, almost crazy factor pioneered by Lynn Margulies is symbiosis, the concept that some early creatures evolved as aggregations of one cell or relatively simple structures.

The investigation of the origin and nature of life in different astronomical environments is known as astrobiology. What was fringe science not long ago is becoming more mainstream by the month. In the next decades we can look to a number of places to find signatures of life beyond earth including meteors, planets and satellites like Mars and Europa, and the atmospheres of extra-solar planets. As a result, progress in astrobiology is and will be tightly linked to progress in space as indicated in Figure 1. On the more theoretical side one can assay the planetary, galactic, and even cosmological habitable zones. This subject has been covered recently in the book *Rare Earth*[4]. Rare Earth is a useful book but I take issue with the word "rare" and find their conclusions sometimes pushed aside by new developments. I suspect a book "Not So Rare Earth" might be just as interesting and perhaps closer to the ultimate truth.

A direct link between particle physics–cosmology and biology is posited by the anthropic principle first suggested by Brandon Carter[5]. Bjorken discusses the anthropic principle in a recent article[6]. In brief, the anthropic principle states that we have our universe because it is the only universe that works for us. The actual theses for this principle can be considerably more complicated. A particularly telling argument for the anthropic principle is the bottleneck noted by Fred Hoyle, the famous triple alpha reaction that produces ^{12}C in a star. Bjorken observes that a 0.3% change in the strong nuclear force would have drastically modified the production of carbon and oxygen thereby ruining the chances for life as we know it. For myself I wonder if this point of view does not suffer from some of the problems with *Rare Earth*. "Life" out there may be considerably more diverse than we credit it with. We need to ask the question posed by Schrödinger, namely "What is Life?"

6. Computing

One more pillar of science and technology that has emerged in the last sixty years is shown on the lower right in Figure 1, namely computing or more generally computing, mathematics, and electronics.

For a decade I oversaw technology transfer at Fermilab. More than 800 technological developments touched by Fermilab work were cataloged. The most important one, the World Wide Web, was totally missed. Granted the development occurred at CERN but some diligent heroes at Fermilab like Ruth Pordes worked to establish one of the first three transoceanic links for the Web. One can safely say that Tim Berners-Lee's development of the World Wide Web at CERN to facilitate collaboration in large particle physics experiments utterly changed the world.

A second "computing" wave is the advance of Moore's law, the shrinking of an electronic gate on a silicon chip with time. Electronics technology is sweeping over civilization at a rate such that it may outdistance us intellectually before our hundred years is over. The next Fermi may be a computer!

It is interesting to investigate the size of some data storage devices and databases[7]. The human genome "database" consists of three billion base pairs. The Fermilab tape robot handling colliding beam data and Monte Carlo calculations is 10 to 100 million times larger. The Sloan Digital Sky Survey database occupies only a small part of the robot. The robot could handle several years of the world's printed material. A typical graduate education might be subsumed in 1 to 10 Gbytes while a lifetime of images requires roughly 1000 times more space. The knowledge base in a human brain is in the 0.25 to 2.5 Gbyte range (stuffed in 10^{11} neurons). $6*10^{9}$ people gives 1.5 to $15*10^{9}$ Gbytes to profile everyone on earth. The lesson here is that the capacity for these databases is within range in the next decades.

Kurzweil[8] estimates the speed of the human brain is $200*10^{3}$ Giga computations/s or 10^{5} times the speed of a typical personal computer. Kurzweil calls the crossover point when computing speed exceeds our brain computing speed a "singularity". "Phase change" might be a better term. At the present rate of "Moore's Law" development the crossover point is 2020 to 2030. Actually the computing rate for processors used to study quantum chromodynamics, 1000 Giga Flops, is getting in range. Over the horizon is another startling possibility, quantum computing.

Computers will be an even bigger part of the future. Some of the problems and challenges will include the virus mess, the approach of the Kurzweil "singularity", qualifying Internet and World Wide Web information, and indeed qualifying programs. Finally we need working theories of computing, knowledge, and mind. Computing needs to be made into a science!

Computing then, is deeply linked to DNA. It has already been applied to untangling the human genome. Even more important, we have seen biology go from being wet, disagreeable stuff to irritating computer code. Life is mathematical. The future possibilities are vast. Computing is also tied to particle physics and cosmology as a faithful servant and is absolutely necessary for space exploration as we now know it.

7. Summing Up

In summary, particle physics and cosmology link and explain much of the universe. Biology has become a mathematical science driven in part by developments stemming from quantum mechanics. With these insights in biology we may soon understand the origin of life. Scientific experiments have moved into space. The development of sophisticated satellites and robotic explorers may shed more light on cosmology, extra-solar planets, and astrobiology on planets and moons in our solar system and beyond. We may find extraterrestrial life. Computers and computer science are reaching the scale of human capabilities. This may have even more profound impacts.

Many of these threads are embodied in the NASA *Origins* program. Similar grand programs are underway in Europe and the world in general. Over the next several decades *Origins* and other programs will look at cosmology, solar planetary science, extra-solar planets, and astrobiology. This is an important agenda for us as particle physicists, astronomers, and cosmologists.

We must continue to be aware that our sciences are linked to these other great developments in space science, biology, and computing. Science is no longer a confederation but a unified subject relying on the same types of resources, techniques, and ideas. Progress in each area depends on developments in all the others.

The future ahead promises that the period 1944 to 2043 will be one of the most interesting 100 years in the history of man.

References

1. See http://www.artmyth.org/Nancy_Jean_Carrigan_1.htm.
2. Abstracted, in part from recent lectures by C. Quigg, Fermilab.
3. C. De Duve, *Life Evolving. Molecules, Mind, and Meaning*, Oxford, New York, (2002).
4. P. Ward, D. Brownlee, *Rare Earth*, Copernicus, Springer-Verlag, New York., (2000).
5. B. Carter, Phil. Trans. R. Soc. Lond. A 310, 347 (1983).
6. J. D. Bjorken, Phys. Rev. D67, 043508 (2003).
7. Taken, in part, from discussions with Alex Szalay (Johns Hopkins) and Michael Lesk (Rutgers).
8. R. Kurzweil, *The Age of Spiritual Machines*, Viking, New York, (1999).

SUMMARIES OF THE POSTER SESSIONS

Organized by:

YVONNE BECHERINI
Dipartimento di Fisica, Università di Bologna and INFN Sezione Bologna
Viale C. Berti Pichat 6/2, 40127 Bologna, Italy
becherini@bo.infn.it

ZOULEIKA SAHNOUN
Astrophysics Department C.R.A.A.G.,
B.P.63, route de l'observatoire, Bouzareah, 16340 Algiers, Algeria
sahnoun@bo.infn.it

and

ALEXANDER ZAKHAROV
Institute for Theoretical and Experimental Physics (I.T.E.P.)
B. Cheremushkinskaya ul. 25 117218 Moscow, Russian Federation
zakharov@itep.ru

Three afternoon Poster Sessions have been organized during the 7th School. The Poster Sessions covered physics and astrophysics items, both experimental and theoretical. For each poster a short presentation was made, usually followed by a discussion. We collect here for each poster: the title, the list of authors, a short abstract and one relevant reference.

NUCLEAR DOUBLE BETA DECAY

R. CHANDRA[1], P. K. RATH[1] and P. K. RAINA[2]

[1]*Physics Dept., University of Lucknow, Lucknow-226007, India*
[2]*Physics Dept., I.I.T., Kharagpur-721302, India*
ramesh_ dbd@yahoo.co.in

The two neutrino double beta decay of ^{94,96}Zr, ^{98,100}Mo, ^{104}Ru and ^{110}Pd nuclei for $0^+ \rightarrow 0^+$ transitions are studied in the PHFB model in conjunction with the summation method. In the first step, the reliability of the intrinsic wave functions has been established by obtaining an overall agreement between a number of theoretically calculated spectroscopic properties and the available experimental data for ^{94,96}Zr, ^{94,96,98,100}Mo, ^{98,100,104}Ru, ^{104,110}Pd and ^{110}Cd isotopes. Subsequently, the PHFB wave functions of the above mentioned nuclei are employed to calculate the nuclear transition matrix elements $M_{2\nu}$ as well as the half-lives $T_{1/2}^{2\nu}$. Further, we have studied the effects of deformation on the $M_{2\nu}$.

Ref. R. Chandra et al., *nucl-th/0405074*.

THE DEGREE OF POLARIZATION OF THE TAU LEPTON PRODUCED IN CHARGED-CURRENT (CC) NEUTRINO-NUCLEUS SCATTERING

K.M. GRACZYK, J. A. NOWAK

Inst. of Theor. Phys., Univ. of Wrocław, pl. M. Borna 9, 50 - 204 Wrocław, Poland, kgraczyk@ift.uni.wroc.pl, janow@ift.uni.wroc.pl

In new long-baseline experiments the neutrino oscillation can be studied by the detection of the tau neutrinos produced in the $\nu_\mu \rightarrow \nu_\tau$ oscillations. The tau produced in the charged current ν_τ scattering off the nucleus can be partially polarized which affects the distribution of its decay products. We calculate the tau's degree of polarization and discuss its dependence on the nuclear effects. Within the framework of the relativistic MFT we show that the tau is highly polarized and only in the case of the forward scattering and neutrino energy around 4.5 GeV it may be unpolarized. Taking into account the effective mass the RPA corrections significantly influence the tau's degree of polarization.

Ref. K.M. Graczyk, Accep. in *Nucl. Phys.* **A**, *hep-ph/0407275.*

SPIN LIGHT OF NEUTRINO IN GRAVITATIONAL FIELDS

M.DVORNIKOV, A.GRIGORIEV, A.STUDENIKIN
Moscow State University, Dept. of Th. Phys., 119992, Moscow, Russia
maxim_dvornikov@aport.ru, alex.grigoriev@aport.ru,
studenik@srd.sinp.msu.ru

We developed a quasi-classical theory for description of a neutrino spin evolution in a gravitational field and obtained the corresponding neutrino spin evolution equation. Our study was performed in the framework of the quasi-classical approach to describe the spin evolution of neutrino interacting with general types of external non-derivative fields. On this basis we derive the expression for the probability of the corresponding spin oscillations and investigate the spin light of neutrino ($SL\nu$) in gravitational fields. We also consider the combined effect of neutrino interaction with matter, electromagnetic and gravitational fields on the neutrino spin and derive the spin evolution equation accounting for these neutrino interactions. As an application we considered the $SL\nu$ radiation in the case when a neutrino is moving in the accretion disc of a neutron star and along the relativistic jet from a quasar. It is also shown that the $SL\nu$ photon energy in these cases can span up to gamma rays.

Ref. M. Dvornikov et al., *hep-ph/0406114*.

GENERAL BIG BANG NUCLEOSYNTHESIS CONSTRAINTS ON NEUTRINO OSCILLATIONS

D. KIRILOVA, M. PANAYOTOVA
Institute of Astronomy, Bulgarian Academy of Sciences, Sofia

We provide numerical analysis of the Big Bang NucleoSynthesis (BBN) production of He-4, Y_p, in the presence of $\nu_e \leftrightarrow \nu_s$ oscillations, effective after neutrino decoupling, for the case when ν_s is partially filled initially, i.e. $0 < \delta N_s < 1$, $\delta N_s = n_{\nu_s}/n_{\nu_e}$. We account for all known oscillations effects on cosmological nucleosyntesis. We have calculated the cosmological constraints on oscillation parameters corresponding to different initial populations of the sterile state and He-4 overproduction $\delta Y_p/Y_p = 5\%$. The cosmological constraints corresponding to $\delta N_s > 0$ are relaxed in comparison to the $\delta N_s = 0$ case and the relaxation is proportional to the level of initial population of ν_s, i.e. to δN_s.

Ref: D. Kirilova et al., *ICTP report*, **IC/IR/2004/13** (2004).

ON SCALAR CONDENSATE BARYOGENESIS MODEL

D. KIRILOVA, T. VALCHANOV

Institute of Astronomy, Bulgarian Academy of Sciences, Sofia

We provided precise numerical analysis of the scalar field condensate baryogenesis model (Kirilova & Chizhov, *MNRAS*, **314**, 256 (2000)). We updated the model parameters range according to the current observational cosmological constraints and analyzed numerically ϕ post-inflationary evolution until its decay $\phi \to q\bar{q}l\gamma$. During that period the ϕ amplitude decreased due to Universe expansion and particle production processes by the oscillating field. Particle creation processes were shown to play essential role for B evolution and its final value. They may lead to a considerable decrease of the field's amplitude for large ϕ couplings or/and large H_I values, which finally reflects into strong damping of the baryon charge carried by the condensate. Hence, the observed small baryon asymmetry can be naturally obtained in the discussed model of baryogenesis.

Ref. D. Kirilova et al., *ICTP report* **IC/IR/2004/12** (2004).

MICROSCOPIC NUCLEAR STRUCTURE EFFECTS ON DGT NUCLEAR MATRIX ELEMENTS FOR $0^+ \to 0^+$ TRANSITIONS OF ^{78}Kr AND ^{106}Cd NUCLEI

A. SHUKLA, P.K.RAINA, P.K. RATH

Department of Physics and Meteorology, IIT Kharagpur-721302, India
Physics Dep., University of Lucknow, Lucknow-226007, India,
ashukla@phy.iitkgp.ernet.in

The Double Gamow-Teller matrix element $M_{GT}^{2\nu}$ for two neutrino positron double beta decay e$^+$DBD($\beta^+\beta^+$, β^+EC, $ECEC$) of ^{78}Kr and ^{106}Cd for $0^+ \to 0^+$ transition has been studied in the Hartree-Fock-Bogoliubov model. The wave functions of initial and final nuclei have been first tested for yrast spectra, $B(E2{:}0^+ \to 2^+)$, $Q(2^+)$ and $g(2^+)$ values within the same framework. The calculated e$^+$DBD half-lives predict that $T_{1/2}^{2\nu}$ for β^+EC transition in ^{78}Kr and ^{106}Cd nuclei is the most favored case for presently undertaken/planned experiments. Further, the effect of deformation is found to be important in case of ^{106}Cd but not for ^{78}Kr.

Ref. A. Shukla et al., *nucl-th/0405066.*

COSMOLOGY OF PRIMORDIAL BRANEWORLD BLACK HOLES

A.S. MAJUMDAR

S. N. Bose National Centre for Basic Sciences, Block JD, Sector III, Salt Lake, Kolkata 700098, India, archan@bose.res.in

We consider the evolution of primordial black holes formed in the high energy radiation dominated phase of the braneworld scenario. These black holes have modified Schwarzschild geometry and hence Hawking radiate at a different rate compared to standard 4-dimensional black holes. We show that accretion of radiation from the surrounding radiation bath is a significant effect which dominates over evaporation. Thus primordial braneworld black holes grow in size through evaporation and can survive up to several cosmologically interesting eras. For suitable parameters, some of these black holes could survive up to present times, hence acting as candidates of cold dark matter.

Ref. A.S. Majumdar, *Phys. Rev. Lett.* **90**, 031303 (2003).

NUMERIC SIMULATIONS OF STELLAR COLLAPSE AND FORMATION OF REISSNER-NORDSTROM SPACETIMES

C.R. GHEZZI, P.S. LETELIER

Instituto de Matemática, Estatística e Computação Científica, Univ. Estadual de Campinas, Campinas, São Paulo, Brazil, ghezzi@ime.unicamp.br, letelier@ime.unicamp.br

We simulated the collapse of a supermassive star and the formation of a charged black hole using a general relativistic code. It was assumed a polytropic equation of state and an initial uniform distribution of charge and mass. It is found a very different behavior on the dynamics of the collapse of a charged star compared to the collapse of a neutrally charged one. For a charge to mass ratio $Q/\sqrt{G}M > 0.1$, a shell-like structure surrounding an interior region of lower density and charge is formed. It is possible to form highly charged black holes in stellar collapse (up to $Q/\sqrt{G}M < 1$). This result could have profound implications for astroparticle physics, since highly charged compact objects can act as potential accelerators (selective accretion of charges onto charged black holes remains to be proven).

Ref. C. R. Ghezzi et al., *gr-qc/0312090.*

MIRAGES AROUND KERR BLACK HOLES AND RETRO GRAVITATIONAL LENSES

A.F. ZAKHAROV[1,2], A.A. NUCITA[3], F. DE PAOLIS[3], G. INGROSSO[3]
[1]*Inst. of Th. and Exp. Phys., Moscow, 117259, Russia, zakharov@itep.ru*
[2]*Space Research Centre of Lebedev Physics Institute, Moscow*
[3]*Dip. di Fisica Univ. di Lecce and INFN Lecce, Italy*

Recently Holz and Wheeler (2002) considered an attracting possibility to detect retro-images of the Sun by a Schwarzschild black hole (BH). We discuss glories (mirages) formed near rapidly rotating Kerr BH (KBH) horizons and propose a procedure to measure masses and rotation parameters analyzing their forms. In some sense that is a manifestation of gravitational lens effect in the strong gravitational field near BH horizon and a generalization of the retro-gravitational lens phenomenon. We analyze a KBH rotating at arbitrary speed for selected positions of a distant observer with respect to the equatorial plane of a KBH. Falcke et al. (2000) suggested to search shadows at the Galactic Center. We present shadow boundaries calculated numerically. We also propose to use future radio interferometer RADIOASTRON facilities to measure shapes of mirages and to evaluate the BH spin as a function of the position angle of a distant observer.

Ref. A.F. Zakharov et al., *astro-ph/0411511.*

A NEW POSSIBLE WAY OF PRIMORDIAL BLACK HOLES DETECTION

S.E. SIAHLO, V.V. TIKHOMIROV
Inst. for Nucl. Problems of Belarus State Univ., Bobruiskaya str. 11, Minsk 220050, Belarus, svetaju@inp.minsk.by

We show that primordial black holes (PBH) are able to absorb white dwarfs (WD) for the Hubble time. WD matter absorption is accompanied by neutronization and neutrino emission, with a total energy which exceeds by more than ten orders of magnitudes that of PBHs' Hawking radiation. The most dense WD absorption is accompanied by neutrino burst carrying away 10^{51}erg and more during a tenth of a second. Such a burst taking place at several kiloparsecs from the Earth will be detected by existing neutrino detectors. WDs absorbtion by PBHs and accompanying processes allow to improve constraints on abundance of weakly emitting PBHs by several orders.

Ref. S.E. Juralevich et al., *astro-ph/0202445.*

MATHEMATICAL POST-TREATMENT OF SERIES FOR SOLUTIONS OF THE LANE-EMDEN EQUATION

ROGEL MARI D. SESE[1,2], JOSE PERICO H. ESGUERRA[2],

[1]*Inst. of Math. Sci. and Phys., Univ. of the Phil., Los Baños, College, Laguna 4031, Philippines,* [2]*Nat. Inst. of Phys., Univ. of the Phil., Diliman, Quezon City 1100, Philippines*

rmds@physics.uplb.edu.ph

We obtained approximate analytic solutions to the Lane-Emden (LE) equation for polytropes for values of the index n from 0 to 5 by mathematical post-treatments of power and Hunter series (HS) solution. The technique involves the Laplace transformation (LT) of the truncated series, a diagonal Pade approximation in the transformed variables and the inverse Laplace transform. The result is a compact approximate analytic solution to the LE equation. We validate the method by comparing the solution with previous results and analytic solutions. The values of the first zero are accurate, having an average error or 3.2%, 0.12% and 0.13% for the treated power, 2-term and 5-term treated HS respectively. The HS has an accuracy >40 terms of the original HS.

Ref. C. Hunter, *Mon. Not. Roy. Astron. Soc.* **328**, 839 (2001)

GRAVITATIONAL WAVES FROM WHITE DWARFS

A. SADOYAN

Yerevan State University, Alex Manoogian 1, 375025 Yerevan, Armenia

asadoyan@www.physdep.r.am

Rotating white dwarfs undergoing quasi-radial oscillations can emit gravitational radiation in the frequency range 0.1-0.3 Hz. Assuming that the energy source for the radiation is the deformation energy, the strain amplitude is found to be 10^{-25} at a distance about 50 pc. We calculated the thermal energy losses through magneto-hydrodynamic mechanism to compare with energies emitted in Gravitational Waves (GW) band. When differential rotation energy is feeding GW, the strain amplitudes are again found to be less than 10^{-25} for a white dwarf at 50 pc. Nearby oscillating white dwarfs may provide a clear enough signal to investigate their interiors through gravitational wave astroseismology.

Ref. M. Benacquista et al., *Astroph. J. Lett.* **596**, 233 (2003).

THEORY OF POINT STRUCTURAL PARTICLE

LL. URANGA, A. MARTINEZ, M.Y. BALLESTER

Instituto Superior de Tecnologias y Ciencias Aplicadas,
Quinta de los Molinos, Ave S. Allende y Luaces, C. Habana, Cuba
llinersy@uci.cu, aliezer@uci.cu

The theory of the point structural particle (p.s.p.) is presented. A Schrödinger-like equation describing the state evolution in a six dimensional space is derived. The hamiltonian of the system agrees with the non relativistic limit of Dirac's equation but it remains valid for any particle. An expression for the rest energy is introduced. Spin-orbit and spin-spin interactions are straightforwardly derived from the model. Darwin's contact term is obtained with a clear physical meaning: quadrupolar interaction of each p.s.p. with the external electric field. A new interaction appears which doesn't have analogous in relativistic quantum mechanics.

Ref. Ll. Uranga et al., *Rev. Cub. Fis.* **21**, 1 (2004).

NEW RESULTS FROM THE CAKE EXPERIMENT

S. CECCHINI[1], T. CHIARUSI[1], E. MEDINACELI[1,2]

[1]*Dip. di Fisica, Univ. di Bologna and INFN Bologna,* [2]*Inst. Invest. Fisicas, La Paz, Bolivia*
cecchini@bo.infn.it, chiarusi@bo.infn.it, medinaceli@bo.infn.it

The CAKE balloon experiment measured the composition of the primary charged cosmic rays and searched for exotic particles. It used nuclear track detectors CR39 and Lexan. The effective area was about 1 m^2 and the acceptance was 2.3 m^2 sr for $Z > 30$. The balloon flew from the ASI Trapani Milo launch base in Sicily to central Spain in 22 hours, at an altitude of about 40 km. Once recovered, the detectors were etched in 6N NaOH at 70 °C for 30 hours and then measured with optical microscopes. The position and the geometric characteristics of the measured tracks were recorded and the charge numbers Z of the ions were determined on the basis of the detector calibrations. The presented charge spectrum came from the analysis of 5% of the exposed detector and was obtained tracking the particles through the stacks. Nuclei from the CNO group (Z = 6,7,8) up to the Nickel (Z = 28) were observed.

Ref. T. Chiarusi et al., *Radiat. Meas.* **36**, 335 (2003).

OPERA: AN APPEARANCE EXPERIMENT TO SEARCH FOR $\nu_\mu \to \nu_\tau$ OSCILLATIONS IN THE CNGS BEAM

M. COZZI, L.S. ESPOSITO, G. SIRRI FOR THE OPERA COLLAB.

Dip. di Fisica, Univ. di Bologna and INFN Bologna
cozzi@bo.infn.it, esposito@bo.infn.it, sirri@bo.infn.it

The data on atmospheric neutrino are well described by a dominant $\nu_\mu \leftrightarrow \nu_\tau$ oscillation, with oscillation parameters $\Delta m^2_{23} = 2.4 \times 10^{-3}\ eV^2$ and $\sin^2 2\theta_{23} = 1.0$. The primary goal of OPERA is to measure the ν_τ appearance in a pure ν_μ beam. OPERA is a hybrid detector with a modular structure made of electronic detectors and nuclear emulsions. The basic emulsion structure is the Emulsion Cloud Chamber (ECC) composed by two emulsion films each 50 μm thick placed on either side of a plastic base, 200 μm thick interleaved with 1 mm thick lead plate (to provide the large target mass). The ECC is used as a tracking device. The detector is now under construction in the Gran Sasso underground lab and will detect ν_τ in the CNGS long baseline ν_μ beam from CERN, starting from May 2006.

Ref. P. Migliozzi, *Int. J. Mod. Phys.* **A 18**, 3877 (2003).

AUTOMATIC SCANNING OF EMULSION FILMS FOR THE OPERA EXPERIMENT

M. COZZI, L.S. ESPOSITO, G. SIRRI FOR THE OPERA COLLAB.

Dip. di Fisica, Univ. di Bologna and INFN Bologna
cozzi@bo.infn.it, esposito@bo.infn.it, sirri@bo.infn.it

Nuclear emulsions provide three-dimensional spatial information with excellent resolution, and are ideal to detect short-lived particles as τ leptons. Technological progresses in scanning instruments and application of automated systems allow the analysis of a very large emulsion target as the OPERA experiment. An automatic scanning system consists of a computer driven mechanical stage, an appropriate optical tube, a camera and its associated electronics. The readout is performed by grabbing and processing images of the emulsion at different depths and measure the position of each grain along the tracks. Track reconstruction is done by fitting aligned grains. The system has an angular resolution of 2 mrad, a spatial resolution of about 0.5 μm and an efficiency of 90 %.

Ref. N. D'Ambrosio et al., *Nucl. Phys.* **B 125**, 22 (2003).

SEARCH FOR INTERMEDIATE MASS MONOPOLES AT HIGH ALTITUDE IN THE HIMALYAN REGION IN NORTHERN AREAS OF PAKISTAN

M.I. SHAHZAD[1], I.E. QURESHI[1], S. MANZOOR[1,2], G. SHER[1], L. PATRIZII[2]

[1]*Physics Research Division, PINSTECH, Nilore, Islamabad, Pakistan, ikram@pinstech.org.pk,* [2]*INFN Sez. Bologna, Italy*

An experimental set-up for the search for magnetic monopoles ($10^5 - 10^{12}$ GeV) similar to that in the ref. is described; it consists of CR-39 and Lexan nuclear track detector modules (82 m^2 area) installed on the false ceiling of a building at Koksil (4265 m a.s.l.), in the Himalayan region of Pakistan in May 2002. The environmental radon level measured using CR-39 detectors placed in box type dosimeters for the period from May 2002 to May 2003, was (14 $\pm$ 7) Bq m^{-3}. The density of the recoil tracks produced by cosmic ray neutrons in CR-39, exposed for one year at that altitude was measured to be (55 $\pm$ 8) cm^{-2}.

Ref. S. Cecchini et al., *Il Nuovo Cim.* **24C**, 639 (2001).

SEARCH FOR A VERY RARE PROCESS: SPONTANEOUS EMISSION OF MUONS FROM HEAVY NUCLEI

L. CONSIGLIO[1], M. GIORGINI[1] AND V. POPA[1,2]

[1]*Dip. di Fisica, Univ. di Bologna and INFN Bologna*
[2]*Inst. for Space Sciences, R-76900 Bucharest, Romania*
consiglio@bo.infn.it, miriam.giorgini@bo.infn.it, popa@bo.infn.it

A possible search was proposed for a very rare process, consisting in the spontaneous emission of muons from heavy nuclei. As a first step, we intend to look for this exotic radioactivity from lead nuclei, using the base elements ("bricks" composed by lead and nuclear emulsion sheets) of the OPERA experiment. We present a Monte Carlo simulation of the expected event topologies and we discuss their detectability. A further step would consist in the substitution of Lead with Uranium in the OPERA bricks. Using few bricks, we could reach a good sensitivity level, considerably better than what has been measured until now.

Ref. M. Giorgini, Proc. of the Vulcano Workshop 2004, SIF (Bologna,2005), in press.

NEW CALIBRATIONS OF CR39 AND MAKROFOL DETECTORS FOR THE SLIM-EXPERIMENT

S. MANZOOR[1,2], A. KUMAR[1,3], M. FRUTTI[1], L. PATRIZII[1], V. TOGO[1], D. DI FERDINANDO[1], Z. M. SAHNOUN[1,4] , T. CHIARUSI[1], E. MEDINACELI[1,5]

[1]*Dip. di Fisica, Univ. di Bologna and INFN Bologna,*
[2]*Physics Research Division, PINSTECH, Islamabad, Pakistan*
[3]*Phys. Dept., SHSL Inst. of Engg. and Tech. Longowal - 148 106 India*
[4]*C. R. A. A. G, Algiers, Algeria,* [5]*Inst. Invest. Fisicas, La Paz, Bolivia*

We present calibrations of the Makrofol-DE nuclear track detector (NTD) using CERN Pb ions of 158 AGeV energy. With new etching conditions with ethyl alcohol we improved the quality of the post-etched surfaces, reducing background tracks and enhancing the sharpness of the tracks. Special efforts were made to keep the etching tanks tight and to control the stability of the etching conditions. The detection threshold of Makrofol-DE is $Z/\beta \sim 50$. The charge resolution close to the Lead peak ($Z = 82$) was improved by measuring the heights of the etched pit cones. The peaks are well separated, and a charge can be assigned to each.

Ref. G. Giacomelli et al., *Nucl. Instr. Meth.* **A 411**, 41 (1998).

SOLAR ENERGY. A PHOTOVOLTAIC ROOF

A. BA[1], D. MATTEUZZI[2]

[1]*Dép. de Physique Faculté des Sciences et Techniques Univ. de Bamako, BP E 3206, Bamako, MALI*
[2]*INFN Sez. Bologna, Italy*

Solar energy is converted into electric energy by a photovoltaic system installated since year 2001 on the roof of the Physics Department of the University of Bologna, in the context of the 10.000 photovoltaic roof project of ENEA (Italy). We measured the solar radiation and the electric power produced daily. We present the daily data obtained from year 2001 to July 2004. We found in average yearly AC produced electric energy of about 2100 kWh, which would be sufficient for the needs of a small family.

Ref. D. Matteuzzi et al., *in preparation*

SEARCH FOR NEUTRINO RADIATIVE DECAYS DURING THE 2001 TOTAL SOLAR ECLIPSE (TSE)

S.CECCHINI, V. POPA FOR THE NOTTE COLLABORATION
Physics Dept. Univ. Bologna and INFN Bologna; also ISS, R-77125, Bucharest - Magurele, Romania

In June 21st, 2001 we observed a TSE, from a location situated North of Lusaka, Zambia, looking for visible photons produced by possible $\nu_{2,3} \rightarrow \nu_1 + \gamma$ decays of the mass eigenstates of the solar neutrinos. From the analysis of the digital pictures of the Moon dark disk no signals compatible with the expectation were seen for both possible decays. We could estimate 95% C.L. neutrino lifetime lower limits. For the $\nu_2 \rightarrow \nu_1 + \gamma$ decays the proper ν_2 lifetime limit is strongly dependent on the ν_1 mass assumption; considering $m_{\nu_1} \simeq 0.1$ eV, $\tau_0 \geq 10^8$ s for lefthanded or Majorana neutrinos, and $\tau_0 \geq 10^6$ s for righthanded ν_1. For the $\nu_3 \rightarrow \nu_1 + \gamma$ decays, the limit depends also on the unknown mixing θ_{13}; assuming $\sin^2\theta_{13} = 0.1$ and $m_{\nu_1} \simeq 0.1$ eV, we obtain $\tau_0 \geq 10^4$ s for lefthanded or Majorana ν_3 and $\tau_0 \geq 100$ s for righthanded ν_3.

Ref. S. Cecchini et al., Astropart. Phys. **21**, 35 and 183 (2004).

DOWNGOING MUON CALIBRATION OF AMANDA

S. MOVIT FOR THE AMANDA COLLABORATION
Dept. of Astronomy and Astrophysics, The Pennsylvania State University, 525 Davey Lab, University Park, PA, 16802 USA, movit@astro.psu.edu

Relative timing for AMANDA (http://amanda.uci.edu) is calibrated with a laser system comprised of fibers and diffuser balls in the ice. However, this method requires many austral summer man-hours, and thus is performed once yearly. A new method has been devised that uses atmospheric muons and Monte Carlo simulations to calibrate timing. Recent improvements to the algorithm include allowance for varying ice properties with depth and automation of the process. A small speed-of-light error (.5%) has been detected in the laser calibration using the muon method, confirming the accuracy of the new technique.

Ref. D. Cowen et al., Proc. 27th ICRC **2**, 1133 (Hamburg, 2001).

THE ANTARES NEUTRINO TELESCOPE

Y. BECHERINI FOR THE ANTARES COLLABORATION
Dip. di Fisica, Univ. di Bologna and INFN Bologna
Yvonne.Becherini@bo.infn.it

The ANTARES Collaboration is building a neutrino telescope in the Mediterranean Sea with the main purpose of looking for high energy neutrinos from astrophysical sources. The detector is a 3D array of phomultiplier tubes aiming to capture Cerenkov photons emitted by the passage of relativistic particles in seawater. The Collaboration has developed, deployed and tested a prototype line with 15 photomultipliers and an instrumentation line equipped with devices for environmental measurements. During the present year many laboratory tests were carried out and it is planned to deploy two test lines in the next months and the first full line in June 2005. The complete 12-lines detector will be operational in 2007.

Ref. L. Moscoso for the ANTARES Coll., *Proc. of 32nd ICHEP 2004*, Beijing.

ACOUSTIC DETECTION STUDIES IN MARSEILLE

V. NIESS FOR THE ANTARES COLLABORATION
Centre de Physique des Particules de Marseille (CPPM)
163, avenue de Luminy, case 902, Marseille cedex 09, France
niess@cppm.in2p3.fr

Ultra high energy neutrinos (UHE, $\geq 10^{18} eV$) coming from astrophysical sources could be detected from the pressure wave generated as they shower in sea water. Whereas sound can propagate on kilometric ranges with little attenuation it also results in high ambiant noise conditions as regard to the expected level of signal. The main issue left in order to know if acoustic detection can become competitive is a practical knowledge of ambient sea noise at depth (2500 m) and in the ultra sonic frequency range ($\geq 20kHz$). Therefore, CPPM aims to integrate a low-self noise ITEP hydrophone ($2\mu V/Hz^{1/2}$) with embedded electronics in the instrumentation line of ANTARES during the second half of year 2005.

Ref. G.A. Askariyan et al., *Nucl. Instr. and Meth.* **164**, 267 (1979).

OBSERVATION OF EGRET GAMMA-RAY SOURCES

M. KHAKIAN GHOMI, ALBOPRZ OBSERVATORY

Physics Dept., Sharif University of Technology, P.O. Box 11365-9161, Tehran, Iran, khakian@mehr.sharif.edu

Ultra-high-energy (E>100 TeV) Extensive Air Showers (EASs) have been monitored for a period of five years (1997 - 2003), using a small array of scintillator detectors in Tehran, Iran. The data have been analyzed taking into account the dependence of source counts on zenith angle. During a calendar year different sources come in the field of view of the detector at varying zenith angles. Because of varying thickness of the overlaying atmosphere, the shower count rate is extremely dependent on zenith angle which has been carefully analyzed over time. High energy gamma-ray sources from EGRET third catalogue where observed and the data were analyzed using an excess method. Upper limits were obtained for a number of EGRET sources, including 5 AGNs, 1 probably AGN and 4 unidentified sources.

Ref. M. Khakian Ghomi et al., *Astron. and Astrophys.*, accepted.

EXTENSIVE AIR SHOWERS AND THE PIERRE AUGER OBSERVATORY (PAO)

M. J. TUEROS FOR THE AUGER COLLABORATION

Departamento de Fisica, Universidad Nacional de La Plata, Argentina tueros@fisica.unlp.edu.ar

The PAO is the first to use fluorescence telescopes and particle detectors simultaneously. The data from both methods can be combined for cross calibration, systematic error reduction and increased energy and angular resolution. Light detected by the telescopes is related to the shower particle content and the primary energy. The track left and the signal timing gives the shower direction. The Cerenkov detectors measure the particle density profile and the primary energy. Timing of the signals gives the shower direction. All PAO detectors store the time profile of the signal, giving additional information on composition, asymmetries and much more.

Ref. The Pierre Auger Coll., *Nucl. Instr. Meth.* **A 523**, 50 (2004).

COSMIC RAY SPECTRUM FROM 3×10^{14} TO 3×10^{19}eV

ANA VASILE[1,3], PETER L.BIERMANN[1,2]
[1]*Max-Planck Institut für Radioastronomie, Bonn, Germany*
[2]*Dep. of Physics and Astronomy, University of Bonn*
[3]*Institute for Space Sciences , Magurele, Bucharest*
plbiermann@mpifr-bonn.mpg.de, avasile@venus.nipne.ro

We discuss the recent success of a proposal made by Biermann et al. over 10 years ago to explain the spectrum of cosmic rays in the energy range 10^{14}-10^{19} eV. This success allows some strong conclusions to be made on the physics of supernovae: in this proposal which explains the origin of cosmic rays, the mechanism for exploding supernovae of high mass should be that proposed by Kardashev (a combination of the effects of rotation and magnetic fields which explodes the star). This leads to further suggestions which may be useful for the study of gamma ray bursts and the search for a new standard candle in cosmology, brighter than supernovae of type Ia.

Ref. P.L. Biermann et al., *astro-ph/0302201*.

THE GLAST MISSION

OMAR TIBOLLA FOR THE GLAST LAT COLLABORATION
Dip. di Fisica, Univ. di Padova and INFN Padova, Via Marzolo 8, 35131 Padova, Italy, tibolla@pd.infn.it

The Gamma-ray Large Area Space Telescope is an international (France, Germany, Italy, Japan, Sweden and USA) space mission that will study the cosmos in the photon energy range 10 keV-300 GeV. GLAST will have a new generation imaging gamma-ray telescope (Large Area Telescope: LAT) and a second instrument for the detection of Gamma Ray Bursts (Glast Burst Monitor: GBM). GLAST has many scientific purposes in several fields of research: AGN/Blazars, Unidentified EGRET Sources, New Particle Physics, Extragalactic Background Light, GRBs, Pulsars, Cosmic Rays and Interstellar Emission, Solar Flares. Other important features are the complementarity with Ground-based Gamma-ray Telescopes and the plan to study the Universe in unexplored regions of EM Spectrum.

Ref. J.E. McEnery et al., *astro-ph/0406250*.

HIGH ENERGY EAS MUON DETECTION WITH THE LST DETECTOR OF KASCADE-GRANDE EXPERIMENT

A. BERCUCI FOR THE KASCADE-GRANDE COLLABORATION

National Institute of Physics and Nuclear Engineering – Horia Hulubei, POB MG 6, Ro-76900 Bucharest, Romania, alex@muon1.nipne.ro

The high energy muons from Extensive Air Showers (EAS) are first candidates for low generation secondaries of the interaction of the primary Cosmic Rays (CR) with the atmosphere. Their number and spatial distribution at the Earth surface can thus discriminate on the primary mass and energy. A Limited Streamer Tube (LST) detector system has been installed in the KASCADE-Grande experiment to study the EAS muon component above a 2.4 GeV threshold. The extended area of 255 m^2 of the LST and the fine granularity of detection (4.5 cm) makes the study of the muon spatial distribution possible in its very central zone. A dedicated, fractal based analysis tool has been developed. The results on CR mass and energy discrimination will be presented obtained from 1.5 years of data collection by the KASCADE-Grande experiment.

Ref. T.Antoni et al., *Nucl. Instr. Meth.* **A 533**, 387 (2004).

GLOBAL ANALYSIS OF THE SOLAR NEUTRINOS HAVING TRANSITION MAGNETIC MOMENT

D. YILMAZ, A.U. YILMAZER

Faculty of Engineering, Ankara Univ., 06100 Tandogan Ankara, Turkey, dyilmaz@eng.ankara.edu.tr, yilmazer@eng.ankara.edu.tr

A global analysis of the solar neutrino data from all solar neutrino experiments is presented, assuming that the solar neutrino deficit is due to the the matter-enhanced spin-flavor precession (RSFP) effect. We used the Wood-Saxon shape of magnetic field profile throughout the entire Sun. We investigated how the RSFP effect change the allowed regions in the neutrino parameter space. We show that the allowed region in the LMA region disappears at 95% CL as μB value increases.

Ref. A.B. Balantekin et al., *Phys. Rev.* **D 41**, 3583 (1990).

POSSIBLE NEUTRINOLESS COMPONENT IN THE ^{50}Cr DOUBLE BETA DECAY

N. TODOROVIĆ[1], I. BIKIT[1], J. SLIVKA[1], M. VESKOVIĆ[1], M. KRMAR[1], Lj. ČONKIĆ[1], D. MRDJA[1], E.VARGA[1]

[1] *Physics Dept., Univ. of Novi Sad, Novi Sad, Serbia and Montenegro, zikic@im.ns.ac.yu*

The intensity of positron annihilation radiation emitted by a sample of natural chromium trioxide, CrO_3, containing 4.7g of ^{50}Cr, was measured for 30 days. The background was measured with samples of iron, copper, and no sample at all (air), during the same time. From the net intensity of the 511 keV line in the HpGe coincidence spectrum the partial half-life for the double beta decay of ^{50}Cr, $T_{\frac{1}{2}}(\beta^+EC, 2\nu + 0\nu) = 1.3(12) \times 10^{18}y$ at the 68%CL was deduced. Alternatively, at 95%CL the effect may be considered absent, the resulting lower limit for this half-life of $1.3 \times 10^{18}y$ representing an order of magnitude improvement over existing limits.

Ref. V.I. Tretyak et al., *ADNDT* **80**, 83 (2002).

ENERGETICS OF SUPERNOVAE FROM MASSIVE STARS. SN/GRB CONNECTION

GABRIELA E. PAVALAS[1], PETER L. BIERMANN[2,3]

[1]*Institute for Space Sciences, Bucharest, Romania,* [2]*Max-Planck Institute for Radioastronomy, Bonn, Germany,* [3]*Department for Physics and Astronomy, University of Bonn, Germany, gpavalas@venus.nipne.ro*

The predictions of the model developed by P.L. Biermann for cosmic rays accelerated by the SN shocks of very massive stars fit well the CR data across the knee. This leads us to conclude that massive stars, with masses above 25 solar masses, end their life with a collapse: a SN explosion which occurs through the magneto-rotational mechanism proposed by Bisnovatyi-Kogan and whose emitted energy is 10^{52} erg. Similar values of the energy have been determined for SN 1998bw and SN 2003dh, which were also associated with gamma ray bursts, possible products of a collapse leading to a black hole.

Ref. P.L. Biermann, *New Astron. Rev.* **48**, 41 (2004).

LIST OF PARTICIPANTS

Abou El Dahab, Cairo, Egypt
Ahriche Amine, Jijel, Algeria
Al-Salti Nasser, Muscat, Oman
Aliouane Said, Algiers, Algeria
Athar M. Sajjad, Aligarh, India
Bâ Abdramane, Bamako, Mali
Bayar Melahat, Trabzon, Turkey
Becherini Yvonne, Bologna, Italy
Benfdila Arezki, Tizi-Ouzou, Algeria
Bercuci Alexandru, Bucharest, Romania
Boruah Kalyanee, Assam, India
Chandra Ramesh, Lucknow, India
Chiarusi T., Bologna, Italy
Chita Sabina M., Bonn, Germany
Consiglio Lucia, Bologna, Italy
Cozzi Michela, Bologna, Italy
Das Chitta R., Chennai, India
Di Ferdinando Donato, Bologna, Italy
Dutta Ruma, Stennis, MS, USA
Elkharrim A., Oujda, Morocco
Esposito Luigi S., Bologna, Italy
Frutti Massimiliano, Bologna, Italy
Fu Yongming, Hangzhou, Rep. of China
Ghazghouz Sameh, Trieste, Italy
Ghezzi Cristian R., Campinas, Brazil
Giorgini Miriam, Bologna, Italy
Graczyk Krzysztof M., Wroclaw, Poland
Grigoriev A.V., Moscow, Russian Fed.
Gutierrez R. A.B., Zacatecas, Mexico
Horvat Dubravko, Zagreb, Croatia
Hounkpe Mathias, Porto Novo, Benin
Ilijic Sasa, Zagreb, Croatia
Khakian Ghomi Mehdi, Tehran, Iran
Khlopov Maxim, Moscow, Russian Fed.
Khoi Dinh Phan, Vinh, Viet Nam
Kirilova Daniela P., Sofia, Bulgaria
Kumar Ashavani, Longowal, India
Liao Wei, Trieste, Italy
Luca Melisa, Villeurbanne, France
Majumdar Archan S., Kolkata, India
Mandal Abu S., Kolkata, India
Manzoor Shahid, Islamabad, Pakistan
Martinez Aliezer, Habana, Cuba
Mastropas E., Rostov, Russian Fed
Medinaceli Eduardo, La Paz, Bolivia
Mehta Poonam, New Delhi, India
Morales T. H.A., Mexico City, Mexico
Movit Steven, Univ. Park, PA, USA
Niess Valentin, Marseille, France
Nowak Jaroslaw, Wroclaw, Poland
Oliveira L. F.A., Rio de Janeiro, Brazil
Ozden Asli, Ankara, Turkey
Pavalas G.E., Bucharest, Romania
Reyes Carlos M., Santiago, Chile
Saharian Aram A., Yerevan, Armenia
Salamida Francesco, L'Aquila, Italy
Sarelo Konstantin, Gomel, Belarus
Sese Rogel M.D., Laguna, Philippines
Shahzad M.I., Islamabad, Pakistan
Sahnoun Zouleika F., Algiers, Algeria
Sheng Z.-M., Hangzhou, Rep. of China
Shivamoggi V.B., Cambridge, USA
Shnir Yakov M., Oldenburg, Germany
Shukla Amritanshu, Kharagpur, India
Siahlo Svetlana, Minsk, Belarus
Sioli Maximiliano, Bologna, Italy
Sirri Gabriele, Bologna, Italy
Tibolla Omar, Padova, Italy
Todorovic Slavko, Novi Sad, Serbia
Todorovic Natasa, Novi Sad, Serbia
Tueros Matia J., La Plata, Argentina
Van Elewyck V.J., Brussels, Belgium
Vasile Anna, Bucharest, Romania
Vishwakarma R.G., Zacatecas, Mexico
Xu Yupeng, Zurich, Switzerland
Yilmaz Deniz, Ankara, Turkey
Yue Chong-X., Dalian, Rep. of China
Zakharov A.F., Moscow, Russian Fed.